NORTH WEST WETLANDS SURVEY **1**

THE WETLANDS OF

Merseyside

R W Cowell
J B Innes

With a contribution by

C E Wells

GIS mapping by

John Dodds

LANCASTER IMPRINTS 2

1994

Published by
Lancaster University Archaeological Unit
Storey Institute
Meeting House Lane
Lancaster LA1 1TH
Phone: 0524 848666, *Fax*: 0524 848606

Distributed by
Oxbow Books
Park End Place
Oxford OX1 1HN
Phone: 0865 241249, *Fax*: 0865 794449

Printed by
The Short Run Press, Exeter

ISBN 0 901800 40 6
ISSN 1343-5205

Editors
B Coles, G Friell, E Huckerby, R Middleton, R M Newman, C E Wells, A Wilmott
Copy editor
J K Lambert
Design
R Middleton
Layout
V Tomlin

Front Cover: View of Orrell Hill, Ince Blundell (Photo R Cowell)

LANCASTER IMPRINTS is the publication series of Lancaster University Archaeological Unit. The series covers work on major excavations and surveys of all periods undertaken by the Unit and associated organisations.

Contents

List of illustrations

Figures

Plates

Tables

Abbreviations

EH	English Heritage
LUAU	Lancaster University Archaeological Unit
NERC	National Environmental Research Council
NWWS	North West Wetlands Survey
LPFB	Lower Peat Forest Bed
LRO	Lancashire Record Office
OD	Ordnance Datum
OS	Ordnance Survey
PRO	Public Record Office
SMR	Sites and Monuments Record
SSSI	Site of Special Scientific Interest
UPFB	Upper Peat Forest Bed

Palynological sites

AV	Alt Valley
BM	Bidston Moss
FMW	Flea Moss Wood
IB	Ince Blundell
KP	Knowsley Park Moss
NC	Newton Carr
PMB	Parr Moss
PRM	Park Road, Meols
PM	Prescot Moss
SMB	Simonswood Moss B
SMC	Simonswood Moss C
SW	Sniggery Wood

Contributors

R W Cowell
National Museums and Galleries on Merseyside, Liverpool Museum, William Brown St, Liverpool L3 8EN

J B Innes
Dept of Geography, University of Durham, Science Laboratories, South Road, Durham DH1 3LE

C E Wells
Archaeological Unit, Lancaster University, Storey Institute, Lancaster LA1 1TH

J T Dodds
Archaeological Unit, Lancaster University, Storey Institute, Lancaster LA1 1TH

Acknowledgements

Our thanks go to the Trustees and the Director, Richard Foster, of the National Museums and Galleries on Merseyside, and the Director of Liverpool Museum, Eric Greenwood, and Ed Southworth for their continuing support and funding for the majority of this project. English Heritage also assisted in the funding of fieldwork and archive and thanks go to the Regional Inspector, Gerry Friell, for his assistance. The many farmers and landowners who allowed access to their land have been vital to the success of the survey and Lord Derby of Knowsley Hall and Mr G Washington of the White Moss Peat Co, Simonswood, are thanked especially. The White Moss Peat Co also kindly donated the funds for one radiocarbon date for a peat sample from Simonswood Moss, and a further date from the site was provided by the Merseyside Archaeological Society. Funding for the other dates was provided by NERC, English Heritage, and Liverpool Museum.

The archaeological fieldwork was assisted by Gary Ashby, and archive and documentary assistance was provided by Jan Grove and Susan Nicholson. Philippa Tomlinson was jointly responsible for some of the palaeoecological sampling in 1979-80, and Colin Wells, assisted by Mike Peace, of Lancaster University Archaeological Unit (LUAU) carried out the stratigraphic survey of the central mosslands as part of the NWWS. The majority of the artwork has been undertaken by Richard Danks and Bob Middleton with contributions by Ruth Parkin (LUAU), and Mark Faulkner of Liverpool Museum. Elizabeth Huckerby designed and produced the pollen diagrams. John Dodds' predecessor Deirdre Winstanley (LUAU) is thanked for the original transcription of my fieldwork notes onto the GIS maps reproduced in this volume, and Janet Lambert (LUAU) is thanked for her patience with the copy editing. The Trustees of the Vyner estate are thanked for their permission to reproduce Plate 2.

While the fieldwork for the project has largely been the remit of Liverpool Museum, LUAU was responsible for the production of the volume. John Coles and Adrian Olivier are thanked for helping to shape the early versions of the report, and Rachel Newman, Colin Wells, and Elizabeth Huckerby (LUAU) all greatly assisted in the emergence of the final version. The external readers, Charly French and Peter Moore, are thanked for their views on the text, and Bryony Coles and staff of the Ancient Monuments Laboratory also commented helpfully on various aspects of the final version. Great help was afforded by Bob Middleton in technical matters and liaison in the final production of the report, as well as much discussion on regional prehistory and fieldwalking results. Chris Cox of Air Photo Services provided information relating to the aerial photographic evidence, and Rob Philpott of Liverpool Museum also kindly provided information from his recent flying programme, in advance of publication, and discussion of the Romano-British evidence from the county. Grateful thanks go to Ray Kenna for commenting on the Wirral chapter and the provision of unpublished data relating to his work on the palaeoenvironment of North Wirral for use in this report. Dave Bedlington of Durham University is also thanked for providing unpublished information from his work on sea-level change in North Wirral.

Summary

This Merseyside volume is the first of six detailing the results of the landscape surveys associated with the NWWS in the counties of North West England. The archaeological and palaeoecological fieldwork presented here is a combination of work undertaken intermittently during the 1980s under the auspices of Liverpool Museum, and between 1990-92 under the auspices of the NWWS. Most of the evidence recovered from fieldwalking and pollen analysis relates to prehistoric settlement and landuse, while aerial photography and documentary study have also been used to present a balanced view of the exploitation of wetlands in the county up to the post-medieval period.

The wetlands in Merseyside began forming in some areas as early as the immediate post-glacial period, but the main period of accumulation began in the seventh millennium BC, resulting in raised mires in the central part of the county and in more limited form along the coasts. The coastal peats were subject to the effects of repeated changes in sea-level over 3000 years from *c*6000 BC.

It is in these coastal areas that the greatest density of sites has been found, mostly belonging to the hunters and gatherers of the Mesolithic period and the first farmers of the Neolithic. It is in these periods that the archaeological and palaeoecological evidence complement each other most effectively, suggesting extensive use of the wetland areas for a range of wild resources. Inland, the archaeological evidence is biased to an unmeasurable degree by modern, extensive quarrying of late glacial sand, the most favoured subsoil for early prehistoric settlement elsewhere in the county. The pollen evidence provides a good framework for the more limited archaeological evidence in this part of the county, suggesting that the pattern of exploitation inland may have differed at this time from that near the coast.

Forest remained the dominant vegetation throughout the prehistoric period in the county, and it is not until the late Bronze Age (*c*1000 BC) that the first evidence for relatively significant woodland clearance and mixed farming appears. Archaeological evidence is slight around the mosslands for this and the subsequent Roman period, which may herald a trend for the relative decline of the wetland areas as a primary location for settlement, and for the spread of farmsteads in well drained locations at some distance from the mosses. This pattern becomes more pronounced in the historic period, when the wetland areas are almost all given over to common land, often used for grazing. They are frequently shared by several townships and so are found generally on their margins. The incidence of archaeological finds and sites is accordingly very slight for these periods, while the peat is largely truncated beyond the Roman or early medieval period by cutting and later erosion.

The intensive farming seen in other areas of wetland in the country has not taken place to the same degree in Merseyside, and so the peat is relatively deep and the water-table relatively high within the less archaeologically rich inland areas. The thinner peat of some coastal areas means that the greatest threat is present in the richest archaeological areas. It is in these coastal areas that the prehistoric sites with the most archaeological and palaeoenvironmental potential have been located, and these will provide the greatest advance in understanding the earliest local settlement and landuse.

Organisation of the Merseyside volume

The aim of this report is to provide an interpretation of the archaeology of the wetlands in the county, within a closely integrated framework of changing vegetation and landscape patterns. It is presented in a form intelligible to the interested layperson, yet it contains evidence of value to research being carried out within the region and nationally by archaeologists and palaeoecologists.

The field survey of the mosslands was organised on topographical criteria which have been followed in the description of the results. The evidence is presented in four stand-alone chapters, representing the coastal wetland areas of the Wirral (Chapter 3) and Sefton (Chapter 4), and the main areas of raised mires in the central and eastern mosslands (Chapter 5) and the southern mosses (Chapter 6). The central and eastern mosslands are discrete areas, separated by a distance of about 2km (Fig 5). These have been grouped in Chapter 5, as they share many common characteristics, particularly for the historic period, in the wider settlement and landuse context. Chapter 6 is concerned with the last remnants, through industrial and urban development, of a series of widely scattered small mires to the south of the main study area. As these have suffered quite heavily through development their archaeological potential is seriously limited, but they are included as they have provided palaeoecological evidence that is of wider importance in interpreting the settlement and landuse, both in the main areas of mossland to the north and in the dryland areas to the south. Chapter 7 brings together and summarises the detail from the four preceding chapters, where possible showing the wetland evidence in the wider landscape context of the county or the region.

The internal contents of the chapters are ordered in a standard way. An introductory topographical description is followed by a section on previous archaeological and palaeoenvironmental research, allowing the reader to judge to what extent the recent archaeological and palaeoecological evidence from the NWWS has advanced our understanding of settlement and landuse. The main section in each chapter presents the archaeological evidence integrated with a summary of the specialist palaeoecological evidence, arranged chronologically within the context of the changing landscape. The detail from which the palaeoecological summary has been drawn is presented in the following section. Finally the main points of importance are summarised and applied to the potential and threats to the resource of each wetland area, putting forward the academic reasons for further research or management policies in selected areas.

The evidence presented in Chapters 3 to 6 is complemented by large-scale foldout maps for the Wirral (Figs 69-72), Sefton (Figs 73-76), the central mosslands (Figs 77-84), and the southern mosslands (Figs 85-86). These maps show the survey results relating to peat survival, archaeological and palaeoecological sites, as well as the field conditions under which the survey was carried out, so that the reliability of the results and their interpretation can be judged objectively.

Appendix 1 consists of a gazetteer which contains more archaeological and topographical detail for the sites mentioned in the text. The numbering system also highlights the advances made during the NWWS fieldwork in comparison to the pre-existing distribution of sites and finds. The latter group are prefixed with an A, while new sites from the NWWS have a county prefix (Me) followed by the NWWS regional archive number. The third part of the gazetteer provides a correlation between the pollen sampling sites (P1-13) marked on the large-scale maps and nomenclature used in the text.

Most of the artefactual evidence recovered was of prehistoric date, other than thin scatters of late medieval or post-medieval pottery, probably the result of manuring, which have not been included in the gazetteer. The prehistoric material in the gazetteer is dated according to the meagre local framework, which means that many finds are difficult to assign to a specific period with confidence. On the maps such difficulties are avoided somewhat by dividing the evidence into only two main periods: Mesolithic/earlier Neolithic, and later prehistoric (essentially late Neolithic / Bronze Age). More explanation of this can be found in the relevant parts of Chapters 3 to 6.

The main body of the report has been written by Ron Cowell, and the palaeoecological survey sections (in Chapters 3 to 6) were written by Jim Innes, with contributions by Colin Wells.

Foreword

The Trustees of National Museums and Galleries on Merseyside (NMGM) are delighted to present *The Wetlands of Merseyside*, the first comprehensive study of its kind for a county area. The concern to make up for the long-term neglect of such a valuable archaeological resource in North West England had seen a limited response by Liverpool Museum using the former Manpower Services Commission resources in the early 1980s. Then, in 1987, English Heritage assessed the potential for a major survey covering Cumbria, Lancashire, Greater Manchester, Merseyside, and Cheshire and, in 1989, an eight-year programme was commenced. The Merseyside survey, undertaken by NMGM's Liverpool Museum, is the first of these and provides, for the first time, an accurate assessment of the potential of wetlands to preserve evidence of past settlement, landuse, and environments.

The results of the survey have been very encouraging in throwing new light on some areas of the county where fieldwork reveals an excellent chance of a number of sites of prehistoric date surviving in their wetland context. This in turn will form a reconsideration of the significance of settlement at this date within Merseyside.

The Trustees would like in particular to thank Ron Cowell, Curator of Prehistoric Archaeology, who undertook the fieldwork, for preparing the report and to acknowledge the contribution of Jim Innes who provided the detailed palaeoecological sections. Lastly, Trustees are grateful to English Heritage and Lancaster University Archaeological Unit for arranging for the publication of the volume.

Richard Foster
Director, National Museums and Galleries on Merseyside

Preface

The North West Wetlands Survey is the third major wetland project established by English Heritage in the past two decades. Work in the Somerset Levels, where quarrying posed a substantial archaeological threat, demonstrated the wealth of environmental and cultural organic evidence contained within peat deposits. Contemporary work in the Fenlands of East Anglia and Lincolnshire exposed the abundance of ancient settlement scatters and industrial debris within and adjacent to the organic and inorganic soils of the Fenland basin. These two wetlands contained substantially different and complementary evidence, not surprising due to their different environmental contexts. The archaeological methodology devised for the two regions had broad similarities in terms of fieldwalking and recording, and in the essential associated palaeoenvironmental investigations that accompanied the field surveys. Differences in the two investigations were mainly concerned with the scientific studies directly linked with the range and type of evidence recovered. Each of the two projects has now completed its primary survey work, published its basic discoveries, assessed the threats to the surviving archaeological remains, produced a synoptic volume, and deposited its archive and artefacts in appropriate institutions. Work building on the surveys now continues in both areas, concerned with preservation of particular areas as well as investigations of sites identified to be of special interest and importance.

By 1989, the Somerset Levels Project had completed its programme, and the Fenland Project was about to finish its surveys. Various discussions were being held about the low-lying wetlands of North West England, in the five counties of Cumbria, Lancashire, Merseyside, Greater Manchester, and Cheshire. The existence of well-preserved and abundant archaeological evidence in the region was known, but widely scattered and inaccessible to the wider world. Environmental studies in the mosses were equally numerous and more informative. Many of the wetland areas were damaged or under serious threat so English Heritage funded a desk-based survey of the existing evidence, and its results, published in 1988 as *Peat and the Past* (Howard-Davis *et al* 1988) set out the state of knowledge and made recommendations for future work.

In 1990, the full North West Wetlands Survey began an ambitious eight-year programme, and the region was expanded to include Shropshire and Staffordshire. Field Officers and Assistants were appointed and archaeological survey and environmental investigations commenced. Attention was first directed to the north, in Cumbria and Lancashire, and ongoing survey in Merseyside was incorporated into the North West Wetlands Survey design. In 1993, the survey moved to Greater Manchester and is now underway in Cheshire. By 1998 the survey will be complete and the programmed series of publications will be well underway. In all this, the support of English Heritage has been unwavering.

A long-term and ambitious survey of this character relies upon a firm adherence to the Project Design, with, however, some flexibility to take advantage of new sources of information. Success also depends upon an adequate and responsive organisation for archival and analytical work, as well as in-house publishing. Above all, however, the ultimate value of this survey, as for all other surveys, depends on the quality of the fieldworkers and their assistants.

This volume is the first of the North West Wetlands Survey monographs, and it is a pleasure to me, as retiring Chairman (1990-94), to see the first book advance to the press. I welcome the new Chair (Bryony Coles) to the project and extend thanks and good wishes to all the members of the Survey for their past and anticipated future achievements.

J M Coles
September 1994

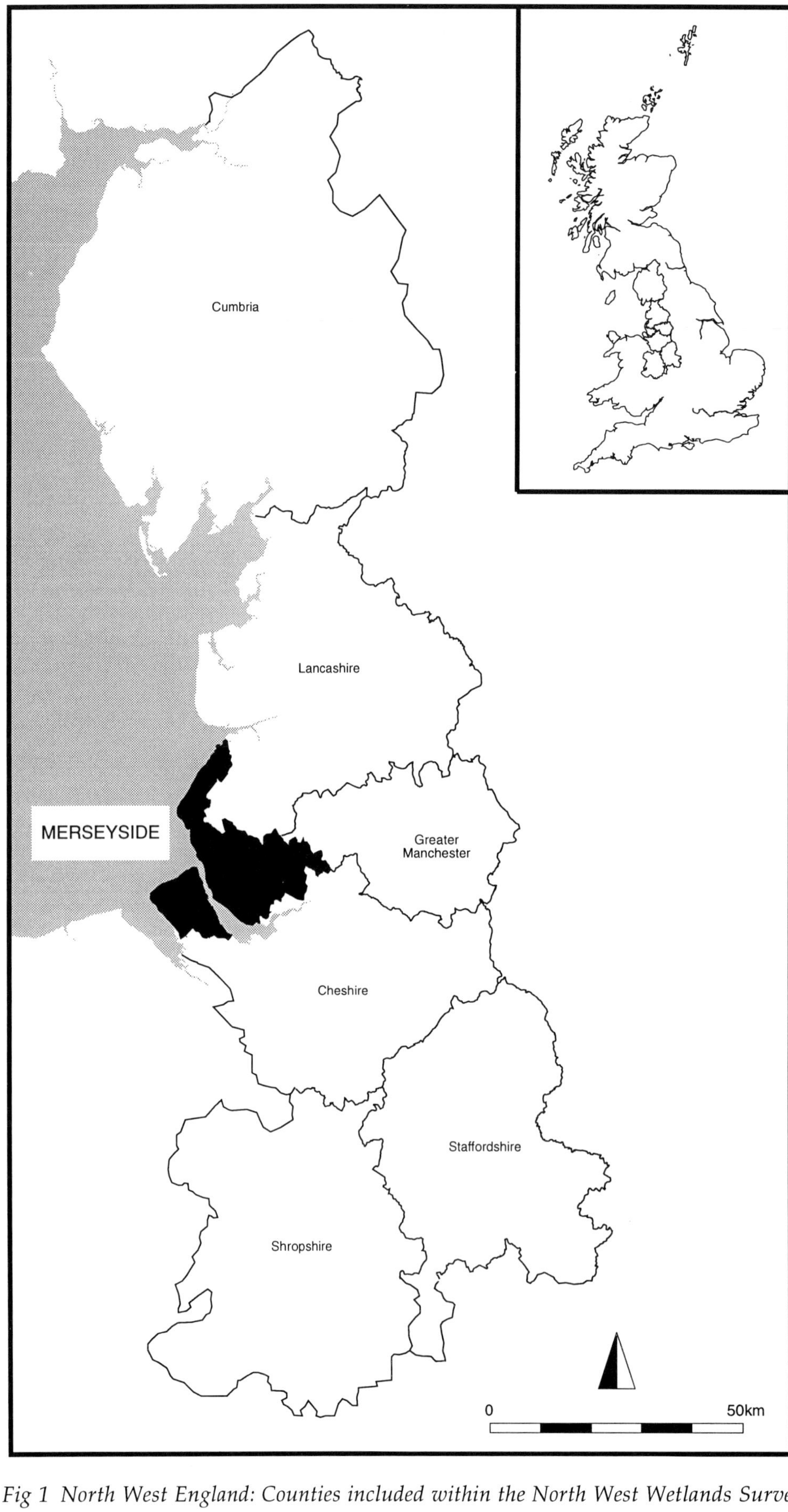

Fig 1 North West England: Counties included within the North West Wetlands Survey

1

THE WETLANDS OF MERSEYSIDE: BACKGROUND

The county of Merseyside was created in 1974 by the amalgamation of the south-west part of the historic county of Lancashire and the north-west part of the original county of Cheshire. At the time of its creation knowledge of the new county's archaeology was sparse, the greater part, mainly that formerly in Lancashire, long having been an area that had received little archaeological interest. The bounds of the new county did, however, include the south-western part of the second most extensive area of lowland peat deposits in England (Shimwell 1985), and it may be the particular difficulties associated with identifying archaeology in such a landscape that partly explains the past lack of attention paid to the area. In more recent times, however, mires have been seen as a source of great archaeological and palaeoenvironmental potential. Systematic work since the 1970s in lowland peat areas such as the Somerset Levels (Coles and Coles 1986), and latterly the Fenlands of eastern England (Hall and Coles 1994), has increased appreciation of the archaeological importance of wetlands. These studies have revealed many instances of settlements of all periods which were associated with wetlands in view of their economic potential for the exploitation of natural resources, such as fish, wildfowl, and plants. These studies also emphasised the almost unique opportunity afforded by wetland archaeology to record and recover well preserved organic artefacts in combination with palaeoenvironmental evidence. As time passes, however, the quality of archaeological preservation in these wetland areas is steadily being degraded by the accelerated rate of erosion of the remaining peat areas. Factors such as drainage, farming, peat cutting, and urban development lead to the loss of the characteristics that preserve the evidence so well.

It was recognised that the long-term neglect of such a valuable archaeological resource in North West England needed to be rectified quickly if its potential was to be realised. Some limited work designed to alleviate this problem was undertaken in Merseyside in the early 1980s through Liverpool Museum (Chitty 1981; Cowell 1981, 1982; Lewis 1982; Innes and Tomlinson 1983a, 1983b), but it was not until the concerted field programme associated with the North West Wetlands Survey (NWWS) began in 1991 that the problem was addressed in a systematic manner.

The North West Wetlands Survey

In 1987 English Heritage initiated the NWWS project with the aim of surveying, analysing, conserving, and managing the wetland archaeology of the region. The first phase in this project was the collation and assessment of existing archaeological and palaeoenvironmental evidence relating to the lowland peat of North West England (Fig 1), and an assessment of the threats that faced this evidence. The findings were presented in the form of a publication, *Peat and the Past* (Howard-Davis *et al* 1988), which emphasised the potential for archaeological discovery in the wetlands of Cumbria, Lancashire, Greater Manchester, Merseyside, and Cheshire. It also identified large gaps in knowledge of the scale and nature of the prehistoric and later occupation of the region, and in the associated development and exploitation of mires and their surrounding landscapes.

Following on from the recommendations made in *Peat and the Past*, the second phase of the project was initiated. In late 1989, an eight-year programme of intensive landscape survey of the lowland peat areas of the North West counties (now expanded to encompass Shropshire and Staffordshire) was begun. The current report outlines the results of the systematic survey of peatlands in Merseyside, the first of the counties to be comprehensively studied. The archaeological fieldwork was carried out during 1990-92 by Liverpool Museum on behalf of the NWWS. Field survey, palaeoenvironmental analysis, and collation and reappraisal of extant evidence, have allowed an accurate assessment of the potential of the wetlands to preserve evidence for past settlement, landuse, and environments. This has been carried out with the intention of identifying the most valuable landscapes and features, so that a sound base of knowledge is available on which to base future management decisions. The results of the survey will also allow the nature of the local wetland archaeology to be placed within its regional context, so that the archaeological record of the North West can, belatedly, play its full part in understanding the development of national patterns of human settlement from the earliest times.

Geographical background

The political unit of Merseyside encompasses the south-west part of the Lancashire plain, and incorporates most of the Wirral peninsula to the west of the River Mersey. It became one of the six Metropolitan counties of England in 1974, and has as its focus the conurbations of Liverpool, St Helens, Southport, and Birkenhead. Since the abolition of the Metropolitan counties in 1986, the five former Metropolitan districts of Wirral, Sefton, Knowsley, Liverpool, and St Helens now function independently within the former geographical area of the county (Fig 2).

Merseyside is the smallest county encompassed by the NWWS, covering an approximate area of *c*890km^2. The main body of the county measures approximately 30km by 20km with a northward extension along the coastal strip towards Southport (*c*18km by 4km in area), and it also contains most of the Wirral peninsula (*c*17km by 13km in area). Much of the county is dominated by urban and industrial use, with agricultural land forming *c*47% of the total. Over 50% of the arable land is of the highest quality (Grades 1 and 2), compared with the national average of *c*18% (Lees 1979); much of this high grade agricultural land is concentrated in the northern part of the county. This includes the peat areas, locally called mosslands, which form the southern limits of the extensive peat deposits of West Lancashire, accounting for *c*3950ha in Merseyside. The main belt stretches across the northern half of the county, from the coast in the Metropolitan Borough of Sefton, to Rainford in the Borough of St Helens. There are additional isolated patches on the North Wirral and in the built-up areas in the southern part of the county.

Geology and topography

Merseyside is on the low-lying Lancashire plain, which stretches from the Pennine uplands in the east to Wales in the south-west. The north-western limit of the plain is the Irish Sea, although the present coastline does not indicate its true boundary, as it continues far out beneath the sea (Fig 3).

The landscape of the county can be divided into four main geological units that each provide a distinctive topography (Fig 4). The earliest deposits are represented by solid rocks, which were laid down in Triassic and Carboniferous times, forming the main relief in Merseyside, with relatively low sandstone ridges and hills distributed across the central and north-eastern part of the county. Much of the character of the area, however, was produced by events associated with the last (Devensian) ice age. Glaciation was responsible not only for producing the major river channels that flow across the area, but also for the second characteristic landscape, formed from the deposition of a thick sheet of boulder clay. This smoothed existing contours and produced a landscape of undulating plains and low hills, which lap against the upstanding ridges produced by the earlier solid rock. These two landscape types dominate the southern two-thirds of the county.

The remaining two landscape areas are mainly found within the northern third, where the topography grades from an almost featureless plain in the west to a more undulating landscape in the east, terminating in the major Carboniferous sandstone ridges that approximately define Merseyside's eastern boundaries. The earlier of these landscapes is formed from superficial deposits which result from the period of Devensian ice-retreat, culminating *c*10,000 years ago, and is largely represented by relatively thin, but extensive, bands of windborne sands that were laid down over the glacial clay from the retreating ice-cap. The final series of deposits are the most recent, and the subject of this survey. They date to the post-glacial period, known as the Flandrian, and particularly to the last 8000 years. These consist of peats, clays, and silts that formed in hollows and low-lying plains in response to changes in the water-table, resulting from factors associated with sea-level and climatic changes.

The sandstone ridges

The 'upland' in the county is represented by a series of low, mainly Triassic ridges, which average between 60-70m OD (Wray and Cope 1948). They were formed partly as a result of the Devensian glaciation, when deep channels were cut in the rock by ice escaping by the most direct route from the Irish Sea to the low-lying Cheshire/Shropshire plain. These channels run from north-west to south-east and the main drainage pattern follows this

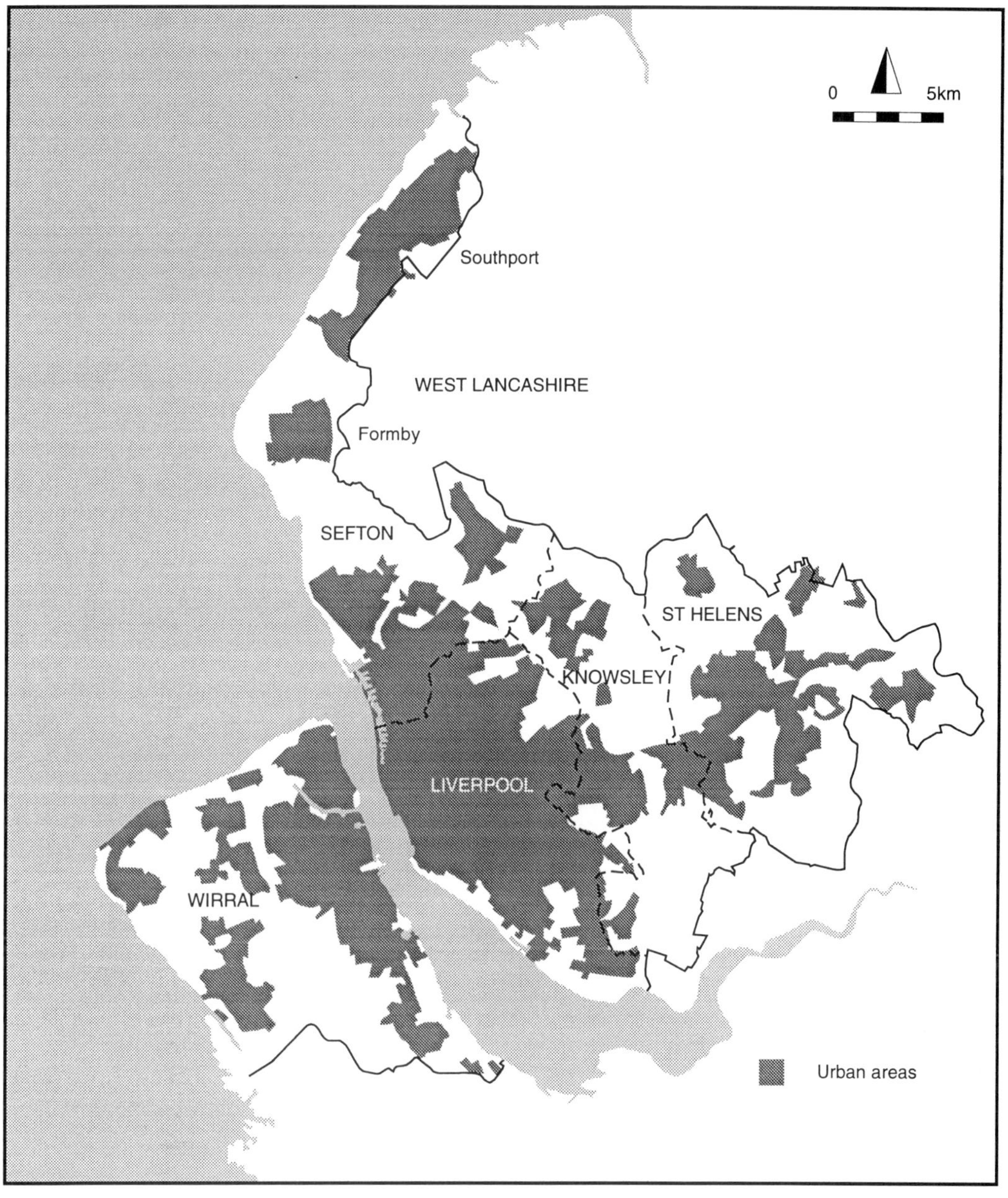

Fig 2 Merseyside: Modern landscape

orientation. The estuaries of the Dee and Mersey flow in the two main channels, the River Fender drains the mid Wirral groove, and the Rivers Alt and Ditton flow at either end of the fourth main former iceway. The River Sankey flows through the major valley across the Carboniferous areas along the eastern borders of the county.

The sandstone ridges are mainly found in the southern part of the county, in four main locations. Along the west and east sides of the Wirral, and east of the Mersey, forming a plateau between Liverpool and Woolton, the geology of these ridges comprises fine-grained, soft, red, mottled sandstones. These are separated by the coarse-grained, often buff or yellow pebble beds of the Bunter series, or the harder, coarser, and less homogeneous Keuper series (Johnson 1985). The fourth area of Triassic age ridges is situated to the east of Woolton. These are more restricted in area, but the main outcrops are located between Prescot and Knowsley. Smaller outcrops are found to the south, extending towards the Mersey across northern Cheshire.

To the north-east of St Helens, between Parbold and Billinge, the upland ridges are mainly of Carboniferous age. In general they have proved more resistant to erosion than the Triassic rocks, and consequently form the highest ground in the county, reaching heights a little over 100m OD along the

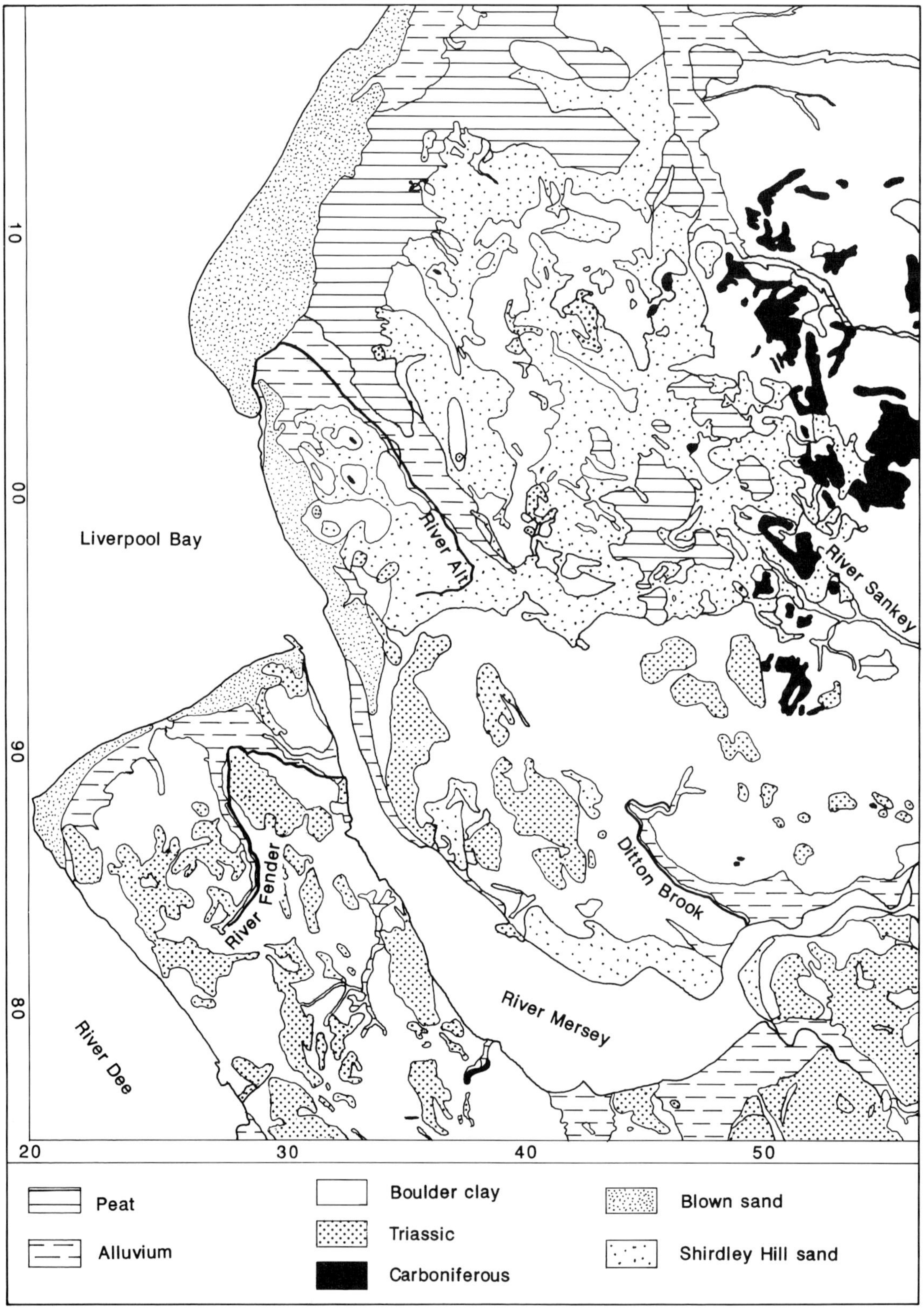

Fig 3 Merseyside: Geology

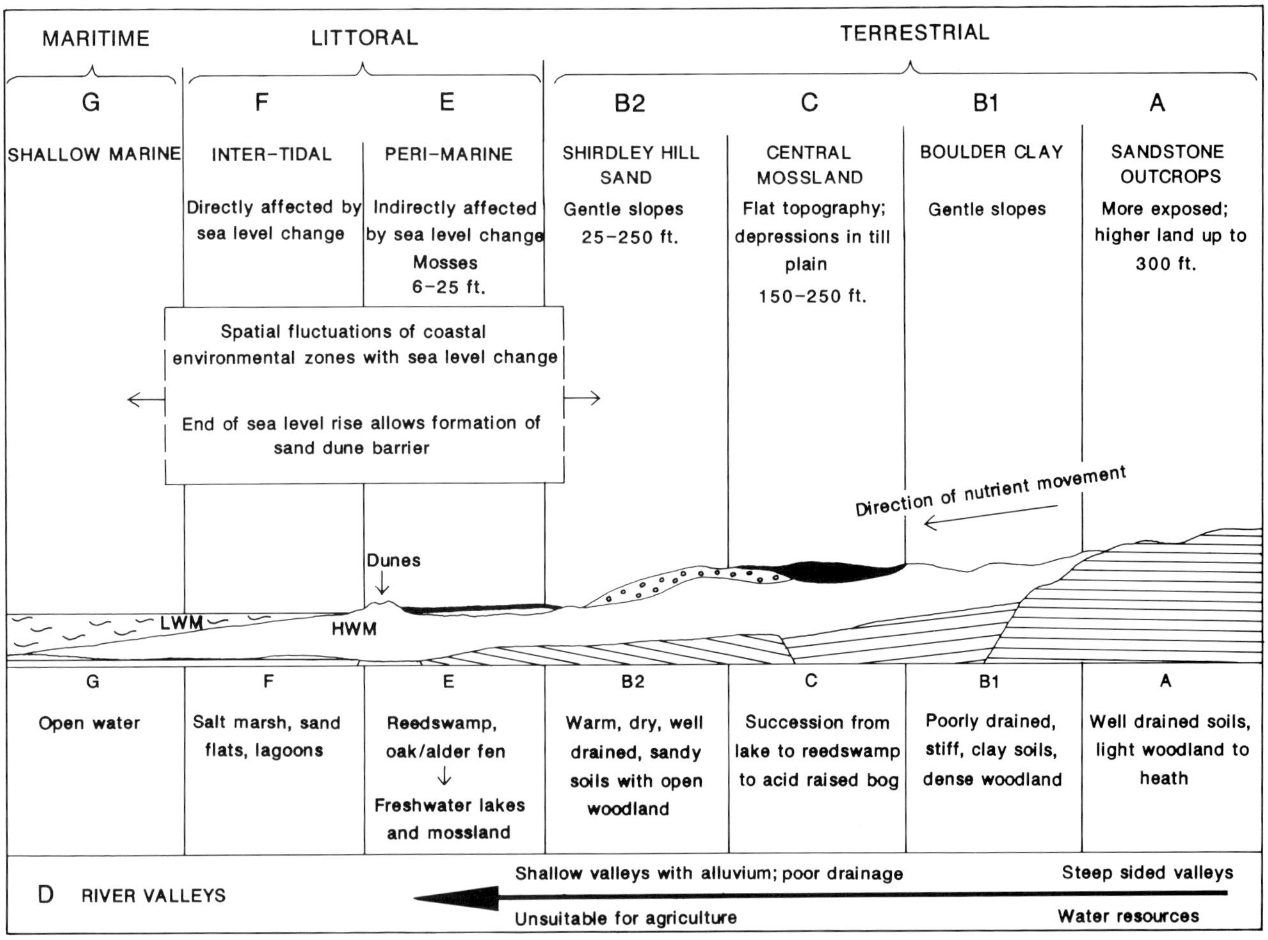

Fig 4 Merseyside: Idealised geological transect across the county

Billinge ridge. This ridge extends north-westwards into Lancashire and marks the eastern edge of the Lancashire plain.

The glacial till lowlands

The dominant relief in the county is, however, mostly independent of the pre-Devensian solid geology. The topographical effects of a large number of faults have been obliterated by erosion and smothered by glacial drift deposits. The southern two-thirds of the county mainly comprise a gently undulating till landscape, derived from deposition by the retreating glaciers (Johnson 1985), although in some areas the boulder clay is interspersed with areas of slightly higher relief formed by Triassic and Carboniferous sandstones. The boulder clay varies in texture and composition (depending on its source and mode of transportation) from extremely stiff or very sandy, almost stoneless clay, to gravel, containing frequent pebbles set within a clay matrix. In Merseyside it is a reddish-brown, slightly calcareous clay, which is medium to fine textured and rarely completely stoneless (Hall and Folland 1970). Some of the pebbles encountered in the clay are of flint, or very rarely, chert, which have been eroded by ice from Cretaceous deposits under the Irish Sea and deposited in clay across the lowlands of North West England. These pebbles provided a local source of moderate to poor quality raw material for the manufacture of stone implements in the prehistoric period.

The late glacial coversand lowlands

The other main geological feature of the area occurs more widely in the northern third of the county and extends northwards into West Lancashire. It comprises a widespread deposit of wind-blown sand, Shirdley Hill Sand, of fluvio-glacial origin. For the most part this forms a plain in the west, with the occasional low ridge, such as at Little Crosby and Ince Blundell in Sefton, gently rising from *c*6m OD to 10-15m OD. In the east of the county the sand is found at heights of *c*75m OD, where it caps or skirts a number of sandstone and boulder clay ridges. The sand was first deposited *c*9000-8200 BC, during the culminating stages of the

Devensian glaciation (LDe III) and redistributed by wind at subsequent periods during the early to mid Flandrian (Tooley 1978a; Wilson *et al* 1981; Innes 1986). It appears to have remained unstable and sparsely vegetated until Flandrian Ib, dated *c*7800-6900 BC at the type site of Red Moss (Hibbert *et al* 1971), although with large standard errors on these dates. At this time, a decrease in climatic continentality and increase in humidity, allied to a marine transgression of the coast, contributed to its stability and encouraged the formation of vegetation cover (Tooley and Kear 1977). The sands are rarely more than 5m thick, and are often less than 2m thick in the survey area. The Shirdley Hill Sand is often closely interspersed with areas of rock, boulder clay, and peat, providing landscapes of great environmental diversity.

Flandrian deposits

Superimposed on this glacial drift geology are deposits of Flandrian date (after *c*8000 BC), which mostly comprise the areas of wetland in the county. These include peat, marine and alluvial clays and, along the coastal strip, blown sand, derived by wind action from sea-sorted beach sand.

The peat

The mosslands in the county divide into two main types. The first is represented in those areas away from the modern coast where peat formed in local hollows. This is found both in the south, where impermeable boulder clay areas dominate, and in the northern third of the county where the Shirdley Hill Sand occurs as a capping, of varied depth, to the boulder clay (Fig 3). The first stage was the collection of water in these hollows due to a combination of factors associated with increasing rainfall, after *c*6400 BC (Tooley 1980, 83). These factors included a rising water-table, probably partly due to increasing sea-level, and the impermeable nature of the extensive areas of boulder clay. Even in the northern areas where the sand is extensive, the boulder clay was close enough to the surface to prevent easy percolation of the water.

The peat stratigraphy of these mires reflects their evolution from fens to bogs. Fens are minerotrophic systems fed mainly by ground water from beyond their margins. They are floristically diverse, and are often characterised either by open landscapes dominated by sedges, reeds, and other grasses, or by carr habitats, where trees such as alder, birch, and willow are prevalent over an understorey of monocotyledons and bryophytes. The waterlogged conditions produce a tendency towards anaerobic conditions, discouraging bacterial decomposition of organic material. This allows the build-up of old plant material as peat, and as it accumulates the surface of the mire is raised further from the groundwater influence. The vegetation may then become more dependent on atmospheric precipitation for its nutrients, and the whole system becomes more acid. The vegetation changes in response to this alteration of water quality, forming poor fens and eventually bogs.

The water for ombrotrophic mires, or bogs, comes almost solely from the atmosphere through precipitation. The highly acid nature of the system allows a specialised flora to dominate such sites, with sedges such as cotton-grass, dwarf shrubs such as bog myrtle and, especially, *Sphagnum* mosses. Mire expansion may continue as long as the conditions of high precipitation, low evaporation, and impeded drainage continue, although there may be limits to such expansion due to physical constraints (Clymo 1984). The resulting bog is known as a raised mire. These are traditionally characterised by a domed cross-section, where the centre of the bog can be several metres above the height of its margins. In some cases the peat may expand laterally, spreading over low ridges of mineral ground and coalescing with peats originating from separate loci of peat initiation. The composite bogs may then be termed 'ridge-raised' or 'intermediate mires' (*sensu* Moore and Bellamy 1974; Lindsay 1989); in Merseyside raised and/or intermediate mires are found in the central areas of the county, stretching from Kirkby to Rainford (Fig 5). Little palaeoenvironmental work had taken place in the central mosslands prior to the present survey, so that the date for their origin is not clear. In localised areas, organic accumulation began before the deposition of the Shirdley Hill Sand, as at Clieves Hill just outside the county boundary in West Lancashire (Tooley 1978a), or at Moss Lake, Liverpool, where it was succeeded by late glacial lake deposits (Godwin 1959). The NWWS Merseyside survey shows that a few of the mosses date to the early post-glacial, but most of the central mosses began to form in the Flandrian II period (*c*6190-5720 Cal BC to 3990-3640 Cal BC), or at least spread during that period from localised foci. The mosses to the east of Southport (Fig 2) have also evolved through similar stages of development.

The second variety of mossland is found in the remaining coastal areas of Sefton and Wirral (Fig 4), where most of the mires remained as fens throughout their existence. Work associated with the NWWS shows that these formed slightly earlier than some of the central mosslands. Low gradients, poor drainage, and high water-tables provided the conditions for formation of the peat, but in this case

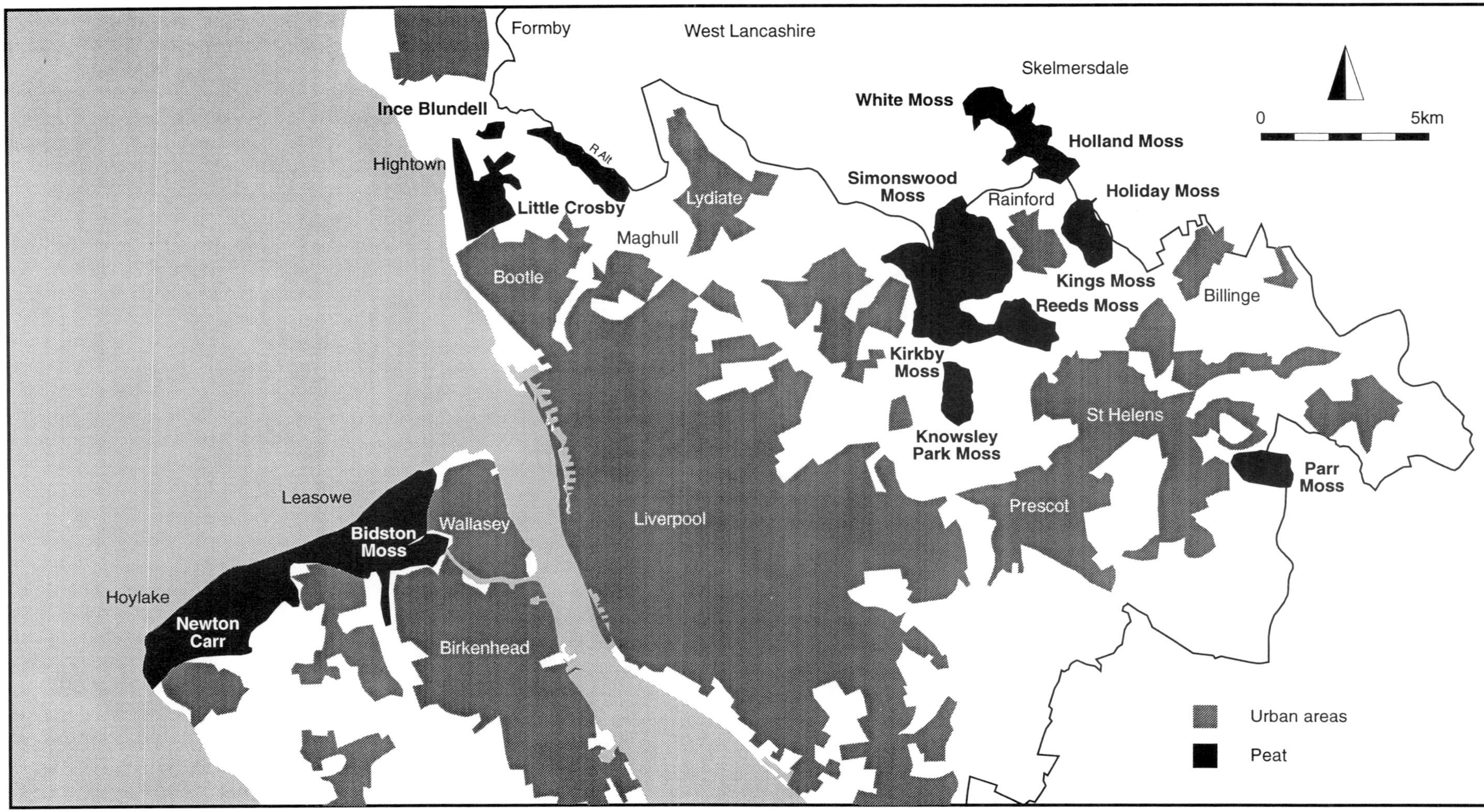

Fig 5 Merseyside: Peat distribution

groundwater levels are probably linked to the oscillations of sea-level. This means that peat growth along the coastal plains was primarily dependent on groundwater influence and so rarely attained raised mire status.

Marine Flandrian deposits

Since prehistoric times, the relevance of wetlands to the history of human settlement in the present coastal areas of the county has been conditioned by the effects of changes in sea-level. This is seen most explicitly in the peat beds along the present intertidal zones on the Wirral, at Hoylake, and on the Sefton coast at Hightown. They were known by earlier workers as 'forest beds' because of the horizons of tree stumps that occur within the peat. These former woodland areas, and the peat that subsequently engulfed them, are essentially a continuation of the inland mosses, and reflect lower sea-levels than obtain today.

Merseyside was subjected to fluctuating sea-level movements between the seventh millennium BC and the fifth millennium Cal BC (Tooley 1978a, 1985; Kenna 1986). An important feature of successive fluctuations has been the burial of former landscapes by layers of clay, silt, and blown sand. Therefore evidence for human occupation is often buried under, or within, these deep deposits. These sediments also contain evidence for the changing environment of the present coastal areas, involving diverse landscapes including sand dunes, intertidal sand and mudflats, saltmarshes, marine and freshwater lagoons, and swamps, meres, and fens. These environments, which varied in distribution through time, will have played a physical role in limiting or encouraging the location and development of human activity in the present coastal areas of the county. The waterlogged nature of many of these sediments also creates the potential for the preservation of organic remains, and for better precision in the dating of buried archaeological sites by means of radiocarbon dating.

Sea-level change

The evolution of Merseyside's coastline and the sequence of post-glacial coastal deposits are the result of the relative changes in sea-level which have occurred since the last ice age. Changes in relative land and sea-level result from two principal causes. Firstly, local upwarping or downwarping of the earth's crust has been brought about by the deformation or displacement of rocks. This may be due to changes in the load of ice, water, or sediment on the crust (isostatic), or to tectonic movements. Secondly, worldwide changes in sea-level may be due to actual increases in volume of sea-water, or to a change in the shape of the sea basins (eustatic). The main eustatic effect over the last 12,000 years has been an overall rise in sea-level caused by an increase in volume of water from melted ice sheets. The combined effect of eustatic and isostatic changes has apparently caused relative sea-level to rise in an oscillating manner, although the data have also been presented in a more simplified manner in terms of a smoothly rising sea-level (Kidson and Heyworth 1973; Tooley 1978a, 1985). Between *c*8000 BC and *c*3700 Cal BC, when sea-level was rising rapidly, it is possible to identify three transgressions of the coastline in the northern Merseyside area (Tooley 1978a). The recognition and understanding of the nature of these transgressions is crucial to the interpretation of the prehistoric archaeology of the region, and is dealt with in detail in the relevant chapters of this volume.

History of environmental and archaeological research

Previous palaeoenvironmental research

Scientific interest in Merseyside has long been more concerned with the formation of its Flandrian deposits than with its history of human exploitation of those environments. This interest in the post-glacial sediments of South West Lancashire and the Wirral has been evident since the mid nineteenth century, although the 'submerged forests' along the intertidal zone attracted attention even earlier, from James in 1636 (Corser 1845) and Anon (1796). The first scientific observation of the lowland peat mosses was undertaken by Binney and Talbot (1869) who excavated on Downholland Moss, just outside the northern county boundary. However, during this early period the attention of these workers was directed mainly to the coastal mosses, as their approach was essentially a geological one, investigating the stratigraphic relationships of the peat to determine its age relative to the other main strata of the region, in particular the glacigenic drift deposits. Chief among these early workers were the prolific T Mellard Reade (*eg* 1872a, 1878a, 1878b, 1908) and C E de Rance (*eg* 1870, 1877). Others who provided evidence on coastal sections, particularly for the Wirral, were C B Travis (*eg* 1908), W G Travis (1922), Potter (*eg* 1894), Jones (1919), Greenwood (1910), and Maidwell (1920). These writers provided a rich body of stratigraphic observation and description and established the primary relationships of the coastal peats with the clay, outwash sand, and blown sand successions. Few workers attempted to relate past land surfaces to human culture, except perhaps Jackson (1919), Potter (1876), and Smith (1866a) who recorded the stratigraphic context of archaeological material within the coastal successions.

Building upon the accurate stratigraphic work of the nineteenth century investigations, the development of palaeobotanical techniques in the early twentieth century caused the research emphasis to move towards the analysis of plant macrofossils and microfossils within the peat deposits. W G Travis studied macrofossil remains from organic inclusions in sand formations at Wallasey, Wirral, and Aintree, Liverpool. C B Travis (1926, 1929) and Erdtman (1928) undertook the first pollen analytical studies of the major natural intertidal exposures of peat around Liverpool Bay, from the Altmouth near Hightown in Sefton, and from Leasowe, North Wirral. These pollen studies dated the formation of these intertidal peats to the mid post-glacial forest period by analogy with other work in Britain and Scandinavia, and were the first stage in the investigation of the vegetation history of Merseyside.

A reappraisal of the stratigraphic sequences of the belt of coastal lowland mosses between Liverpool and Southport was undertaken by Gresswell (1953, 1958). He was the first to explain correctly the alternating bands of peat and clay, for example in Downholland Moss near Formby, in terms of a fluctuating sea-level since the end of the last glaciation, involving a rise which carried marine conditions well inland of the modern coast before falling back to its present level. Many have followed his lead (*eg* Hall 1955; Hall and Folland 1970), although more recent research has shown many of his conclusions regarding coastal evolution in the area to be in error.

With the advent of modern palaeobotanical techniques the capacity to reconstruct the past landscapes and human impact on them has been immeasurably improved. A number of major pollen studies have been carried out in the region, mostly outside the county, but all providing a context for local material. Work by Godwin (1959) has provided a vegetational history of the late glacial to mid post-glacial period from former lake muds in Liverpool. Work on the development of peat mosses throughout the post-glacial period has taken place along the Mersey terraces in Greater Manchester and Cheshire (*eg* Birks 1964, 1965; Hibbert 1977) and on the northern fringes of the county by Cundill (1981). These studies have developed a framework of vegetation history which shows the spread of forest conditions across the region, and the increasing impact of man in deforestation and agriculture.

The work of Tooley on the stratigraphic sequences of coastal mosses and dunes around Liverpool Bay has made a substantial contribution to knowledge of the environmental history of the area, with particular reference to the history and effects of post-glacial sea-level change. Sites within Merseyside, around Altmouth and Formby, have been linked with work carried out in South West Lancashire to provide seminal papers on the evolution of this coastline (*eg* Tooley 1978a, 1982, 1985). Extensive stratigraphic research by Kenna (1978, 1979, 1986) has also done much to explain the impact of coastal changes on North Wirral. The work of these authors has shown that the coastal sequences mapped by the nineteenth century pioneers in the field result from the effects of a fluctuating sea-level, which over several thousand years alternately caused the deposition of marine and freshwater sediments over a foundation of glacial clays and sands. They also indirectly governed the emplacement and mobilisation of the blown sand formations which characterise the area today.

In the late 1970s, work undertaken through Liverpool Museum, based on pollen analysis of several areas of peat in the county (Sheppard *et al* 1978), represented the first systematic attempt to recreate the palaeoecology of the mosslands and the surrounding vegetation. The peat was specifically targeted for a programme of palaeoecological work, undertaken by Jim Innes and Philippa Tomlinson, who were responsible for providing a palynological (pollen) framework with which to relate the human settlement of the county. In this context they systematically sampled most of the main mires and monitored development threats to known or newly discovered peat deposits elsewhere (Innes and Tomlinson 1983a, 1983b).

Previous archaeological research

In contrast to the environmental research, there has never been a similar tradition of either amateur or professional archaeological involvement in the county, although North Wirral is to some extent an exception. Observation of stratigraphic sequences of post-glacial deposits in the nineteenth century also produced archaeological evidence, as sea erosion exposed organic deposits and associated artefacts. Several antiquaries were active in this area over many years, the most important being Hume (1863). He recorded one of the most important groups of medieval metalwork in the region from the beach at Meols, near Hoylake, alongside finds of prehistoric flintwork and coins, and early and later medieval pottery. Today this might be recognised as being in a wetland context, and although Hume treated the nature of the wetland deposits in his report he never made explicit the relationship between the finds and the natural deposits. Other antiquaries satisfied themselves merely with recording the discovery of finds from the coastal area (*eg* Smith 1871, 1872; Shone 1911), and little excavation took place during this early period. The most significant were of two Neolithic sites on the Wirral shoreline, at New Brighton and Hoylake (Roeder 1900; Glenn 1914), although the reports provide little detail of the techniques or results.

Elsewhere in the county there was less interest in, and observation of, artefacts associated with these environments, so that, for example, there are no records of archaeological material coming from the central peat areas of the county until the late 1970s. The view, expressed by Camden in the early seventeenth century, that the peat areas of southern Lancashire were 'swampy, unhealthy places' seems to have set the tone for subsequent archaeological interest in them (Camden 1610). This lack of an archaeological tradition in the county continued into modern times, when despite the presence of archaeologists of the calibre of T G E Powell at Liverpool University, little local work was done.

The first real progress in recording and analysing the archaeology of the area did not come until the national growth of regional Archaeological Units, which were created in response to the perceived needs of rescue archaeology in the early 1970s. In Merseyside this paved the way, approximately four years after the creation of the Metropolitan county in 1974, for systematic archaeological survey coverage within the county, associated with the development of the county Sites and Monuments Record (SMR). The field survey responsibilities were invested in the Liverpool Museum Field Section, who are also responsible for the Merseyside SMR. The limited amount of work associated with the wetlands of the county, prior to the commencement of the NWWS, therefore dates largely from this time. This comprised mainly fieldwalking, aerial photographic survey, and associated pollen sampling programmes, in the late 1970s and early 1980s. Fieldwalking coverage, mainly undertaken by one of the present authors (RC), was fairly representative of the county's main landscape types, including the boulder clay areas in the districts of St Helens, Knowsley, and the Wirral; the western sandstone ridges in Wirral and a limited area of Knowsley; and the Shirdley Hill Sands in the northwest of the county. Field coverage in the latter area included examination of the boundaries of numerous peat areas in the central mosses, although the peat itself was not walked. The major areas not covered in this fieldwalking were the Sefton coastal area and the north-eastern mosslands.

At about the same time, the first regional rescue units were established in the North West, at Liverpool and Lancaster Universities. The Liverpool University Archaeological Unit undertook no work in wetland contexts, and apart from excavations at Bronze Age barrow sites on the eastern borders of the county with Cheshire, at Winwick (Freke and Holgate 1990), its work in the county was largely associated with historic sites in urban centres such as Liverpool and Prescot, and on moated sites mainly in St Helens (Freke 1981; Davey and McNeil 1985; Holgate 1989; McNeil 1989; Lewis in prep). Since approximately 1986, the Field Section of Liverpool Museum has also been involved in excavations, at first research based, but since about 1990 and the advent of developer funding, including commercially sponsored work.

Substantial results have come from a number of excavations during this time. They have included six Mesolithic hunter-gatherer sites in the districts of Wirral, Knowsley, and Liverpool (Cowell 1991a, 1992c, in prep b), a late prehistoric farmstead in Knowsley (Cowell and Philpott 1994), three Romano-British farmsteads in Wirral (Philpott 1993), Knowsley (Cowell and Philpott 1994), and in Cheshire (Philpott *et al* 1993), and six medieval rural or village sites in Wirral, Knowsley, Liverpool, and St Helens (Cowell and Philpott 1994; Philpott in prep). In 1986, a one-day seminar was held at Liverpool University, reviewing ten years' progress in recording the archaeology of the county since its creation. The results of this seminar were subsequently published in an updated form in 1991, in a collection of papers in the Journal of the Merseyside Archaeological Society, providing detailed accounts of the archaeology of each period in the county. This acts as a useful summary of the situation prior to the commencement of the NWWS which, apart from periodic aerial photography undertaken by R Philpott of Liverpool Museum, is the only new survey work to have taken place since 1990.

Survey and settlement evidence in the North West region

This section deals mainly with the earlier periods of settlement in the area as they relate to their discovery and interpretation from ploughsoil evidence; the major evidence for the discovery of sites in the NWWS. The settlement of the early medieval period is notoriously difficult to discover from fieldwalking survey alone (Schofield 1991a, 4), but the pattern can sometimes be recreated using some of the techniques applicable to the interpretation of settlement of the medieval period. For the latter period, aerial photography, maps, and documents, existing buildings, and landscape interpretation, can provide a reasonably accurate picture of the former settlement and landuse pattern in the county. This can be augmented by the evidence of pottery distributions in the ploughsoil to recover evidence not available from the other techniques (Cowell 1982). Although fieldwalking can be useful for the discovery of historic sites, as will be seen in subsequent chapters, such material was seldom discovered in close association to wetlands in the county, and so has less relevance to the following discussion.

In the North West, material by which the presence of prehistoric activity can be recognised is overwhelmingly dominated by the discovery of lithic items. Pottery is rarely found for this period, with only three pottery sherds of possible prehistoric date having been found in the *c*1200ha previously fieldwalked in the county. This is partly because prehistoric pottery is notoriously fragile and decays very quickly once exposed in the ploughsoil, and partly because of the general paucity of artefactual material associated with sites found in the region (*see below*). Although Romano-British pottery is a little more common, such sites similarly produce little surface pottery, and that which is found is generally relatively soft and fragile and susceptible to rapid decay. Interpretation of Romano-British sites, and probably of local late prehistoric sites, can sometimes be aided by aerial photographic evidence, which is generally not the case for prehistoric sites in the region.

Chronology of local prehistoric material

One problem in the interpretation of surface lithic material in the North West results from the lack of a chronological framework that allows flint material to be easily assigned to the relevant periods. Most of the material previously found in the county has been interpreted as being of Mesolithic date, but the absence of widespread finds of microliths, the most diagnostic type for the Mesolithic, makes this interpretation a little less secure. In general, therefore, identification of lithic material of this period has been based on technological features associated with the occurrence of blade cores and blades with thin butts and diffuse bulbs derived from soft hammer knapping techniques. Cores are mainly single platform, and are usually prepared by having one flake taken off at right angles to the striking axis. Several are only worked part way round the pebble leaving the back as cortex (Fig 20).

For the Neolithic period the criteria used are similarly rather limited, as recognisable implements are generally rare. The criteria used for this period relate to thin, moderately large flakes, or blades, larger than in the Mesolithic, some of which have fine retouch. Soft hammer percussion techniques produce butts slightly wider than on the Mesolithic blades. It is accepted that, in the absence of a firmer chronology, there may be a small degree of confusion in the separation of these two periods from flintwork alone, and that some sites identified as Mesolithic may be a little later and *vice versa* for the material interpreted as early Neolithic. As will be seen in subsequent chapters, this may not be a serious drawback, as the archaeological and palynological evidence suggest that there may be little difference between these periods in the location and type of settlement, and the nature of landuse impact on the vegetation.

For the period after the middle Neolithic, general technological characteristics differ from those of the material already discussed, in terms of waste flake proportions (Pitts 1978) and tool types (Richards 1990, 24), although reference collections are still heavily biased towards southern England. In a number of regions, or at individual sites, it is possible to illustrate technological differences between late Neolithic and early Bronze Age assemblages (Pryor 1978, 142-3; Healy 1984). In North West England such distinctions are more difficult, as hardly any significant assemblages of this period exist, and these are mainly surface sites (Middleton 1993, 5). The main diagnostic pieces previously found in the county for this period are early Bronze Age barbed and tanged arrowheads, but few other tool types occur regularly in surface scatters. Therefore, little distinction can be made confidently within this general period for material recovered from the NWWS. The main characteristic used to assign material to a general late Neolithic/early Bronze Age horizon has been waste material that is not blade orientated, and that is larger, thicker, and less carefully knapped than material discussed previously. A large assemblage from NWWS fieldwork in Little Crosby has, for the first time, provided significant material, including many distinctive tool types, that can be assigned generally to this period (Fig 22).

In southern England flintworking seems not to have continued into the later Bronze Age/Iron Age of the first millennium BC (Ford *et al* 1984). As with earlier periods in the North West, there is no evidence from the region to set against this. A number of characteristic flint pieces were recovered during earlier fieldwalking in the county, and more were located during the NWWS fieldwork. These pieces reflect an expedient flint technology whereby natural, suitably shaped chunks, or sometimes irregularly shaped knapped waste, have been modified by limited retouch along the edge or tip. The lack of formal technological traits for this material suggests that it may fall into a period after early Bronze Age flint technology had ceased. This material may therefore be of middle or later Bronze Age date, rather than representative of an alternative functional technology contemporary with material derived from an earlier period.

The nature of the evidence

Lithic material from the previous systematic fieldwalking surveys in Merseyside and North Cheshire (Cowell 1982, 1992b) has shown that the local evidence for the prehistoric period differs strongly from field surveys in other parts of the country. There is little of the pattern seen elsewhere of overlapping, chronologically mixed, dense spreads of material occurring over several hectares (Richards 1990; Clark and Schofield 1991, 104). Instead, large areas of the landscape are devoid of prehistoric and Romano-British artefactual evidence. Even where concentrations of artefacts occur there are large intervening areas where not even 'background' material is found. This latter material reflects use of the landscape outside habitation and specific activity locations, relating to loss or discard of material during, for example, transit or visiting, which might occur close to the home base or at some significant distance from it (Foley 1981, 5). In this way human use of the landscape can be seen as an almost continuous distribution, with certain loci representing more concentrated activity.

In order to interpret the pattern of artefact distribution recorded during the NWWS, and from previous non-wetland survey in the county, the issue of the definition of a 'site' as opposed to off-site activity has to be addressed. This is a complex problem and the subject of much debate (*eg* Schofield 1991b; Middleton 1993). Other surveys have treated varied surface densities of lithics as part of a continuous distribution across the landscape. Within this distribution, significant patterns in the makeup of the flint collections, across various scales of landscape, are looked for to reveal settlement patterns, rather than using the criteria of subjectively defined 'sites' (Schofield 1991b). This approach has not been used for the description of material from the NWWS, for a number of reasons. One of the main aims of the NWWS is to identify the most likely locations where habitation or associated activity has taken place in the landscape, in order to assess the potential of waterlogged sites for either management or rescue purposes. As continuous flint distributions do not occur in the county, and concentrations of material are generally reasonably discrete, 'site' definition was therefore judged to be both necessary and possible.

There are still problems in interpreting sites and background material from field survey. Human artefactual discard patterns vary and there are a whole series of post-depositional factors, such as natural soil and landscape changes, or displacement of material through modern agriculture, that can affect interpretation (Schofield 1991a). Work by the writer in the Merseyside area, outside the scope of the NWWS, is complementing the field survey of the dryland areas with sample excavation to test the nature of the surface evidence as an aid to interpretation. Although this work is in its early stages a few tentative patterns are emerging that allow some criteria to be suggested for the definition of a 'site' for the purposes of the NWWS.

Previous survey in the county has produced a few locations with 100-500 surface lithic pieces, found over a hectare or more, with an average density of 1-3 per m^2. This material appears to be early Mesolithic in date. Such densities are, however, anomalous in the area, the more normal pattern being for total numbers, and consequently densities, of recorded lithic material to be low compared with other parts of Britain. A number of localities have produced a pattern of lithic distribution whereby *c*5-20 struck pieces are found over areas of between 50-500m^2. The fall-off rate for the surface finds distribution is additionally often quite sharp. By analogy with those locations where trial excavation has taken place, such low density sites in the landscape can generally be interpreted as loci of activity of various kinds, probably equated with habitation. These might fairly easily be defined as 'sites' in the traditional sense.

The most common flint distribution previously found, however, is of single or widely spaced 'concentrations' of two or three lithic findspots. This material might more generally be seen as 'background', representative of 'off-site' activity. This suggests that low densities of finds indicating prehistoric loci of activity in the county are mirrored by large areas with only a very sparse distribution of potentially 'off-site' material.

Excavations that have taken place at locations with varying surface flint densities in the county provide some detailed evidence with which to compare the general surface distribution located by systematic fieldwalking (Cowell in prep a). This suggests that settlement features may exist at locations with as few as ten surface pieces, at a density of *c*0.5 per $10m^2$, while there are some indications that even 'concentrations' of as low as three pieces (depending on morphology and distribution factors) can reflect sites where recognisable activities took place. One such site lies in Croxteth Park, *c*3km south-west of Kirkby and Knowsley Park Mosses in the central mossland belt (Fig 81), which was excavated in 1992 (Cowell in prep b). The distribution of surface struck flint would probably classify as 'background' material in other regions (Foley 1981). The greatest density was seven flint pieces spread over an area of *c*2.5ha, although the 'background' distribution of 20 pieces over an area of *c*18ha was the most extensive in the area, without providing obvious, discrete foci. On excavation in the area of highest density (*ie* one per 0.4ha), although no settlement features were recognised in the sandy subsoil, apart from an amorphous hollow, burnt flint suggests that some activity had taken place there during the late Mesolithic, requiring a fire. The activity was not intensive, as only 535 struck pieces of flint came from a sample programme of sieving through a 2mm mesh over an area of *c*$250m^2$. The material was distributed as two or three weak concentrations within the subsoil of the field, suggesting re-use of the area over a period of time. The thin scatter of surface flint, most of which is probably Mesolithic, may reflect similar episodes over a long period of time, for which the term 'background material' is not suitable.

In summary, therefore, the local evidence has certain characteristics which have dictated the approach to its discovery, treatment, and interpretation. 'Sites', defined as locations representing occupations of a specific area for variable periods, dedicated either to single tasks such as animal butchery or raw material procurement, or to a combination of tasks such as would be found on a more long stay habitation site, may be represented by only a few finds. Even single findspots may, depending on factors such as post-depositional conditions or field retrieval techniques, reflect such small occupation loci rather than always reflecting 'off-site' activity. This is of particular relevance for those locations in the NWWS survey area where individual pieces have been found on peat or associated wetland surfaces. These may reflect the location of much larger collections of material, essentially protected, and hidden, by the build-up of the Flandrian deposits. Additionally, much of this evidence is likely to be missed wherever close interval fieldwalking has not taken place, so under-representing the number of potential sites in the landscape. When found, its interpretation as 'off-site' material may also under-represent the frequency of prehistoric living or working sites in the North West. Consequent rescue excavation of such surface sites is also likely to be neglected on account of a perceived lack of archaeological value. In order to counterbalance such potential bias inherent in the identification of sites from sparsely distributed surface material, individual lithics have been given equal weight in the depiction of the evidence on the archaeological distribution maps, and the interpretation of the evidence in the relevant chapters. In the interpretation of flint distribution as a whole some attempt has been made to discuss the variability of prehistoric subsistence activities across the survey area within a general chronological framework.

2

THE NORTH WEST WETLANDS SURVEY

The aims of the NWWS have been set out in a submission to English Heritage by Lancaster University Archaeological Unit (North West Wetlands Survey Project Design 1993) which also states the methods to be followed in each of the seven counties covered by the survey (Cumbria, Lancashire, Merseyside, Greater Manchester, Cheshire, Shropshire, and Staffordshire). These objectives can be summarised as follows:

i) To survey the extent and nature of settlement, from prehistory to the medieval period, within and on the margins of areas of lowland peat
ii) To place archaeological remains within an environmental context through palaeoecological studies
iii) To achieve an understanding of the development of the individual mires through palaeoecological analysis, and identify those mires with palaeoecological potential
iv) To identify gaps within the palaeoecological record
v) To identify the distribution of peat, its depth, and degree of preservation
vi) To assess the degree of preservation of archaeological remains and associated peat, and to determine the degree and timescale of any threat to the deposits
vii) To establish an archive at a scale of 1:10,000, which will contain the locations of sites and the extant areas of peat and associated features
viii) To make recommendations for the future management and conservation of the archaeological and palaeoecological resources

The Merseyside survey

Within North West England there are a number of mire types found in the lowlands including raised/intermediate mires, valley mires, floodplain mires, and basin mires. The definition of wetland in the context of the NWWS refers to areas of peat occurring within these ecosystems, all of which occur or occurred in Merseyside. The peat identified in the county is subjected to a variety of current landuses including farming, recreation, and commercial peat extraction. In addition small areas remain unreclaimed and are covered in semi-natural vegetation communities. Peats are found both as surface exposures (as in the case of inland raised/intermediate mires), or covered by later deposits (*ie* silts, clays, and blown sand), as in the case of some mires or portions of mires situated in coastal locations.

The identification of these wetland areas for the survey has relied on two sources. The starting point has been the distribution of surface peat shown on the 1:50,000 OS Soil Survey map, and former or existing peat deposits implied by place and field names. This has been augmented by archaeological field survey in these areas, which has plotted the peat in some detail from its occurrence in bare fields and ditch sections. In some instances, such as in Rainford, the field survey has identified surface peat that is not shown on the soil map. The second source is derived from borehole information, which can be used to identify peat that has been buried deeply by later deposits. This information has been kindly provided by private and local authority contractors, the Mersey Docks and Harbour Board, and through the personal research of Ray Kenna, copies of which are held in the Merseyside SMR.

The modern wetland landscape

There are three main areas of wetland in the county, which are shown in some detail on Figure 5. Two are found on the coast, forming a narrow strip in North Wirral and a similar but wider strip along the Sefton coast (which includes a belt in the Alt Valley with which it is almost contiguous). These areas of peat are largely covered by later silts, clays, and sand deriving from past changes in sea-level. The third area is the largest, stretching across the north-central and north-eastern part of the county, comprising extensive surface exposures of peats from raised and intermediate mires. These lie in the boroughs of Knowsley and St Helens, and for the purposes of this survey have been termed the central mosslands.

The few areas left today, where unreclaimed mossland is as yet undisturbed, have a distinctive vegetation of coarse grasses and sedges (particularly *Molinia caerulea* and *Eriophorum* spp), rushes, bracken, birch, and heather. These are mainly confined to small areas in the central mosslands, mainly in the higher parts of Rainford in the north-east of the county. Most of the central and Sefton coastal mosses have been reclaimed for agriculture or drained for game shoot, and over the last 30 years several of the central raised mire complexes have become subject to commercial peat cutting and subsequent landfill. Small outliers to the south of the main mire system have been largely subsumed within post-medieval and industrial development, centred on St Helens and adjoining towns such as Prescot. This has led to the peat here being heavily affected by buildings and waste tipping, although this has not precluded the retrieval of some useful palynological evidence from these former mires. The Wirral coastal mosslands are the most extensively developed in the county, with large areas of housing interspersed with small areas of market gardens, and around Newton, the main agricultural land in the coastal belt.

The North Wirral mosslands

The area of mossland in North Wirral comprises *c*1880ha in a strip *c*1km wide from the high tide line, with embayments at the north-east at Bidston, and at the south-west at Newton. Almost everywhere the peat lies below later deposits of silts, clays, and sands, often several metres in depth. This is a result of changes in sea-level since *c*6200 Cal BC, culminating in the modern, narrow (*c*300m wide), coastal dune belt which overlies peat almost the full length of the coast.

Much of the mossland zone is given over to landuse other than arable farming, although small areas of arable occur on the dryland soils flanking the wetland in the south-west, around the village of Newton. More typical of the North Wirral wetland zone are the large expanses of urban development at Hoylake, Leasowe, and the edge of Wallasey village, all of which have, at least partly, been built upon buried peat. The growth of urban Wirral and associated road developments has led to pressure on open areas of land for domestic and industrial refuse disposal. Parts of both Bidston Moss, particularly, and Newton Carr, have consequently been used for such purposes. Much of the rest of the North Wirral wetland area is a combination of market gardens flanking the dunes, and pasture land, with intervening areas given over to recreational use, including golf courses and school playing fields. Open areas of land along the coastal strip around Leasowe are being used increasingly for small-scale house building projects.

The Sefton mosslands

The area of peat in Sefton comprises *c*470ha, almost 90% of which lies beneath later marine, dune sand, and alluvial deposits. It is only found on the surface in small patches along the Alt valley, in a small area around Lady Green hamlet in Ince Blundell, and in a narrow channel north of Little Crosby. Along the width of the coastal plain to the west of Little Crosby it is covered by blown sand, although the peat outcrops on the beach at Hightown, which appears to mark its approximate northern limit. Inland of Hightown, however, the peat becomes impossible to trace towards the loop of the Alt as significant depths of marine clay are encountered.

The belt of dunes which overlies the western edge of the coastal peat, in a band *c*1km wide, is a zone designated a Site of Special Scientific Interest (SSSI) and is managed by the National Trust and Sefton Borough Council. Away from this coastal fringe the northern part of Sefton affords a flat, wide aspect as it stretches northwards across the county border into the large wetland complex centred on Martin Mere in West Lancashire. Here the Merseyside peat is capped by deep marine clays. The southern part of the Sefton wetland landscape has more variety. A few low, wooded eminences at *c*8-15m OD project into the peat and the other wetland environments. These have attracted small villages such as Thornton, Little Crosby, Ince Blundell, and Lydiate. To the east, the valley of the Alt also provides gentle slopes adjacent to wetland deposits. The peat areas provide mainly high-grade arable land with dispersed farms. This agricultural landscape is broken by the main urban development of Formby and, at the northern end of the coastal wetland zone, by the expansion of Southport. Both these centres, neither of which is in the present survey zone, have spread over peat buried beneath later wetland silts and clays.

The central and eastern mosslands

In the central and eastern part of Merseyside the peat occurs in the form of raised mires, covering an area of 1600ha. These mosslands lie at a higher altitude than those in the coastal belt, between *c*40-75m OD. The central mosses are the core of the Merseyside wetlands lying in the townships of Kirkby, Simonswood, Knowsley, and Rainford. The western part of this area has the typical wetland landscape, which in winter appears desolate and bleak, with low grey skies meeting flat, black, arable fields stretching for miles into West Lancashire. Only the outlines of occasional planted copses, and the flanking Carboniferous sandstone Billinge ridge, break the vista. Flocks of geese and other migrating birds are common in these skies at this time of the year, flying to and from the rich winter feeding grounds around Martin Mere and the Sefton coastal estuary areas. These wetland environments are amongst the most important in the country for wintering birds.

The core of this area, totalling *c*120ha, is being commercially mined for peat, but around this lies an agricultural landscape of dispersed farmsteads. In the eastern part of the central mosslands the landscape becomes more undulating and varied, as low boulder clay and Triassic sandstone hills and ridges break up individual mires. These hills mark the beginning of the long ridge, reaching *c*110m OD, which stretches from Billinge to Ormskirk, effectively bounding the Merseyside mossland system on the east. This, the most easterly part of the Merseyside wetlands, is still basically an agricultural landscape, although the historic village of Rainford has now expanded along one of the lower ridges to become a large town at the southern edge of the main mossland belt.

To the north-east, just over the border in Lancashire, the new town of Skelmersdale was built in the 1960s along the foot of the Billinge ridge and onto some of the skirting mossland, to house some of the population of Liverpool after inner city clearance. To the south, the proximity to the central mosslands of St Helens with its large glass industry has caused widespread disruption to many of the fields adjoining the mosses, particularly along the main routes through the area. Here the sand subsoil has been extracted, and the reconstitution of the fields has left not only an archaeologically sterile area, but fields with a distinctive hummocky topography, effectively resulting in an artificial landscape.

The southern mosses

The mosses in this group are separated from the central mosslands by the southern edge of the Triassic sandstone ridges in the eastern part of the county. They lie in areas where the boulder clay subsoil is found extensively on the surface, which also coincides with the most intensive historic settlement in the county. There are only two mosses included in this group: Prescot Moss and Parr Moss, Burtonwood, which are widely separated and have both suffered heavily from historic and industrial development.

Prescot Moss, formerly *c*50ha in area, is largely built over by the post-medieval spread of the town. A few areas are still, or have recently become, open ground which is given over to car parks and public gardens. In contrast, Parr Moss, formerly *c*65ha in area, still retains some surface exposures of peat. This lay on the edge of the post-medieval settlement of Sutton. Since the early twentieth century and the growth of adjacent coal mining, the moss has been used as a tip for colliery spoil. Other associated industrial processes have also been placed either on or adjacent to the moss, so that only 12ha of it is now available for study. On its western and northern sides the landscape is totally urban; to the east, lying in Cheshire, it retains a semi-natural vegetation cover, and on the south the nearest arable land, 800m away, is separated from the moss by industrial buildings.

Survey area

The Merseyside peat deposits are to a large extent a continuation of the West Lancashire system, and certain wetland areas cross administrative boundaries. A degree of rationalisation has therefore taken place in determining the respective responsibilities of the various survey teams regarding these border areas. Some small and self-contained mosses (*ie* White Moss and Holland Moss in south-east Lancashire), parts of which run over the border into north-east Merseyside (Fig 5), were included as part of the present survey. Conversely, the areas of peat along the coastal plain around Formby, which are an integral part of the large area of moss centred on Martin Mere in West Lancashire, will be included in the West Lancashire survey to be carried out by the Lancaster University Archaeological Unit.

A small area of dryland fringe in Merseyside, north-east of the Alt valley, around Lydiate, which abuts onto the southern edge of the large expanse of West Lancashire mossland, was surveyed during 1991, but the results will be included in the West Lancashire survey. The exposures of the 'peat and forest beds' along the intertidal areas of Wirral and Sefton have not been directly treated in this survey, although it is recognised that they are of great potential. The evidence gathered from them by others, however, has been used to set the surveyed areas of the coastal fringe into a wider context.

Archaeological survey techniques

The NWWS is an integrated and multidisciplinary project relying on the co-ordination of a number of institutions and specialists to locate, analyse, and interpret evidence of relevance for an understanding of the past importance of wetlands for human exploitation. The project design (1993) sets out the main methodologies followed by workers in each county, ensuring that standardisation of results and recording systems is achieved.

Fieldwalking

The main technique for the recovery of archaeological data has been fieldwalking of ploughed fields, although pasture fields were also walked to identify earthwork sites. In Merseyside this was undertaken by a field officer from Liverpool Museum (RC) during the winters of 1990-1 and 1991-2, mostly between late October and late March. An assistant was employed to help with finds processing, archiving, and fieldwork. In the first season *c*75% of the field survey completed was conducted by the two workers, while in the second season this was done solely by the field officer, except for a small area to the south of Holiday Moss which was fieldwalked by the assistant survey officer.

The individual mire has been the unit of survey, although as much as possible of the physical relationship between wetland and the surrounding dryland margins has been studied. The boundaries defining the limits of the survey of the dryland margins have been drawn subjectively. This has usually included the slopes of the nearest significant change in relief adjacent to the wetland. Often a variable area beyond this has also been surveyed, limited to some extent by land ownership and field conditions. This has been necessary to provide a wider context for activity associated with the wetland areas. It was judged that in the absence of the discovery of artefacts within deep peat deposits because of ground conditions, the potential importance of the wetland might be gauged from the degree of activity taking place in the immediate hinterland.

Recovery techniques and biases are absolutely crucial in determining how much of the prehistoric settlement pattern is located. Given that some types of site may be represented by as few as one or two lithic pieces, it is highly likely that many of these isolated occurrences will consistently go unobserved in this region. In order to retrieve as full a distribution of sites as possible, the sparse nature of the local prehistoric material dictated that the approach to the field survey was based on intensive field coverage. In the NWWS fieldwork, more emphasis has been placed on single artefacts or extremely sparse scatters of material than is usual for surveys such as this. As overall densities of lithic material are so low in the county, and the field coverage fairly detailed (*see below*), it is possible to record quite accurately the relationships between concentrations of material and the more diffusely distributed material. The survey distribution maps for prehistoric activity around the Merseyside wetlands accordingly reflect the particular nature of the North West evidence.

The method involved walking fields along transects, at the relatively time consuming interval of *c*12m. The nature of many early sites discovered prior to this survey in Merseyside would suggest that even this interval may on occasions be too wide. It does, however, have several advantages, apart from allowing a realistic timescale for completion of the work. The vast majority of cereal fields are sown leaving two *c*0.4m wide tractor tyre 'tramways' of bare soil, each pair being 12m apart. This both allows a consistently spaced transect interval to be maintained for finds plotting purposes, particularly in large and /or irregularly shaped fields, and provides reference points within the field if return visits are necessary. It also offers a slightly better scanning efficiency once a crop becomes established. Finds were individually recorded by counting paces along each transect, and marked onto a sketch plan of the field which was built up as the field was walked. These findspots were scaled off onto individual large-scale OS plans of the field in the processing stage. If a number of finds occurred in close proximity then they were recorded in 10m segments along the transects. All areas around isolated find spots were rewalked at 5m, 3m, or 1m intervals to test if the limited recovery pattern was due to the interval distance being too large, or some other factor. Locations where finds were more common were also rewalked at a reduced interval in order to identify the limits of the scatter, to clarify its relationship to topographical or soil features, and to collect enough struck flint to provide some indication of the date and potential function of the site. In all such cases in the survey areas, all material visible within the confines of the rewalking interval was collected, in order to have a large enough assemblage to be able to consider questions relating to date and function of the sites. All topographic features and soil boundaries were also plotted onto the field plan, regardless of whether there were associated finds.

The only variation from this method was on areas of featureless deep peat and alluvial or marine clays. Peat was walked at 30-50m intervals, depending on the depth of the deposit. This was to identify potential mineral soil areas within the peat (although these were often visible from the edge of the field, if present). Alluvium and marine clay was walked at variable intervals from 25-50m, depending on the depth of the deposit and local conditions. The absence of any return on many such fields walked on the northern edge of the Sefton peat areas determined the cut-off point for the survey in this area. Where the peat or clay was found to be less than *c*0.5m deep, particularly where areas of mineral soil were identified within these deposits, the standard 12m coverage, and less if necessary, was applied.

As such low densities of material are represented at many locations, the total number of 'sites' recovered from the NWWS will be depressed in relation to other areas of the country where artefact visibility is so much higher, and where even the lightest activity will leave more debris than similar activity in the North West. Though the overall number of sites may be depressed, the standardised techniques of recovery, and the intensive coverage of the peat margins, mean that the distribution of lithic material may be broadly representative of differences in the prehistoric settlement pattern. This distribution is probably affected to some extent, however, by the mining of areas of Shirdley Hill Sand around the central and eastern mires.

Monitoring of drainage ditches and peat cutting

Not all parts of the survey area were conducive to fieldwalking of arable surfaces, as a proportion of land is taken up by pasture, woodland, and peat cutting areas. In these environments, and also in the arable deep peat areas, monitoring of ditches provided a guide to the existence, depth, and condition of the peat. It also provided sections that could potentially reveal the mineral subsoil boundary under the peat, where in many cases archaeological evidence is most likely to be found. Unlike other areas of wetland in the country, for instance the Fens (Pryor and French 1985), there are hardly any large drainage dykes cutting through the Merseyside wetland areas. Most ditch sections in the county are narrow and, except for the cut area of Simonswood Moss, often less than 1.5m deep, and usually contain standing water during the winter. The ditch cleaning cycle is also quite long, so that in any one year only a very few ditches provide good uninterrupted sections through the peat. Many others are choked for most or all of their lengths by brambles and other weeds, making observation and recording extremely difficult.

All ditches within the wetland areas were monitored as part of the fieldwalking survey, to record the depth and condition of peat, subsurface features, and the identification of the junction with the mineral soil. Although many ditches were not conducive to rapid scanning, a number could be intermittently scanned to provide at least two points of reference to map these factors.

Peat extraction monitoring

Three main areas of peat cutting within the survey area were monitored to identify archaeological remains and suitable areas for palaeoecological study. The largest, over *c*120ha, is found on the central mosslands at Simonswood. Here access was granted for regular monitoring of the large-scale milling operations, on peat that is in general *c*2-4m deep. This area also provided the deepest and cleanest sections through the peat around the margins of the works, as the White Moss Peat Company, who are responsible for the site, regularly clean them out to a depth below the mineral soil horizon. The second, smaller area of peat cutting is undertaken by hand on a small area of peat (*c*8ha) in the extreme north-east of the survey area in White Moss in West Lancashire. The third area, on Holiday Moss (*c*30ha), has seen the almost complete removal of peat and is now used as a landfill site, although access was not granted to monitor the extent of peat survival within the works area.

Palaeoecological survey techniques

Palaeoecological studies were undertaken with the aim of identifying palaeoenvironments associated with the human activity represented by the archaeological evidence around the wetlands. The main technique used in the current survey was pollen analysis, through which broad changes in the composition of the regional vegetation can be recorded over time. It also allows identification of human activity associated with woodland clearance and farming, aiding interpretation of the archaeological evidence.

The second main technique used was that of gross stratigraphical survey. This allows the ontogeny of the mire itself to be studied, which is of prime importance in understanding the relationship through time between the human communities exploiting the wetland and the nature of the contemporary mire ecosystem.

The sites at which the NWWS conducted research are shown on Figure 35, together with the distribution of surface peat deposits and major previous research in the Merseyside area. These areas were chosen in the light of previous work within the region and the distribution of available peat deposits suitable for sampling. The field areas and palynological sites, and the criteria used in their choice, are listed below.

Sefton

This area was chosen for study as it offered opportunities for increasing our knowledge regarding the development of the coastline and the effects of marine transgression events on the environment. Particular attention was given to the response of man to the littoral environments and resource zones created by sea-level fluctuation. A number of sites in Little Crosby and Ince Blundell were considered, and Sniggery Wood (SD 307015) was chosen as the sample site. Subsequently, a second site was chosen a little further north at Flea Moss Wood (SD 311027), as a means of sampling a site closer to the deposits of marine clay that lie across this part of Little Crosby township. As the archaeological fieldwork of the NWWS progressed, the density of flint sites recovered from the Alt Valley made it advisable to sample a site in the valley. This was done in 1992 at SD 342031.

North Wirral

This area was selected to supplement the work of Ray Kenna, and to correlate results regarding coastal development with information from the Sefton area. Bidston Moss (SJ 285909) was chosen as the research site for this purpose, and to provide data on possible landuse on the adjacent upland areas of Wallasey and Bidston. Two more sites were chosen to complement the Bidston Moss site that might provide a more direct association between the coastal peat and marine clays. Commercial boreholes were sunk by Merseyside County Council as part of an evaluation of the wetland at Newton Carr, Hoylake (SJ 225885). The complexity of the stratigraphy in these samples made them suitable for further analysis and three of the cores were subsequently analysed (Newton Carr (NC) 7, 8, and 11). A final site at Park Road, Meols (SJ 239901) was sampled as it was the most westerly site identified as potentially productive for investigating the history of marine influence on the North Wirral.

Rainford

The area around Kirkby and Rainford has deep peat deposits which contain a long record of vegetation history for this important central area of Merseyside. The work of Baxter and Taylor (1981) has permitted an understanding of much of this environmental history. It would appear that deposits representative of the most recent periods of vegetational history have not thus far been sampled in that area, however, and further work was undertaken on an uncut area of Simonswood Moss (SD 442005) in an attempt to continue the environmental record into modern times.

St Helens

The area to the south of St Helens is poor in peat deposits and pollen work has not previously been carried out there. Two profiles were obtained from Parr Moss (SJ 548939 and SJ 542940) to provide a vegetational history for this area.

In addition to these sites, work was carried out at Prescot when organic deposits were revealed by building development in St Helens Road (SJ 468929).

Knowsley

A site at Knowsley Park (SJ 455961) was also selected, as it seemed that this area, because of its agricultural history, may provide well protected peat deposits. It also lay furthest to the south of any of the central mosslands, and so might be expected to provide palaeoecological evidence for the vegetation in an area of contrasting topography to the northern part of the central mosslands.

The palynological results presented for the Merseyside survey derive solely from the work undertaken between 1979-81 by J Innes and P Tomlinson on behalf of Liverpool Museum. The publication programme of the NWWS allows this evidence to be brought together, and in many cases published, for the first time. In addition, a new pollen core, financed by Liverpool Museum, was taken from the Alt valley in Sefton and analysed specifically for the NWWS.

Collection of material and preparation methods

Samples for detailed study were collected using a variety of tools. Aluminium alloy monolith tins were used for sampling exposed peat faces, and where this was not possible a Russian pattern peat corer was employed (Jowsey 1966). Sediment depths were measured from the levelled mire surface. Immediately after sampling, the cores or monoliths were normally wrapped in a protective 'sandwich' of plastic film and aluminium foil sealed in thick polyethylene. At Bidston Moss and Sniggery Wood, however, sub-samples were taken from the chamber of the corer in the field, and sealed in small glass tubes. All samples were subsequently stored in the dark at a constant temperature of 4°C prior to analysis.

Pollen preparation followed standard procedures (HCl, NaOH, followed by sieving, HF if necessary, and Erdtman's acetolysis; Faegri and Iversen 1989). Samples were mounted in silicone oil and examined microscopically with x400 magnification. Usually a total of at least 150 tree pollen grains were counted at each level, plus all other pollen and spore types identified on the way to that total. At some levels in the Bidston Moss and Sniggery Wood samples pollen was less abundant, and in these cases counting was terminated at 100 tree grains. In levels with low tree pollen content, a total of 500 land pollen grains was used instead. Due to these relatively low pollen counts, there are limitations to the interpretation of changes in the minor components, *ie* taxa whose pollen is poorly represented. Identifications were carried out using the keys of Faegri and Iversen (1989) and Moore and Webb (1978) plus a reference collection. The pollen grains of *Corylus* (hazel) and *Myrica* (bog myrtle) are difficult to separate from one another and so have been grouped as Coryloid throughout the volume. Cereal-type pollen was identified using the criteria of Andersen (1979).

The pollen sums for all the diagrams are the same. They are defined as percentages of trees and trees-plus group. The summary diagrams on the right hand side of each diagram are grouped as percentages of total pollen, *ie* combined trees, shrubs, and herbs. The pollen diagrams have been divided visually into local pollen assemblage zones and brief phases denoting possible periods of forest clearance and regeneration.

Analysis and storage of palynological and macrofossil data was accomplished using computer-based tools, particularly the TILIA software package (Grimm 1991). This was used to categorise data, aid its interpretation, and produce diagrams. As a result of limitations of TILIA-GRAPH it was not possible to include both local pollen zones and forest clearance phases on the pollen diagrams. Therefore at the majority of palaeoecological sites the zones have been drawn on the tree and shrub pollen diagrams and the phases on the herbaceous ones. Where no forest clearance phases have been identified local pollen zones appear on both diagrams.

Data were analysed to ascertain possible human influences on the natural environment, and to attempt to relate this to known and newly discovered archaeological remains where possible. As the major part of the palaeoecological work was completed prior to 1989, the nomenclature of plant names follows Clapham, Tutin, and Warburg (1962). A full list of botanical names occurring in the text, and their English equivalents, is given in Appendix 3.

Biostratigraphic survey

The full-time NWWS palaeoecological staff, based at the Lancaster University Archaeological Unit, limited their involvement in the Merseyside survey to investigation of the gross stratigraphy of the major mire complexes in the central mosslands in 1992. This involved sampling White Moss, Holland Moss, Simonswood Moss, and Knowsley Moss (Wells 1992), and was intended to complement the earlier pollen work of Innes and Tomlinson by identifying patterns of mire development across the county. The broad-scale biostratigraphical studies were undertaken with the following aims:

i) Elucidation of mire ontogeny
ii) Determination of the degree of preservation of the peat
iii) The provision of a contextual framework in which to place detailed (particularly palynological) studies

Fieldwork methods

A 30mm bore Eijkelkamp gouge auger was used to obtain cores of peat for rapid field description. A Zeiss optical level was used for determination of relative surface levels of core positions. The exact location, number, and core intervals were varied according to the different conditions and depositional histories encountered in the various types of mire bodies occurring in the county. The more homogeneous large-scale peat deposits contained in many of the extensive 'raised mire' deposits required relatively wide-spaced cores (*ie* >100m) whereas the more complex seral developments recorded in small systems necessitated close sampling (<30m) in order to achieve a coherent picture.

Analysis

Field descriptions of stratigraphic elements were complemented by frequent sampling from the cores collected. These were examined in the laboratory using a Wild M3 Stereozoom microscope to check and add detail to field determinations of material, and to establish the presence and frequency of any carbonised plant material. A final complete stratigraphic description from each core was synthesised from amendments and additions derived from laboratory examination of peat samples, and transcribed to computer database. This information was then used to draw a series of stratigraphic sections through each major mire system surveyed. The sum total of all the data was then used to interpret mire ontogeny.

Humification values of *in situ* peats were measured crudely in the field using the same modified Von Post determination method as employed by the Soil Survey (Burton and Hodgeson 1987).

Presentation of results

The results are presented as a series of stratigraphic diagrams. Each diagram displays stratigraphic information from an individual transect. Composition of the gross components of the peat and organic deposits are indicated using symbols based on Troels-Smith (1955) with some modifications (*see keys for details*).

The deposit elements recorded are based on a combination of field observation and rapid Stereozoom scans of sieved material from samples of *c*25cm^3 of peat from selected segments of cores. The presence of charcoal was usually detectable only by the microscope examination except when present in exceptionally large quantities. In the charcoal diagrams the density of shading represents the relative abundance of carbonised particles, based on a semi-quantitative scale identical to that used for recording macrofossil remains described earlier.

Core position in the diagrams is drawn to various scales chosen to enhance clarity of presentation. The physical size of the cores displayed in the diagrams is exaggerated in order to enable clear presentation of the results. Heights relative to Ordnance Datum of the surface of the core positions are displayed where known, otherwise relative heights of cores are shown.

Radiocarbon dates

If human activity in and around the wetlands is to be placed in its proper landscape context then it is vital that the dates, and rates, of environmental change taking place within the mires and in the surrounding vegetation are understood. Radiocarbon dating of organic remains is the essential technique for achieving this. Dating within the Merseyside survey has been confined to the biostratigraphic boundaries within the peat profiles, with particular emphasis on those that may reflect human-induced changes to the vegetation. These have largely been provided independently, through Liverpool Museum and NERC funding, between 1979 and 1987 from Birmingham University, for a number of pollen sites in the county. In all, 31 dates have been acquired from these sources for the sites presented in this report. An additional series of seven dates has been provided, under the auspices of the NWWS, to cover the new pollen site in the Alt Valley, and to extend the sequence already acquired for the regionally important pollen diagram from Knowsley Park Moss *(see Appendix 2)*. All NWWS radiocarbon dating was carried out by the Scottish Universities' Research and Reactor Centre at East Kilbride.

Calibration of radiocarbon determinations

Where possible, radiocarbon dates have been calibrated to calendar years using the conventions established at the 1986 Trondheim Radiocarbon Conference (Switsur 1986; Pearson 1987). This uses a calibration curve based on dendrochronology, and all dates thus expressed in this volume have been calibrated using the computer program CALIB (Stuiver and Reimer 1986; University of Washington Quaternary Isotope Lab 1987), and employing a curve based on a 20-year atmospheric record dating to 8100 radiocarbon years BP (Stuiver and Pearson 1986) (*see note in Appendix 2*). Calibrations in the text were established prior to Stuiver and Reimer's 1993 revision.

Calibrated radiocarbon dates are presented at a 2 sigma (96.4% confidence) level. Such dates are denoted by the suffix 'Cal BC' or the prefix 'Cal AD'. Beyond the limits of accurate calibration (currently 7210 BC) dates are presented as raw radiocarbon determinations.

Aerial photography

The use of aerial photography was also applied to the survey, with the primary aim of identifying subsurface archaeological features. A second important result of this technique is the mapping of landscape features such as ancient buried watercourses, which will help to place sites in the context of the changing physical landscape. An aerial photographic consultant, Chris Cox, was commissioned through the NWWS to undertake this work. This entailed a systematic assessment of the existing aerial photographic coverage of the county, and the plotting of relevant data onto a set of 1:10,000 scale archive maps. No new aerial reconnaissance took place in the county to fill identifiable gaps in coverage. In Merseyside the main sources of aerial photographic data come from photographs taken over approximately 15 years. A proportion of these have been taken for specific archaeological purposes, including those held at Liverpool Museum (Philpott 1994), and those held by the RCHME at the National Library of Air Photographs. Good quality False Colour Infra Red vertical coverage, obtained by Merseyside County Council in 1975, was also studied.

Although no good examples of potential early sites were identified in the vicinity of the wetlands, numerous pre-1849 features were located (Cox 1992). The Merseyside aerial photographic programme only started in the second season of fieldwork, and accordingly it was not possible to check all potential sites on the ground, due to time factors or lack of suitable field conditions.

Evidence was also mapped for medieval field systems and early watercourses. No pattern comparable to that of the dense configuration of old watercourses or roddons found in the Fens (*eg* Hall 1987) was located from this source in the Merseyside survey, but stratigraphic and aerial photographic evidence has been used to postulate such a pattern in the estuarine areas of the River Alt in Sefton (Huddart 1992).

Documentary research

Documentary research has been used in the survey for two purposes. One was to identify former distributions of peat from historic maps and field and place name evidence, as a preliminary to the fieldwork programme. The second, which was not intended to form a significant part of the NWWS, was to identify recent processes that may have led to changes in the morphology or stratigraphy of the peat deposits, as an aid to the development of management strategies. A degree of documentary evidence is also necessary if the relationship between the mosslands and the organisation of the different elements of the historic township is to be understood. In Merseyside, much documentary work was done in several of the townships included in the NWWS area as part of the work in setting up the county SMR (Chitty 1978, 1981; Cowell 1981, 1982; Lewis 1982). Where possible, this information has been used to gauge the importance of individual mires within the wider context of the historic landscape. This has allowed a particularly detailed description of the central mosslands in the medieval and post-medieval periods. In other townships, such as in Wirral and Sefton, where historical documentation was not readily available, the treatment of the relationship between the mire and the rest of the township is not as thorough.

The archive

A major aim of the project is to build up an archive of archaeological and palaeoecological information that can be used for all future decisions in relation to research and management in the wetlands of North West England. The records in the archive are standardised for all the counties covered by the NWWS, and are to be held centrally at the Lancaster University Archaeological Unit. Copies of the relevant areas will be lodged with individual county SMRs. The archive consists of a series of 1:10,000 scale OS map transparent film overlays, and written individual field and site records. The map overlays cover four categories comprising land ownership, survey coverage, peat distribution and depth, and archaeological sites and findspots. The conventions used for the latter three categories are reproduced in the pull-out 1:25,000 scale maps accompanying Chapters 3 to 6 in this volume. The written records are held on pro formas, and include details relating to peat condition and perceived threats, as well as the nature of individual sites.

3

THE NORTH WIRRAL MOSSLANDS

The Wirral is a low, narrow peninsula, a little under 15km wide and *c*19km long, joined to the Cheshire Plain to the south. It is defined by the large glacial valleys of the Dee and the Mersey with these two major estuaries discharging into the Irish Sea. To the west lies the upland coast of North Wales, and to the east Liverpool and the flat landscape of the Lancashire plain north of the built-up area. The survey zone lies along the narrow northern coastal strip of the Wirral and includes *c*1880ha of peat with associated silts, clays, and sands deriving from successive changes in sea-level (Fig 6).

The main peat lies in a large embayment on the north-eastern side of the Wirral, at Bidston Moss. The large urban areas of Wallasey to the north-east, and Bidston and Birkenhead to the south, have grown up on the surrounding sandstone ridges from historic village cores. This area includes *c*150ha of peat, although it is difficult to separate it from the peats of the coastal plain to the north and that of the former tidal inlet of Wallasey Pool to the east.

Running westwards from Bidston is a coastal strip approximately 1-2km wide. At each end of the plain lie modern built-up areas: Leasowe to the east and Meols to the west. The latter forms the eastern edge of a large urban area on the north-eastern corner of the Wirral, centred on Hoylake and West Kirby.

To the south and east of this conurbation lies the second former wetland embayment of Newton Carr, lying at the foot of a low ridge on which lies the historic hamlet of Grange. The area of wetland in this embayment amounts to *c*100ha.

Landuse

In the historic period this wetland zone was given over largely to marginal land, with nucleated settlements, such as West Kirby, Moreton, or Bidston, set a little off the wetlands on the elevated sandstone areas to the south. The main landuse for the wetland areas was probably reserved for grazing, with little direct impact from settlement or enclosed farming. Many of these areas remained as common land throughout the post-medieval period.

It was not until the late nineteenth century, when new areas of housing grew up across North Wirral, and the development of the Birkenhead and Seacombe docks, that the first real impact was felt on the mosslands. One key development was the construction, in 1863, of the railway line to the north of the Birket Drain on Bidston Moss. This linked the dock areas of Seacombe with the north-west corner of the Wirral and saw the growth of Hoylake and Meols around the coastal wetlands there. The railway also served to attract industry to the area of Bidston Moss adjacent to the new docks, with an extension to the passenger line to New Brighton and a goods station and engine sheds located on the moss by 1911. Despite this development much of the two mossland embayments remained as farmland, and much of the coastal belt still retained its open aspect centred on the dunes and areas of common land behind them.

The surface of Bidston Moss started to be extensively developed in the period beginning *c*1935. Industrial buildings, including a steelworks, arose along its southern edge, while the major impact came with the housing, landfill, transport, and light industrial developments from the 1960s to the present day. More recently, waste management developments designed to serve the growing population of the area are taking place on remaining areas of wetland at Bidston Moss, which is now extensively developed over most of its area. The Newton Carr embayment has retained its agricultural character better, but it also has suffered from housing expansion and waste tips on its western margins.

Over the last 40 years much of the coastal belt between the two embayments has also seen development, mainly involved with the expansion of

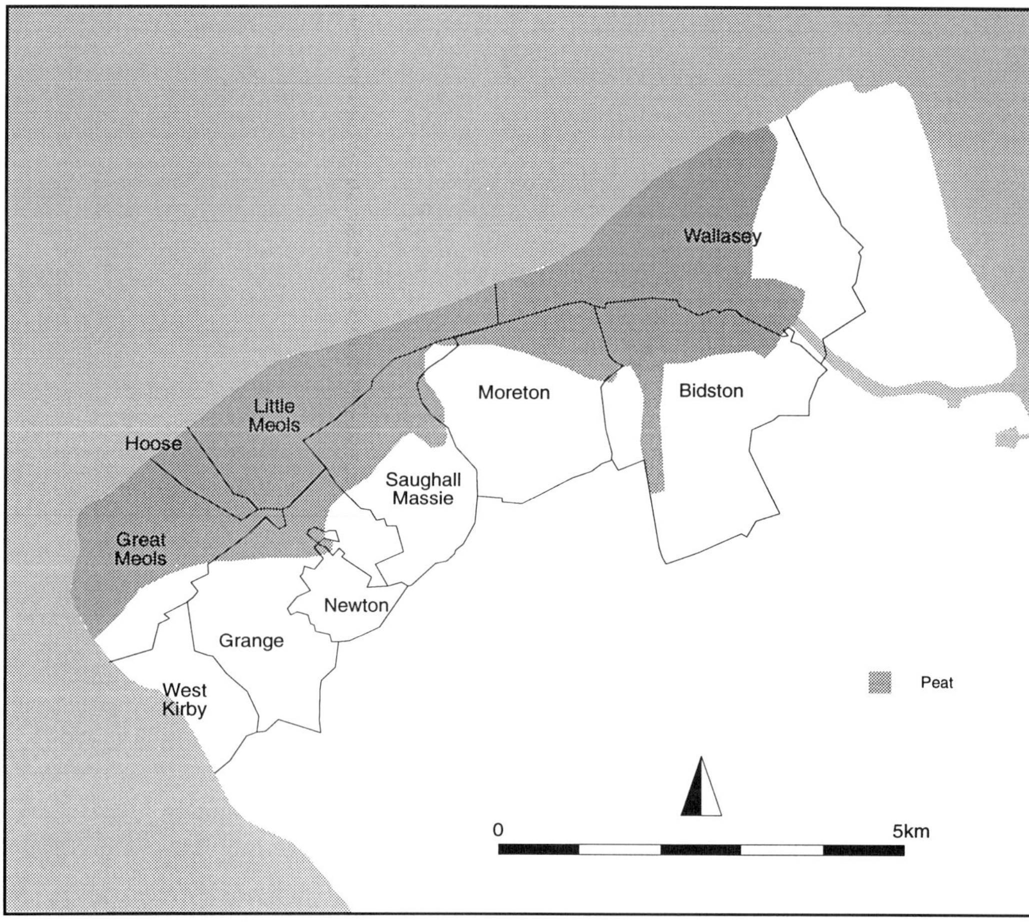

Fig 6 Wirral: Location of survey area

housing estates at Leasowe, Meols, and Hoylake. Intervening areas have remained as open land but much of this is given over to recreation, either for golf courses along the dune belt or grassed over playing fields landward of this belt. A few areas are farmland, although with the exception of a small area east of Leasowe which is dominated by market gardens, most of the rest is pasture land with horse grazing being particularly strong. The only traditional mixed arable farming is found in the north-west corner around Newton Carr, although even here cattle grazing is the dominant economy.

Previous research

Some of the archaeological evidence for the early societies living on the Wirral is likely to be buried within the deep wetland deposits. The waterlogged nature of some of these sediments creates the potential for the preservation of organic remains and for precision in dating of these buried archaeological sites by radiocarbon dating. During the late nineteenth century some of the most important archaeological finds in North West England were discovered in North Wirral, when antiquaries such as Abraham Hume, Charles Potter, and Henry Ecroyd Smith were active along that part of the Wirral which they described as the 'Cheshire Shore'. In 1974 this became part of the Metropolitan county of Merseyside. The material they collected spanned many phases of human occupation of the peninsula, and includes material from settlements of prehistoric hunters and early farmers on rocky outcrops at New Brighton, Hoylake, and the adjacent islands, dating as far back as 7000 BC, as well as more recent medieval metalwork and pottery.

Many of these finds are divorced from their associated deposits and evidence concerning their original environmental setting and date is generally vague. For example, an account of a wooden implement, recovered from a depth of 4m in the peat at Bidston Moss (Ricketts 1866), possibly refers to a bow or spear or digging stick of very early date, although there is now no trace of the object. Some of the other artefacts from the area, such as flint arrowheads, Iron Age coins, Saxon and Viking jewellery, stone tools, and a variety of medieval metalwork, have survived in museum collections, but they provide little evidence concerning the nature of the sites from which they were collected. Some of these artefacts may have come from peat deposits, but the recording is often too imprecise for any firm conclusions to be drawn.

It is a reflection on the nature of subsequent archaeological research in the area, that until the 1980s these finds made North Wirral the only area with significant evidence for early settlement in the county, and indeed, it was one of the more important areas across the whole of the lowlands of North West England. Only recently, with the advent of more systematic survey, has the archaeological record been extended across the county. However, these early finds did not provide the only antiquarian evidence for an understanding of the early settlement and landuse of the peninsula. The coastal deposits were also important for the palaeoenvironmental evidence they contain and also received Victorian investigation.

Many of the archaeological discoveries were made during a phase of coastal erosion in the later nineteenth century. Most notably this revealed the 'submerged forest' which lay over a large part of the foreshore near Meols, representing remnants of an ancient forest that had lain buried by sand and mud for over 5000 years. Attention was also drawn to the thick beds of peat that ran out in both directions from the shore in the north-east around Leasowe (Hume 1863). Although the idea of environmental archaeology did not then exist, many early studies of these sediments and their associated silts, clays and sands were made by geologists and civil engineers (de Rance 1870; Reade 1872a).

Recent work by Kenna, undertaken during a phase of sea wall construction, has been instrumental in providing a base for understanding the age, distribution and character of these deposits (Kenna 1978, 1979, 1986). His work now acts as a guide to the changing pattern of the natural environment, and is vital in any attempt to understand the character of the early occupation of the Wirral.

Geology and topography

The main mossland areas studied in North Wirral lie adjacent to a variety of dryland environments. When taken in conjunction with the effects of fluctuating sea-levels, it can be appreciated that North Wirral would have afforded a continually changing mosaic of environments during the prehistoric period. The stable features in the landscape would have been formed by the ridges of Triassic sandstone, and, to a lesser extent, boulder clay, which flank the wetland (Fig 69). To the north-east of the mosslands, on the ridge where Wallasey is located, the sandstone rises quite steeply to *c*33m OD. To the south-west, forming an embayment around Bidston Moss, lie the slopes of the Bidston ridge, rising to about the same height, which have also attracted much modern settlement.

The second main area of wetland involved in the survey, Newton Carr, occupies a similar inlet at the north-western edge of the coastal plain. The southern margins of the Carr terminate in a low, boulder clay ridge, at a height of *c*22m OD. The south-western edge of the basin meets the northern edge of a high (*c*65m OD), steep, sandstone ridge which follows the bank of the river Dee as far north as West Kirby. The slopes of this ridge would have provided similar soils and vegetation to the high ground at Bidston, but it has likewise suffered heavy modern development which makes archaeological discoveries difficult.

The present coastal dryland fringe lying between these two mosses is formed by a gently rising boulder clay plain parallel to the coast and terminating in the ridge at Newton. This formed the southward limit of marine influence along the coastal plain (Fig 71). A belt of land approximately 1km wide, between this dryland edge and the coastal dune fringe, is formed from the build-up of successive layers and land surfaces, which have accumulated over a period of *c*2000 years during the Flandrian, on a now buried boulder clay plain. These deposits have produced a surface at an altitude of between *c*5-6m OD, with a sand dune belt, effectively marked by the 8m contour, defining its seaward edge.

Flandrian geology

The coastal sediments of North Wirral contain evidence for environmental change resulting from fluctuations in sea-level and ground water during the last *c*9000 years of the Flandrian. It is clear from the archaeological evidence found over many years along the coast that these spatially and temporally changing environments played an important part in the nature and development of early human activity in the area. The deposits reflect the former existence of diverse coastal environments including sand dunes, intertidal sand and mudflats, saltmarshes, marine and freshwater lagoons, swamps, meres, and fens which were widespread in areas that are now covered by recent landuse and urban developments.

Kenna' s (1978, 1986) work provides the main evidence for the palaeogeography of the coastal plain. This shows that the coastal strip between the two main mossland embayments, approximately from Meols to Leasowe (Fig 71), has a very subtle, undulating boulder clay subsurface topography of low gradients. This would have formed the earliest prehistoric land surface prior to the onset of wet conditions after *c*6000 Cal BC (Table 1). In the central part of the plain the subsurface lies generally below *c*0.25m OD with small eminences, such as around Leasowe Castle, reaching *c*2m OD. This subsurface continues in a moderately undulating form out to sea beyond Dove Point (Kenna 1978, 28). To the north-east there is little detail available to identify the pre-Flandrian subsurface topography of the Bidston basin, although three major buried channels have been identified flowing into it (*see below*).

In the north-west of the peninsula the subsurface slopes more dramatically to form an extensive depression, occupying most of the area between Hoylake and West Kirby, including Newton Carr; what has been termed the 'western depression' (Kenna 1986). The sub-surface topography of this depression is not well understood at the present, particularly in the north, but it appears to slope quite sharply to the west and south-west, lying at *c*-6m OD in the centre, perhaps forming a shelving ledge out to sea. The palaeogeography of the southern part of the Carr, to the south and east of the present line of the River Birket (Fig 69), is known in more detail from the work of D Bedlington of Durham University (Innes *et al* 1990; D Bedlington pers comm). This shows that the hillslope of the boulder clay ridge on the south forms two main steps within the Carr. The first runs along the line of the Newton Brook and a little to its north-west. Between this break of slope and the River Birket to the north lay an area of now buried, flatter gradient, although with slight ridges that would have formed gentle promontories within the depression.

This subsurface palaeogeography of the North Wirral coastal area was subsequently masked by a series of wetland deposits deriving from sea-level changes and associated water-table movements. These produced a succession of different land surfaces throughout the prehistoric period which would have played important roles in determining the location and nature of settlement and landuse in this zone. The sequence of Flandrian deposits overlying the earlier land surface has been outlined by Kenna (1978, 1979, 1986), who divides it into seven units below the dune sand, from observations mainly along the central part of the coastal plain. Palynological and stratigraphical work in the main mossland embayments, reported in detail in this volume, provides additional information to extend the sequence more widely across North Wirral (summarised in Table 1).

The earliest deposit is represented by a thin band of laminated peats containing stumps of oak and birch, known as the Lower Peat/Forest Bed (LPFB) (Reade 1872a; Kenna 1978) which have been dated to 5490-5245 Cal BC (6420±60 BP; SRR-1494) and 5439-5194 Cal BC (6460±40 BP; SRR-1496) (Kenna 1986). These represent conditions during the late Mesolithic period, when deciduous woodland was more extensive and lay beyond the present low water mark.

Above this layer, and thickening towards the south-west from the vicinity of Leasowe lighthouse, is a deposit, over 1m thick, of grey and blue clays and silts (the Leasowe Marine Beds). This probably represents, at least partly, upper tidal flat conditions when the prehistoric coastline lay a little further inland than today. The relationship of the deposits associated with sea-level changes on the Wirral coastline to the sequence of deposits in South West Lancashire (*see Chapter 4*) can be correlated to some extent. Tooley (1976) suggests that a relative rise of sea-level on Helsby marsh in South Wirral can be identified in late Flandrian II (dated between 4711-3990 Cal BC (5470±155 BP; Hv-2686), with a sea-level fall at *c*4937-3100 Cal BC (5250±185 BP; Hv-2685). This is probably the same phase that affected the North Wirral coast, as seen at Park Road, Meols, where the end of a marine phase is dated 4234-3980 Cal BC (5250±50 BP; SRR-2694) (Table 5).

A second peat and *in situ* tree root layer, known as the Upper Peat/Forest Bed (UPFB) (Kenna 1978, 27), succeeds the Leasowe Bed marine clays along the coastal section. A black peat shows the transition from saltmarsh to fen carr woodland (identified from pollen evidence) during the mid to late Flandrian II (before *c*5000-4000 Cal BC). The work of the NWWS at Park Road, Meols shows a marine phase in the early fourth millennium Cal BC (Table 5, PRMb). This may equate with a short period noted by Kenna (1986, 13) when saltmarsh lay close to the present central part of the coast during the accumulation of the Upper Peat/Forest Bed, dated by him on pollen to the same period.

In Flandrian III, after *c*3700 Cal BC, conditions dried out in these coastal areas, matching the retreat of marine conditions noted generally in North West England (Tooley 1978a, 35), so that by *c*2560-2295 Cal BC (3910±40 BP; SRR-1493) oak, pine, and birch woodland was growing on the earlier peat sediment in the central part of the coast. The Park Road,

Table 1 Wirral: Sea-level changes summary

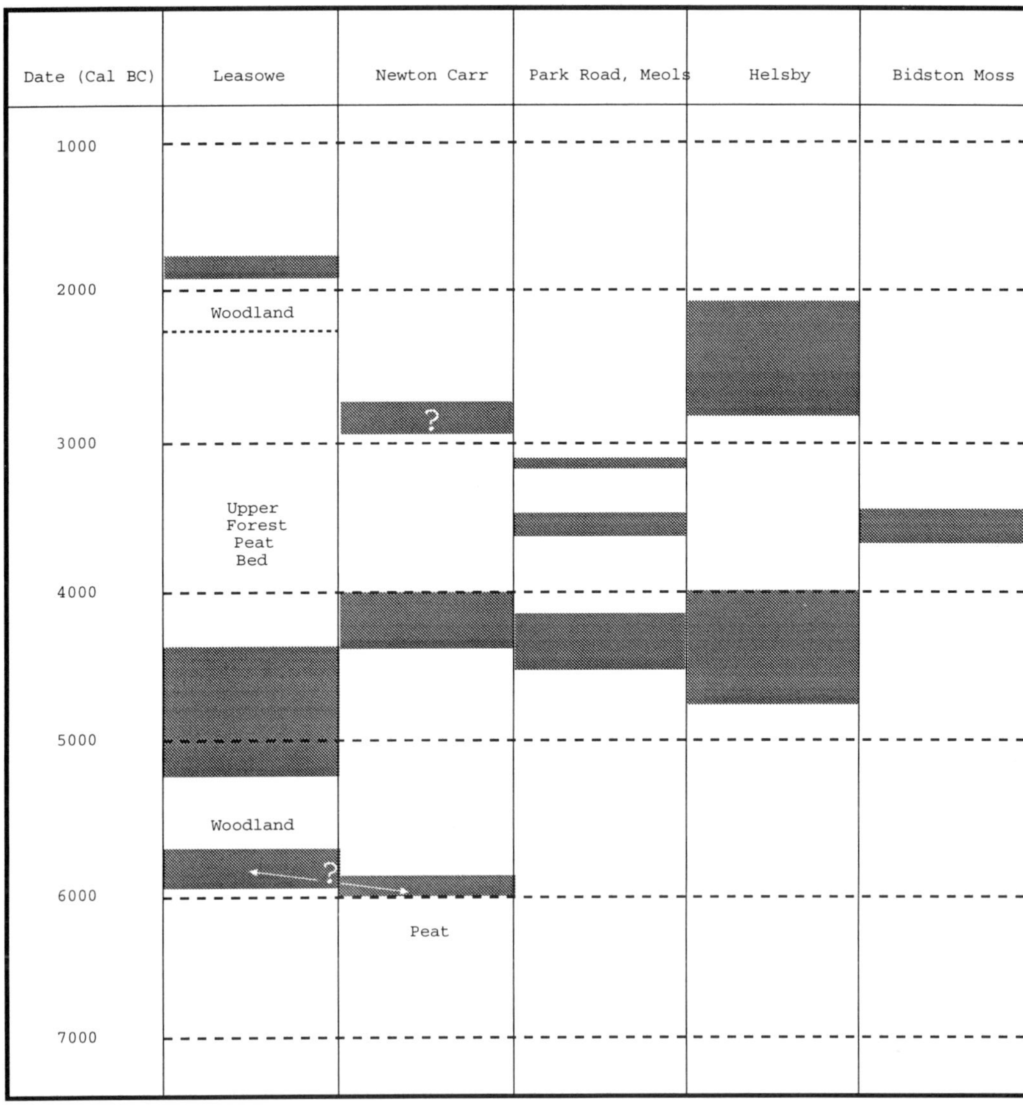

Meols site shows that even in these drier conditions, marine influence was still felt locally with a third marine incursion around the beginning of the third millennium Cal BC (Table 5). The more general wooded conditions lasted until at least *c*2460-1774 Cal BC (3695±110 BP; Q-620) in North Wirral (Kenna 1986, 14). This Upper Peat/Forest Bed, which is not necessarily continuous across North Wirral, has been traced inland to within approximately 1km of Bidston Moss, where it is found at thicknesses of over 1m overlying the silty clay fills of the early Birket and Fender channels (Kenna 1986, fig 6b). A fine sand, with marine-estuarine shells, which postdates 2012-1680 Cal BC (3490±60 BP; GU-1271), is found in a few localities overlying the Upper Peat/Forest Bed (Kenna 1978, 30; 1986).

Bidston Moss seems to have been directly inundated by at least one marine incursion, as a surface alluvial clay overlying a woody peat with tree trunks and stumps is recorded by Smith (1849), who attributed it to estuarine conditions. This was confirmed by Davies and McMillan (1956), who recorded the part of the moss now covered with tipped refuse as having a thin layer of marine *Scrobicularia* clay, which postdates 3690-3360 Cal BC (4740±70 BP; SRR-2924), overlying the peat. Despite a large discrepancy in date, this may equate with Tooley's second marine incursion on the Wirral, dated after 2876-2143 Cal BC (3965±110 BP; Q-620) at Reeds Lane, Leasowe (Godwin and Willis 1962), as the surface peat at Bidston appeared to have been eroded before the deposition of the clay, allowing for the possibility of an older date for the peat/clay contact than originally existed.

At Newton Carr, further evidence is available from commercial boreholes and work undertaken by

Durham University (*see below*). This shows that there are three distinct peat horizons, representing three periods of terrestrial and freshwater depositional environments (Fig 11; Table 6) which correlate with Kenna's sections. The lowest layer of peat, the upper contact of which is found at a depth of *c*-5m OD, is only found at the western end of the site, which the pollen assemblage suggests is likely to be Flandrian I in date (*ie* before *c* 6200-5700 Cal BC), and thus represents the first evidence for marine incursion of the Wirral coast.

This peat layer at Newton Carr is overlain across the basin by a cohesive blue-grey silt, representing an estuarine or marine phase, which spread as far as the present foot of the dryland slopes. This marine phase does not seem to be represented further to the north-east in the central coastal plain where, allowing for the imprecision in dating, woodland or reedswamp is likely to have been prevalent, although the date brackets are rather too wide to be of great value. This marine-estuarine deposition in the Carr gives way to a second peat layer, indicating the development of reedswamp and fen carr across the former marine clay surface, and is Flandrian II in date (Table 6). At this time most of the coastal plain, including the now submerged parts, would have been well studded with pine and birch woodland as represented by the Lower Peat/ Forest Bed.

The peat layer is overlain by clayey silts or silty sands which represent a second phase of direct marine-estuarine conditions in the embayment, beginning sometime before the start of Flandrian III (beginning of the fourth millennium Cal BC). This phase is present in the coastal plain also, as wet, intertidal environments became widespread again. A better idea may be gained from the Park Road, Meols site, where the same marine phase is probably represented terminating 4234-3980 Cal BC. In the centre of Newton Carr a third peat layer, averaging *c*1m in depth, is found, although to the north the marine deposits continue to the surface, suggesting that this part of the site was never far from direct marine influence throughout the Neolithic and early Bronze Age. This upper peat at Newton Carr probably equates to the Upper Peat/ Forest Bed recognised on the coastal plain and should therefore date to around the mid fourth millennium Cal BC. At Newton Carr, the landscape towards the beginning of the Bronze Age would therefore have consisted of a fen carr woodland skirting an open marine-estuarine environment.

The palaeoecological evidence from the main mosslands associated with the NWWS does not continue beyond the early Bronze Age. At Newton Carr a capping of alluvial silty clay of unknown date is found above the upper peat (Fig 11), while at Bidston the marine phase of the late Neolithic completes the sequence (Table 2). On the coastal plain there is more evidence, but it is still incomplete after *c*2012-1680 Cal BC (3490±60; GU-1271) by which time the coastal woodland had been replaced by fen carr west of Leasowe (Kenna 1986, 14).

The upper layers of the coastal sequence may relate to more recent episodes and are mainly represented by what is known as the 'Soil Bed' (Hume 1863), which consists of peaty sand and clayey peat. There appear to be local phases of erosion which may have removed the original boundary between this layer and those beneath it, but the upper part of this bed formed under dune slack conditions. Dates for this have been produced of Cal AD 1010-1220 (925±50 BP; GU-1311) to Cal AD 1304-1438 (540±40 BP; GU-1403) (Kenna 1986), and may equate with late medieval dune formation on the Lancashire coast (Tooley 1978a). A second phase of dune slack and sand dune building conditions then succeeds the Soil Bed phase under the present day dune system.

Natural drainage patterns

The pollen data from the basal peats in Newton Carr show that the northern part of the moss lies in a deep marine channel present from *c*5900 Cal BC onwards, running north-westwards towards Hilbre Island (D Bedlington pers comm), with shallower slopes lying to the north. This channel would have been present through much of the prehistoric period despite the minor fluctuations in sea-level which may be recognised elsewhere at Newton Carr. This would have produced surrounding areas of fen carr woodland or saltmarsh/estuarine conditions at different times, depending on the height of sea-level. In the prehistoric period these areas would have received drainage from the southern hillslopes in a north-westerly direction, in contrast to the present day artificial south-west to north-east drainage pattern.

To the north-east of Newton Carr, the earlier drainage pattern has also been reconstructed from borehole logs, geophysical techniques, and observations (Kenna 1978, 1979, 1986). Several observation points along the present course of the River Fender, to the south of Bidston Moss, show that it follows the earlier course of the river (Fig 72). This has a steeply profiled channel, of probable late Devensian age (*c*13,000-8000 BC), whose base lies *c*9m below present ground level close to the south-western edge of the moss. From this point the channel splits, running northwards and deepening steeply to *c*15m beneath modern ground level to meet the present coast to the east of Leasowe Castle. A second channel runs north-westwards towards Leasowe lighthouse. The low base of the peat reflects a period when sea-levels in the Irish Sea basin were as low as -20m OD, during the eighth millennium BC (Tooley 1978a).

The other main drainage of the wetland zone is formed by the River Birket. Its early channel had an easterly arm which flowed into the tidal Wallasey Pool on the south-western edge of the Wallasey ridge, approximately where the Birkenhead docks are today, before joining the Mersey. The Pool has a shallower gradient and it is suggested that this may have been cut once the northerly arm of the Birket, which flowed to the coast west of New Brighton, had been blocked by dune formation, sometime after *c*2000-2500 BC (Kenna 1986). The modern drainage of the area now differs from that of the prehistoric period in some important respects. In 1665 the Kingston estate map (Plate 2) shows the Pool with a dendritic pattern of small tributaries flowing across the moss, with the Birket joining it at its northern end. This river then formed the main drainage of the coastal area as it ran parallel to the coast joining the other main river, the Fender, west of Bidston Moss. The main difference is that the present profiles lie several metres above what would have been the earlier banks of these rivers and their alignments have also subtly changed, particularly the Birket, which has either been canalised or responded naturally to higher ground levels. One major diversion occurred *c*1849, to facilitate the proposed docks construction along the line of the Pool, when the Birket/Fender was diverted to the south with the excavation of the Great Drain through the centre of the moss (Smith 1849).

Survey techniques

The two main areas of mossland studied within North Wirral lie at Bidston Moss and Newton Carr, Hoylake, on the landward limits of the belt of wetland deposits along the coast (Fig 70). Both occupy large drainage channels or basins which have acted as sheltered inlets to the sea. The natural development and attractiveness for human exploitation of these embayments is linked to the environmental changes wrought by changes in sea-level. In order to study the temporal and spatial relationships for settlement and landuse around the wetland embayments, the survey area therefore also includes the wetland deposits of the coastal zone. The conditions along much of North Wirral are not conducive to systematic archaeological survey. Most of the coastal strip is dominated by housing, golf courses, market gardens, and sand dunes which prevented any extensive coring, excavation, or consistent survey work from being carried out for the NWWS. The archaeology of this area therefore has to be based on the work of the early antiquaries and a few more recent chance finds, in attempts to explain the human activities and exploitation of the mosses proper. Field survey was restricted to a few areas around Newton Carr, where agriculture is the dominant industry, and around Bidston Moss on some of the areas of public land.

The deep wetland deposits, heavy clay soils, landuse, and the built-up nature of the terrain of North Wirral have also made archaeological discovery around the peat margins through aerial photography very difficult. Since 1987 this area has been flown by R Philpott as part of a Liverpool Museum/RCHME programme of aerial survey in Merseyside and the North West (Philpott 1994), but it has yet to provide archaeological evidence from around these mossland areas.

The mosses

Bidston Moss (SJ 285915)

Bidston Moss lies at the head of the former large tidal inlet of Wallasey Pool, now occupied by the Birkenhead Docks complex. Although previously a large area of raised bog and marsh (Smith 1849), the moss has been greatly disturbed in the recent past and very little of the original area of wetland now survives (Fig 71). A small area of 3ha in the centre of the moss has been turned into a wetland nature reserve, while a large proportion of the central part of the moss has for some years been covered by local authority domestic refuse. Other parts of the moss are covered by factories, motorways, and railway lines. Of the undeveloped area, some is given over to pasture or uncultivated ground, while a golf course covers the northern part of the old peatland area.

The nature of this landscape militates against archaeological survey, as nowhere are there ploughed surfaces or ditches that might provide evidence of the extent and nature of the surviving peat and associated archaeology. Ground survey has consequently been difficult, as most of the area is inaccessible and unsuitable for field survey of any practical value (Fig 70). It was possible to survey only 50ha of the area, all of this consisting of fields given over to pasture or left waste, with the occasional area of parkland. The margins of the main golf course around the moss were also walked. The market gardens in the area were not walked because of potential access problems.

It has also been difficult to trace the extent of the moss from the surface, although in some fields wetland type plant cover was used to suggest likely areas of wetland. Documentary sources, borehole records, and various observations, made over a number of years by various workers when the surface has been disturbed, have largely been used to identify the extent of the peat (Kenna's notes, Merseyside SMR; R Kenna pers comm). The continuing survival and present condition of the peat is less easy to verify.

The south-eastern edge of the survey zone is marked by the Poulton Road bridge (SJ 297908), but Mersey Docks and Harbour Board boreholes confirm that in the 1970s the peat extended for at least another kilometre to the south-east (Fig 71), along the line of the former Wallasey Pool (*see below*). The north-eastern boundary of the wetland, and coincidentally the high water mark of the Pool in this area is approximately marked, as far as Bidston Dock, by the main road alongside the north-eastern edge of the docklands. From here the moss runs in a more northerly direction, although its full extent ceases to be known, except in relation to the Triassic ridge flanking it to the east towards New Brighton. The present day vegetation of *Molinia* grass suggests it probably occupies all the open land to the north of the earlier course of the River Birket and probably links, at least intermittently, with the peat along the coastal plain to the west.

To the north-west of Bidston Moss, borehole information shows that as the proto-Birket channel was buried, probably quite quickly, as sea-level rose, the upper silts and clays became capped by a bed of peat, 1m or greater in depth, which is Kenna's (1986, fig 6b) Upper Peat/Forest Bed. It has not been possible to identify the relationship between this peat and that within the moss proper. To the north-west of the peat in the earlier Birket channel, observations have shown peat across the coastal plain and beneath the sand dune belt, as far as Dove Point.

The south-western edge of the moss has been located from commercial borehole data relating to modern industrial developments. One of these developments was monitored in November 1992 and showed that the peat is *c*1.4m thick near the foot of the Bidston sandstone slope. The ground surface here is *c*5m above that in the centre of the moss where the Birket Drain is cut. The upper 1.5-2.7m of the stratigraphy consists of building debris related to the factories and industrial developments on this part of the moss.

When the Great Drain was cut it provided a section through the central part of the moss which showed that the peat was *c*4m deep on average and contained many fallen trees and root systems (Smith 1849). Clay-filled channels postdate the peat at each end, probably marking the later drainage system of the original Pool (Plate 2). In places the peat was underlain by a 'white sand' with shells, *c*1m deep over the pre-Holocene boulder clay surface (Smith 1849).

Palaeoenvironmental knowledge of this mossland relies on the detailed pollen sampling of the moss in 1979. This was prior to the last major light industrial development of the moss area when four peat cores were taken by J Innes and P Tomlinson. Three radiocarbon dates have since been acquired through NERC funding for this site. Table 2 summarises the palaeoenvironmental evidence from the most useful of these cores, BM2, while the detailed report is contained in the palaeoecological survey section.

Newton Carr, Hoylake (SJ 225885)

Newton Carr lies at the extreme north-western edge of the Wirral, about a mile south of the present coast. The area is basically flat with low undulations along the eastern side. The northern and north-western edges of the study area are represented by the built-up areas of Hoylake and West Kirby which border the present coast. Much of Hoylake lies on former coastal deposits, but to the west lie the three islands of Hilbre, Middle Eye, and Little Eye, which form slightly elevated areas of sandstone.

The landuse of Newton Carr is similar in some respects to that of Bidston (Fig 70). The northern and western part of the Carr is given over to a combination of golf courses, market gardens, and a partly grassed over and landscaped council waste tip. This presented the same problems for archaeological survey as at Bidston Moss for over 50% of the area. Unlike Bidston, however, the rest is farmland. Over 85% of the fields on the Carr are under pasture on mainly blown sand or more recent alluvial clay soils. The pasture extends eastwards onto the boulder clay plain skirting the wetland. The main arable area is found on the north-facing slopes of the boulder clay ridge marking the southern boundary to the Carr.

This landuse pattern limited fieldwalking to areas located on the ridge slopes and on a small part of the Carr itself. All arable land was walked, but this amounted to only *c*40 hectares in all. The early sowing regime in the area also meant that the greater part of this had to be scanned in poor recovery conditions, and at best the conditions for the rest could only be termed moderate. As with Bidston, therefore, the archaeological evidence associated with the wetland has to come mainly from existing sites and findspots from around the margins of the Carr.

The existing ditches across the Carr reveal nothing of the early Holocene deposits as they are cut no deeper than the depth of the more recent deposits capping the wetland. As the mapping of the original extent and character of the mossland is made difficult by later soils and current landuse, its extent has had to be defined from commercial boreholes. These have been taken both by the former Metropolitan County Council (Fig 10), and in the course of palaeoenvironmental research by Durham University, which has been under way on the Carr for some time (Innes *et al* 1990). The main thrust of this work has been in relation to sea-level changes, with detailed lithostratigraphical investigations through extended seismic and borehole survey by D Bedlington (Bedlington in prep). The area to the north of the present canalised course of the River Birket has not been investigated in detail because of the presence there of running sand and stiff clay deposits which are very difficult to penetrate with hand coring equipment.

The depression in which the moss lies is shown from commercial logs to be at least 6m in depth, with organic deposits likely to be present across the whole of the north-central part of the basin. The depth of the peat as identified from these borehole data is shown in Figure 11. It is present across part of the basin in three bands alternating with clays and silts. The southern and eastern limits of the peat basin are defined by the pre-Holocene geology of the ridges at Newton and West Kirby. Its western and northern limits, however, extend only as far as the outer sample points, and the peat continues for an unknown distance in these directions. The layers of peat are largely unaffected by the desiccation caused by modern drainage which has damaged surface sediments.

The cores from three of the commercial boreholes (7, 8, and 11) have been selected for more detailed examination by pollen analysis and the results are presented for the first time in this report. Two are summarised in Tables 3 and 4. Unlike the pollen diagrams from the other two wetland areas studied in North Wirral, radiocarbon dates are unavailable and the chronology of the Newton Carr diagram is based on comparison with the pollen zones of the regional Flandrian type site diagram from Red Moss (Hibbert *et al* 1971).

Park Road, Meols (SJ 239901)

No fieldwork took place as part of the NWWS along the coastal peats due to difficult survey conditions (Fig 70). The western zone is taken up by housing, while to the east lies a combination of market gardens and fields given over to pasture and caravan parks. The mapping of the wetland areas for this survey has therefore come from Kenna's work, associated largely with the construction of the sea defences and opportunistic observations during other development works of the 1970s (Kenna 1978a, 1979, 1986). This shows the two main peat beds of the Upper Peat/Forest Bed and Lower Peat/Forest Bed alternating with bands of marine clays and silts to a depth of 4.5m. The sea wall is now continuous along this section and further erosion of these deposits is prevented.

As the suitability and attractiveness for human settlement on North Wirral will have been conditioned by environmental changes taking place along the coastal strip, a pollen site from the edge of the built-up area of Meols is included in this report. It provides a reference point for the wider ecological changes which took place along the coastal zone, so that the two main mossland areas covered by the survey, Bidston and Hoylake, can be seen in context.

The pollen site, at Park Road, Meols (SJ 239901) lies at the landward edge of the zone of peat and clay deposits, at a height of 3.5m OD. The detailed pollen diagram is presented in the palaeoecological survey section, while a summary is presented as Table 5. Four radiocarbon dates were subsequently obtained for this site through NERC funding.

Archaeology and palaeoecology

Early Mesolithic *c*8000 BC – *c*6800 Cal BC

The Flandrian deposits in the area only started forming after *c*6500 Cal BC. Prior to this, in the earliest post-glacial times, the North Wirral landscape was dramatically different from the present one; the coastline lay well to the north-west, where it formed an arc running from Anglesey to the southern Lake District (Tooley 1985, fig 6.1). The now tidal Dee and Mersey rivers may have presented less of a barrier to the adjacent uplands of North Wales than they do today with their high tidal ranges (Tooley 1985, 96), and they may therefore have appeared as major, wooded, river channels at this time. The Birket and the Fender would have flowed out, across a pine and birch-clad plain, to the distant coast over 8km from the present one.

As wetland growth in these coastal areas rested on high water-tables, dependent to some extent on high sea-levels, wetland would probably have been restricted, possibly confined to the margins of the major channel features, where waterlogging is likely to have been prevalent. Throughout the Mesolithic, the inhabitants of the area would have had to contend, over generations, with the gradual landward retreat of the coastline, with the consequent loss of land and the inland extension of perimarine landscapes. The seventh millennium Cal BC was a period of relatively rapid sea-level rise in the region (Tooley 1982). By the late seventh millennium Cal BC (Table 2) the area around Bidston had become part of the perimarine zone of swamps, meres, and fens lying between the river systems of the Fender valley to the west and the inlet of Wallasey Pool to the east.

There is some slight evidence, from collections of flintwork along the present coast, to show that before the onset of these wet conditions Mesolithic hunters were active in the former wooded plain that now lies beneath the sea. These sites were discovered through sea erosion around the turn of the century. Evidence of settlement activity at these sites has not survived, but the existence of a small number of obliquely pointed microliths from two sites on the coast suggests that hunter-gatherers were in the area during the early Mesolithic. The form of the flintwork suggests that this could have been during the eighth/early seventh millennium BC (Jacobi 1980).

One site at New Brighton (A 6) includes within its chronologically mixed flint assemblage some blade debris, which is probably also Mesolithic (Pitts and Jacobi 1979). The site lies at the junction of the now inundated boulder clay plain to the north and a low Triassic sandstone cliff. The buried northerly channel of the Birket probably flowed a little to the west of the site (Fig 72).

Another possible site is found to the south-west, on the present coastal plain. On a small area of sandstone to the north of Newton Carr, the flint scatter excavated in 1914 by Glenn at Red Rocks, Hoylake (A 10; now in Manchester Museum) and characterised as Neolithic, includes some pieces that have Mesolithic blade and blade core affinities. From an assessment of the raw materials used (by RC), one type, a light blue patinated possible chert, is paralleled at an early Mesolithic site at Greasby, *c*4km to the south (Cowell 1992c). This suggests that it is possible that the Hoylake material may be of a date similar to that from New Brighton, although it is not yet demonstrated that this type of material was exploited locally only in the early Mesolithic. The only distinctive topographical feature adjacent to this site is the major, buried, early channel which ran north-westwards through Newton Carr towards the island of Hilbre. To the north-east, at an altitude of *c*0.5m OD, an early Mesolithic flint point may be indicated in illustrations of flintwork from the submerged forest at Dove Point, Meols (A 5). The Flandrian stratigraphy of the site suggests that this is likely to have come from the boulder clay surface before the onset of the Lower Peat/Forest Bed in the late Mesolithic period (A 1).

The paucity of the evidence from these sites makes it very difficult to understand the nature of the activities carried out at these locations, or their precise dating. They were unlikely to have been placed to exploit the coast directly, as this lay well to the north-west, but they probably were within several hours' walk of it. The main reason for their siting, with the possible exception of the Dove Point site, may have been related to the early channels. These would have provided transport, and opportunities to exploit fish, plants, birds, and water for hunters and animals. At New Brighton and Hoylake this may have been combined with the existence of well drained land with, in the case of the former site, strategic views over the plain to the north, which is now covered by Liverpool Bay.

These sites can be seen as part of the wider exploitation of the North Wirral area during the early Mesolithic. A number of better understood sites lie a little to the south of the ridge at Newton Carr (Cowell 1991a), the nearest *c*3.5km away at Greasby. Raw material exploitation at Greasby and the similar, nearby site of Thurstaston (Cowell 1992c), strongly suggests that the north Welsh coast was being used as a supply of chert for the manufacture of implements. These sites appear to be residential camps and are provisionally placed at approximately the beginning of the seventh millennium BC, on the form of the obliquely pointed microliths (Jacobi 1980). The lower sea-levels at this time would have allowed easy access across the now tidal river Dee to the upland areas of the north Welsh coast, where raw materials and complementary wild resources to those found in the lowlands of Wirral would have been available. The present day coastal sites might therefore be seen as part of this wide ranging exploitation of the landscape, representing locations for camps, probably of a temporary nature, of unknown duration and function. The surviving pattern of sites along the present coast may reflect the exploitation of the river valleys in the absence of more widespread wetland environments, while sites of the same period positioned closer to the contemporary early Mesolithic coast are now hidden under the Irish Sea.

Later Mesolithic/earliest Neolithic *c*6800 BC – *c*4000 Cal BC

By the late seventh millennium Cal BC sea-level increase had led to waterlogged conditions along the present North Wirral coastline, and low-lying areas became dominated by wetland vegetation. This date also marks the beginning of a period of about two millennia when, for the first time, the prehistoric archaeology of North Wirral relates directly to fluctuating coastlines and periods of direct inundation within the wetland embayments. As a consequence, the tidal zone seems never to have been too far away from the ridges flanking the coastal plain during the later part of the Mesolithic.

Prior to *c*6400-6080 Cal BC (7360±60 BP; SRR-2926) freshwater clays were being laid down in the Bidston basin, which were succeeded by reedswamp conditions (Table 2). At Newton Carr, from Flandrian II times (before *c*6190-5720 Cal BC) the north-west part of this basin had also ceased to be dryland, as mire conditions spread into the thick pine, birch, and hazel woodlands that had existed previously (NC7a, Table 3). The period of several centuries a little before *c*5500-5250 Cal BC (6420±60 BP; SRR-1494, and 6460±40 BP; SRR-1496) saw the approach of coastal influence. Although birch and oak woodland spread to the north of the present coastline between Dove Point and Leasowe, saltmarsh is indicated in the pollen diagrams (Kenna 1978, 31).

The first direct effects of sea-level rise are seen several centuries later, both along the present coast and at Newton Carr (Table 1). At the latter site, towards the end of pollen zone NC7b (post FII; Table 3), the embayment was directly inundated by marine and estuarine deposits. The succeeding dry phase may not have lasted for long, as a second marine incursion is seen at Newton Carr (NC7c, Table 3; pre-NC11a, Table 4), later within Flandrian II, probably towards the end of the fifth millennium Cal BC, and at numerous places along the coast within the same chronozone (Kenna 1986, 11). The only firm date for this phase comes from Park Road, Meols, where marine clays were laid down before *c*4234-3980 Cal BC (5250±50 BP; SRR-2694), which may correlate with the second marine phase at Newton Carr. There is evidence that this inundation may have also affected the Wirral peninsula as far south as Helsby (Tooley 1976).

During those periods when the coast lay close to its present line, the dryland areas beyond the limit of tidal flat and saltmarsh formation (such as the ridges at Wallasey and Bidston in the north-east, and at West Kirby and Grange in the south-west) would have formed suitable bases from which to exploit the plentiful natural resources of the wetland areas (*see below*). These sandstone slopes would have provided light, well drained soils supporting woodland extending down to the edge of the wetlands of the moss. During periods of high sea-level, saltmarsh or estuarine conditions would have advanced into the present coastal plain and into the Newton Carr basin where the boulder clay ridge would have marked the limit of dryland vegetation. The central part of the Bidston basin seems never to have been directly affected by the sea during the late Mesolithic.

During times of lower sea-level, the moss areas would have carried freshwater fen and reedswamp, or wooded carr, with alder very common and oak and birch present in the carr community. The surrounding dryland areas sustained a vegetation comprising mixed deciduous forest in which elm, oak, and hazel were most abundant, and which was dramatically different in composition and density from the pine, birch, and hazel woodland of the earlier period (*eg* BMa, Table 2). A survivor of those earlier woodlands, pine, though now reduced, was still able to find suitable habitats in the area (*eg* NC7a, Table 3), probably upon thinner sandy soils within the coastal zone. When marine influence was reduced, as during the late sixth millennium Cal BC and the mid fifth millennium Cal BC, conditions within the mire areas became drier. Woodland cover reflected this and changed, with oak dominating at the expense of alder while pine and birch were reduced. The extent of the woodland also increased as it spread down from the high ground and, as during the second half of the sixth millennium Cal BC, beyond the present coastal areas. At Newton Carr similar conditions would have obtained during the late Mesolithic, with the ground to the south of the Newton Carr Brook remaining dry, but heavier clay soils on the slopes probably carried a denser woodland cover than those around Bidston.

Human activity has to be viewed in the context of the continually alternating sequence of environmental changes related to changes in sea-level. This will presumably have led to changing patterns of settlement and landuse around the margins of the mosses and the adjoining plain. However, the lack of suitable fieldwalking conditions in North Wirral means that the main evidence remains that already known from a few extant sites. These, however, are too sparse and ill-dated to allow human settlement to be matched to the changing environmental conditions in any detail.

Human occupation during this time must have been positioned so as to take advantage of natural breaks in the dense woodland canopy that had developed by the late Mesolithic. The main breaks would have occurred along the edges of channels, at the coastal fringe, and around the edges of the mosslands. Most of these potential locations are now unfortunately hidden from standard survey techniques: the former coastal fringe sites lie under the sea, the channels are now mostly buried by Flandrian deposits, and present landuse around the fringes of the mosslands is mainly pasture or developed land. There is consequently a lack of archaeological evidence for the later Mesolithic around these mossland basins in North Wirral. To some extent this is compensated for by the palaeoecological evidence.

The archaeological evidence for this period is limited to the north-western tip of the Wirral and is concentrated on the islands west of Hoylake, although it is not known if they were islands during the Mesolithic. They approximately marked the slightly elevated, western edge of a former channel running out beyond the present coastline to the north (Kenna 1978; D Bedlington pers comm). The attraction of this channel may be suggested by the site at Little Eye (A 7) which has produced a rod microlith and a few small cores alongside a small amount of knapping debris. Vague accounts of flint pieces, that could be interpreted as microliths, may also reflect further Mesolithic activity on the adjoining island to the north-west (A 8).

The wetland basin of Newton Carr, lying 2.5km to the south-east of the islands, with its alder carr woodland, rich fen and reedswamp, would also have been exploitable from this base for plants, birds, wetland animals and, in the surrounding forest, woodland plants and animals. It might be thought probable that the elevated land along the boulder clay ridge to the south would have offered a suitable location for exploiting the wetland in the Newton Carr basin, but in fact this seems not to have been the case to any great extent. The original Mesolithic junction of the dryland and the wetland, on the southern edge of the basin, is partly hidden by later peat and clays and mostly inaccessible to fieldwalking because of the occurrence of pasture. The upper part of the slope, however, where survey conditions are better, has produced only one possible site (Me 001). It reflects a fairly typical site encountered in field survey across the county generally, comprising a blade core and five waste pieces of local pebble flint. Dating of such sites is difficult, but the core may suggest a Mesolithic date, and comparison of the raw material with that from the early Mesolithic sites a little to the south suggests a date later in the period for the Me 001 flints, as the raw material is not similar.

Although the evidence is sparse, it does point to late Mesolithic activity around the present north-western coastal area. Due to the lack of precise dating evidence, it is not clear if the activity was directly dependent on coastal resources or on inland freshwater wetland resources. If the latter, coastal localities were probably almost always close enough to have been exploited from these locations. The flintwork from Little Eye is mostly beach flint, but that from the Newton ridge could have been collected from the immediate vicinity of the site.

In the north-eastern part of the peninsula, Bidston seems to have been less affected by marine incursions throughout this period. Although human activity cannot be recognised by artefactual evidence from the surrounding sandstone slopes, the palynological record hints at its presence. Four small disturbances may be tentatively recognised in the pollen diagram from Bidston Moss 2 (Table 2, Phases 2, 4, 6, and 8). Mixed oak woodland on the surrounding dryland appears to have suffered limited reduction between 6400-6080 Cal BC (7360±70 BP; SRR-2926) and *c*4900-4530 Cal BC (5840±70 BP; SRR-2925). These disturbances may reflect the environmental consequences of hunter-gatherers in the area, although some caution ought to be applied to this interpretation in the absence of contemporary human artefacts in the vicinity. Such evidence does however exist in areas less than 5km to the south, where survey conditions are more conducive to recovery of artefacts (Cowell 1991a). The flint blade debris recovered from New Brighton, and assigned to the earlier Mesolithic on very limited criteria, could include material of this later date as blade technology continued throughout the Mesolithic (Pitts and Jacobi 1979).

The possibility that hunters and gatherers were ranging across the sandstone slopes and adjacent wetlands during the late Mesolithic is strengthened by further pollen evidence from Bidston Moss. At an horizon dated to the beginning of the fifth millennium Cal BC, good evidence for a human presence in the woodland may be recognised for the first time (Table 2, Phase 10). This is before the latest secure date for the end of Mesolithic culture in the region, of *c*4750-4500 BC, although a number of other less secure dates do take this horizon several centuries later (Jacobi 1987, 163). This suggests that the activity is more likely to have been carried out by hunter-gatherers.

Table 2 Wirral: Bidston Moss 2, palaeoecological summary

cms	Date (Cal BC)	Local Pollen Zones		Ecology of Moss	Regeneration/Disturbance Phases	
				Oxidised organic clay topsoil		
32 36	3690-3360	BM-e	Tree pollen decreases. Maximum of grass pollen. Pollen of ruderal weeds and cereal type.	Drier, more acidic environment. Vegetation developing into raised mire communities.	Major woodland recession. Tree pollen 20% of total pollen. Rising grass pollen with cereal type pollen and arable weeds recorded. Temporary Elm Decline early in phase.	16
50		BM-d	Tree pollen *c* 40% of total pollen. Oak major tree type. Alder decreases. Birch, elm, and ash increase as do heather and herb pollen. Bracken and fern spores prominent.		AP rising to 50% of TP. Calluna high.	15
58					AP 30% of TP. Maxima of herb pollen types (mainly grasses). Increase in *Calluna* and birch pollen.	14
66					Regeneration to mature deciduous woodland.	13
74					AP (mainly alder) reduced to 30% of TP. Grasses and ruderal herbs increase. Cereal-type pollen recorded.	12
86		BM-c	AP greater than 70%. Alder dominant type. Other tree taxa low. Low values of wetland and ruderal herbs. Marsh and fenland ferns present throughout.	↑ Mosaic of open water, fen, and carr communities. ↑	Woodland regeneration	11
94					Major forest (alder) reduction. Herb pollen high - increase in ruderals. Cereal-type pollen recorded.	10
98 106	4900-4530				Woodland regeneration	9
118					Significant temporary clearance of alder with increase in ruderal herbs.	8
130					Woodland regeneration with locally dominant alder.	7
146 148				Seasonal flooding (146–198)	Significant, brief episode of local clearance. Grasses and ruderal herb pollen increase.	6
150					Woodland regeneration. AP greater than 70% of TP. Alder dominant. Some wetland herb types recorded.	5
170					Some disturbance. Increase in pollen of plants indicative of open ground.	4
174 178		BM-b	Tree pollen greater than 60% of total pollen. Alder with some birch and oak. Pine values decline. Low hazel. Herbaceous pollen low mainly wetland types.		Woodland regeneration.	3
186					Tree pollen reduced. Possible localised grassy clearings.	2
194 196 198 200	6400-6080	BM-a	Tree pollen greater than 80% of total pollen. Dominated by pine with birch, oak, and hazel.	Reedswamp	Mixed woodland with pine, oak, and birch	1

Table 3 Wirral: Newton Carr 7, palaeoecological summary

cms	Date (Cal BC)	Flandrian Zone	Local Pollen Zones	Ecology of Moss	Dryland Vegetation
600			Blue-grey silty clay	Marine incursion	
693		FL II	**NC7-c** Major tree types oak, alder and hazel. Birch, pine, and elm important. Pollen of grasses and coastal herbs.	Saltmarsh Fen carr and reedswamp	Mixed deciduous woodland: elm, oak, and hazel most abundant.
714			Blue-grey silty clay	Marine incursion	
984	6180-5720*	FL II	**NC7-b** Characterised by alder pollen but other tree types high. Little herb pollen. Coastal-type pollen increases towards top of zone.	Estuarine Saltmarsh Fen carr and reedswamp	Mixed deciduous woodland: elm, oak, and hazel most abundant.
992		FL I	**NC7-a** Characterised by oak, birch, and pine. High frequencies of elm and hazel. High grass pollen values. Alder rise marks top of zone.	Reedswamp	Mixed pine, birch, and hazel woodland
998					

* Date relative to the FLI/FLII boundary at Red Moss, Greater Manchester (Hibbert *et al* 1971)

At this time the mire was a mosaic of open water, fen, and carr communities, while oak-dominated, mixed, deciduous woodland occurred around the margins. The scale of the evidence for disturbance is substantial, with alder reduced and ruderal pollen types indicative of clearance present. One pollen taxon, of cereal-type (*Triticum* sp), present at this horizon may suggest a possible link with cultivation. Using Andersen's (1979) criteria for the identification of cereal pollen, some large grains originate from uncultivated grasses, common in natural grassland. Taxa such as *Glyceria, Agropyron,* and *Elymus,* common particularly in coastal areas, may be contained within the *Hordeum*-type (barley) group. The validity of cereal-type pollen as a true indicator of agriculture cannot, therefore, be taken for granted.

Such occurrences have been noted in fifth millennium Cal BC contexts which predate the elm decline in Ireland and North West Britain (Edwards and Hirons 1984). In the absence of Neolithic cultural material or cultigens (with one exception, *see below*) in secure Neolithic contexts in the country at this early date (Zvelebil and Rowley-Conwy 1986, fig 3; Kinnes 1988, table 1.1), such evidence may be seen as indicating cultivation thinly scattered across North West Britain in the late Mesolithic. There are a number of difficulties, however, with this theory, which have already been mentioned *(and see* Kinnes 1988, 3).

If the Bidston pollen evidence can be used to suggest the early adoption of cereal cultivation in these coastal areas, the radiocarbon date of 4900-4530 Cal BC (5840±70 BP; SRR-2925) for Phase 10 at Bidston suggests that one element of farming was present in the area, predating by *c*500 radiocarbon years the earliest appearance of Neolithic cultural and economic evidence in the country. The one exception to this is the site of Ballynagilly in Northern Ireland, which has a slightly later date than the Bidston evidence for Neolithic cultural material (Pilcher and Smith 1979), although the uniqueness of the Irish site causes some doubt as regards its reliability (Kinnes 1988, 6).

The characteristics of Phase 12 at Bidston (Table 2) are very similar to those of Phase 10, and may point to another occurrence of clearance and cultivation activity in a context that is again likely to predate the elm decline, which may be present in Phase 14 or 16 (Table 2). This clearance phase is also likely to be earlier than the earliest Neolithic material in the country. The most diagnostic Neolithic artefacts found locally are the flint leaf arrowheads, which date at the earliest nationally to the second half of the fifth millennium Cal BC (Green 1984), but there is no indication that the few Wirral examples are as early as this.

Although the Bidston evidence for this potential early cultivation is tenuous, it does raise some potentially important issues concerning the rate and nature of the adoption of a new form of subsistence, and ultimately a new material Neolithic culture (Zvelebil 1986). This is one of the least understood processes in prehistory, making further research in the coastal areas of the county highly desirable. Similar evidence is also present on the Sefton coast (*see Chapter 4*), and therefore a more detailed discussion of its implications is left to Chapter 7.

Earlier Neolithic *c*4000–3200 Cal BC

Current views on the adoption of agriculture in this country suggest that it was a long, gradual, and complex process, whereby in some areas hunting and gathering remained dominant, with domesticates or cultigens providing a useful adjunct to be adopted if and when necessary (Zvelebil and Rowley-Conwy 1986). The pollen evidence for Bidston Moss (Table 2, phases 10 and 12) suggests that palynological and archaeological evidence might exist in these coastal areas (including Sefton) that could bear on this problem. For the moment, however, it must remain only a possibility that locally this process started well before the elm decline.

The local pollen evidence suggests that a slightly more secure identification of the adoption of cereal cultivation might be recognised at Bidston Moss in the early Neolithic. Nationally, the early Neolithic horizon is marked approximately by the elm decline, which regionally falls between *c*3990-3640 Cal BC (Hibbert *et al* 1971). In Merseyside, work reported in this volume shows that the elm decline is present at about the same horizon on the Wirral. At Park Road, Meols it has produced a date of 4034-3790 Cal BC (5120±50 BP; SRR-2929) (Table 5). At Bidston, it is more difficult to identify and could be represented at either the beginning of Phase 16 which dates to before 3690-3360 Cal BC (4740±70 BP; SRR-2924), or possibly by the less clear fall in elm pollen at the start of Phase 14 (Table 2). The wider implications of the elm decline and its associated landscape changes are dealt with in more detail in Chapter 7.

The effects seen at the elm decline at Bidston are unique in the county in their scale and effect on the woodland. Nationally, it is generally associated with woodland clearance and farming (Smith *et al* 1981, 152-167), and at Bidston, both phases 14 and 16 (Table 2) mark major periods of disturbance in the pollen diagram. The latter phase is associated with cereal-type pollen, as in earlier, chronologically late Mesolithic episodes (Table 2, phases 10 and 12). In the case of Phase 16, possible arable weeds are also well represented. The repeated occurrence of *Triticum*-type pollen in three separate phases at Bidston, of which the third is associated with arable weeds, may be held to counter the arguments that are used to question the validity of cereal cultivation being represented by single cereal-type pollen grains (Zvelebil and Rowley-Conwy 1986).

Whatever the interpretation of the pre-elm decline disturbances, by the early Neolithic the landscape had changed significantly over the preceding millennium or so, with tree pollen accounting for *c*20% of total pollen by the end of Phase 16 (Table 2). This suggests that the area around the moss was the scene of repeated inroads into the forest cover, causing major vegetation changes around the mid fourth millennium Cal BC. It is not possible to know from the pollen evidence if the economic balance had tilted in favour of agriculture on the Wirral by this date, or if hunting and gathering still played a dominant role. There is evidence that hunting, and the gathering of wild plants and berries, continued to play an important role through the Neolithic period elsewhere, even in areas where domesticates and cultigens were present (Smith *et al* 1981, 189; Smith 1984, 109).

The identification of early Neolithic material that might provide a human association for these vegetational changes is difficult over the whole county, and indeed over much of the region. The only diagnostic artefacts are the leaf arrowheads and stone axes. Not all of these, however, need belong to the earliest farmers in the area, as their dating range extends into the middle and even late Neolithic (Smith 1979; Green 1984). It is difficult to attribute any other local material, particularly flintwork, to this early period.

Evidence for the settlement pattern of the early Neolithic therefore appears to be sparse in comparison to that of the Mesolithic. This is probably a function of the present lack of criteria by which this material can be recognised locally (*see Chapter 1*), rather than its non-existence. Until early Neolithic flint technology can be identified confidently in the area, most probably from within material currently attributed to the Mesolithic, such sites will remain elusive. Similarities in flint technology nationally (Pitts and Jacobi 1979), continuing exploitation of areas such as Bidston Moss, and similarities in the nature and scale of woodland disturbance across the county as a whole (*see Chapter 7*), point to the conclusion that the pattern and nature of landscape exploitation differed little from that in the Mesolithic to earlier Neolithic.

In the absence of settlement locations, the only evidence for activity of this period comes from previously recorded chance finds in the Merseyside SMR (Fig 72), some of which material is deposited at Liverpool Museum (Cowell in prep c). A number of stone axe findspots are recorded from the sandstone margins of Bidston Moss; the lower slopes to the east yielded an axe made of North Welsh stone (A 2), while two more come from the upper slopes to the west (A 3, A 4). These provide a potential artefactual association with the major impact on the natural vegetation seen in the palynological evidence although, as mentioned above, dating for these types of implement is quite wide. They could, accordingly, belong to a later phase of the Neolithic not covered by the pollen record, which at Bidston is terminated after *c*3700-3300 BC (4740±70 BP; SRR-2924) by a marine clay. The axes may be linked to woodland clearance and farming around the margins of the wetland, although A 3 (Plate 1) is a very fine basalt axe that appears unused and may have fulfilled a prestige role rather than a practical one.

Plate 1 Wirral: Neolithic axehead (Site A3) from the vicinity of Bidston Moss (Photo R Cowell)

There is also a leaf arrowhead included in the material illustrated from New Brighton (A 6) suggesting reoccupation of the Mesolithic site to the north of the moss.

If the clearance at the elm decline horizon took place around contemporary settlements, it argues for quite prolonged and intensifying occupation of the Bidston Moss area throughout the fifth and early fourth millennia Cal BC, probably incorporating the adoption of farming as a feature of the economic base. The location of the associated sites might be expected in two zones, either on the sandstone slopes around the moss or along major river channels. Both these zones are unfortunately not amenable to field survey in the area. The sandstone slopes skirting the moss are quite heavily built upon, although a few small areas around the lower slopes and the moss junction still survive as undeveloped, grassed-over land. The same restrictions apply to the discovery of potential sites along the buried channels of the Fender, the Birket, and that identified running across Newton Carr, which are buried by late Flandrian deposits.

The palynological evidence from elsewhere in North Wirral contrasts strongly with the Bidston evidence. Here the vegetation pattern of the late Mesolithic changed little into the first half of the fourth millennium Cal BC. One small-scale clearance episode may be recognised at Newton Carr, shortly after the elm decline, with forest reduction and the spread of *Pteridium* (bracken), *Plantago lanceolata*, and *Triticum* (cereal-type) pollen (Table 4, NC11b). In contrast to Bidston, however, this phase soon ceased and oak/alder woodland regenerated to former levels and stayed this way throughout the rest of the period.

The archaeological evidence in the north-west of

Table 4 Wirral: Newton Carr 11, palaeoecological summary

cms	Local Pollen Zones		Ecology of Moss	Regeneration/ Disturbance Phases
			Grey silty clay	
135	NC11-d	Tree pollen declines throughout zone. Characterised by alder with birch and oak. Hazel fluctuates. Variety of herbs. Shrubs increase. Rising coastal/ruderal herbs define upper boundary.	Salt marsh ↑ Alder ↑ Birch carr ↑ Drier acidic	Oak, alder woodland dominant with some birch carr around the site.
145	NC11-c	AP *c*80% of TP. Major component birch which declines at top of zone. Lower alder, hazel, and oak values. Low herbaceous pollen.		
185	NC11-b	AP fall to 50% of TP. Characterised by oak (falling at top of zone), alder, birch, and hazel. Range of herbs increases with coastal types present and rising grass values. Aquatic pollen well represented.	Wetter with flooding. Some coastal herb types.	
215				
225				Possible limited clearance with cereal-type pollen recorded. Bracken peak.
245	NC11-a	Tree/shrub pollen 80% of TP. Oak, elm, alder major components. Grass, *Artemesia*, and Chenopodiaceae have significant values. Elm Decline at top of zone.	Reedswamp and fen carr.	Elm Decline at top of zone. Mixed woodland with oak dominant pollen type.
290			Blue-grey silty clay and *Phragmites*	
352			Boulder clay	

the peninsula is also slight, and again consists almost entirely of previously recorded chance finds. It comprises a few leaf arrowheads from the peat beds at Meols (A 5), which may represent hunting in the coastal plain woodlands at a time of lower sea-level. The flint site from Hoylake (A 10), although having a Mesolithic element, reportedly also included stone axe fragments (Varley and Jackson 1940), possibly indicating continuity, or at least re-use, of settlement locations into the early Neolithic. The one piece of evidence from the fieldwalking survey, on the southern slopes of the Newton Carr basin, also suggests that the same areas were being exploited in a similar way in the Neolithic as in the Mesolithic. This is represented by the flint core (Me 002) found adjacent to the presumed Mesolithic flint scatter of Me 001.

The settlement and clearance, farming and hunting activity in North Wirral was, for several centuries leading up to the approximate elm decline horizon, taking place against a background of marine influence in the north-western part of the peninsula (Table 1). From *c*4234-3980 Cal BC (5250±50 BP; SRR-2694), direct marine influence began to decline around the Park Road, Meols site but coastal influence was never far away, as saltmarsh, areas of deep water, and fen reedswamp dominated (Table 5). Possibly at about this time direct marine influence also receded from around the Newton basin, but coastal conditions still lay within a few kilometres of the site, although this also receded during Flandrian III (estimated to approximately the end of the fourth millennium Cal BC) (Table 4, NC11b and NC11c).

Some time after the elm decline an increase in wetness occurred throughout the coastal plain, and the Newton embayment became subject to local flooding as direct marine influence spread into the basin. An inundation took place at Park Road, Meols, which had ceased by *c*3510-3140 Cal BC (4620±50 BP; SRR-2693), although the sampling site probably lay very near the landward limit of this incursion. At Newton Carr the surface topography of the southern part of the basin led to the formation of small estuarine bays at periods of higher sea-level. One such embayment, identified to the east of the Birket-Carr Lane junction (SJ 230882), existed for a few centuries in the middle Neolithic (D Bedlington pers comm). At Bidston, direct marine inundation occurred across the mossland for the first time after 3690-3360 Cal BC (4740±70 BP; SRR-2924).

Although the archaeological evidence is sparse for this part of the Neolithic, the pollen evidence from the survey does provide some potential ecological context for the period. It suggests that there may have been two zones in North Wirral, with the balance of exploitation differing in each. In the north-west corner, which is lower and thus more susceptible to flooding, the land never seems to have been exploited for agriculture to any great extent. Woodland remained dominant outside the flooded areas, and may have led to the area being better suited to exploitation for wild resources around the natural open zones formed by channels, coasts, and mosslands. This type of activity would be difficult to identify from pollen diagrams. Similarly, if farming was established across the whole of the North Wirral in the early centuries after the elm decline, and woodland was being used for pannage for domestic pigs, and for grazing cattle, then woodland clearance may not have been necessary in this case either (Smith 1984, 109).

The less inundated north-east may have had a geographic advantage that led to the development of grassland, which is more amenable to cereal cultivation (Smith 1984). The archaeological evidence is not strong enough to identify whether the settlement associated with these landscape changes was of a more sedentary nature than in the Mesolithic. One attractive model, that might fit the pattern of landscape changes at Bidston around the elm decline, is illustrated in an area of southern England where small-scale, short-lived settlements appear to have been relocated frequently in response to difficult to manage regeneration of weeds and bracken (Smith 1984, 115). This may be one interpretation for the pattern seen at Bidston, where woodland regeneration follows each episode of clearance and subsequent short-lived phases of grassland.

The associated settlement was presumably also located on the lower sandstone slopes around the moss, while the nearby coasts and intersection of several major channels in the area would have provided opportunities for hunting and fishing. As the regularly flooded wetland of the later Mesolithic was becoming ombrotrophic by this time (Table 2, phases 12-16), the reduced capacity of the mire to sustain the earlier biomass, so attractive for a range of wetland flora and fauna, may have led to a greater investment in cultivation. The mossland would have been a useful source of wild plants and animals, as well as a variety of raw materials, adjacent to the cultivated areas which must have existed on the sandstone slopes. The wetland itself could also have provided regularly nutrient-replenished grazing land (Pryor 1978) for either wild, or later, domesticated animals around the margins of the carr woodland.

Late Neolithic/early Bronze Age *c*3500–1650 Cal BC

By the mid fourth millennium Cal BC, the local evidence shows that, as with North West England in general (Tooley 1978a, 35), drier conditions had become prevalent across much of North Wirral, as sea-level fell and forest developed around the mires. At Park Road, Meols this probably lasted several centuries from *c*3510-3140 Cal BC (4620±50 BP; SRR-2693) (Table 5). At Newton Carr this drier phase saw birch carr communities around the margins of the moss, with mixed oak woodlands dominating the dryland areas. These drier conditions were subsequently curtailed by flooding in the coastal strip. At Park Road, Meols marine clays were deposited shortly after 3100-2704 Cal BC (4315±70 BP; GU-1312) (Table 5).

At Newton (NC11), alder carr replaced birch woodland during the earlier part of Flandrian III (Table 4), the pollen record subsequently indicating approaching saltmarsh before the embayment was flooded by marine water. It is not clear whether this equates with the middle Neolithic flooding seen at Park Road and Bidston, or the subsequent late Neolithic/early Bronze Age phase seen at Park Road. At Newton this flooding marked the beginning of the move to estuarine conditions within the embayment.

Table 5 Wirral: Park Road, Meols, palaeoecological summary

cms	Date (Cal BC)	Local Pollen Zones		Ecology of Moss	Dryland Vegetation
			No pollen analysed	Oxidised clayey topsoil	
40			Blue-grey silty clay	Marine incursion	
60 64	3100-2704	PRM-c	AP increases to 70% at top zone alder major type with less oak, birch and hazel. Peak of willow at base of zone. Grass pollen lower. Wetland shrubs and herbs increase.	Fen carr Reedswamp Progressive reduction of marine influence.	Fen carr filters out regional pollen in upper levels. Mixed woodland with no evidence of clearance. Insignificant values of *Plantago lanceolata* pollen and bracken pollen.
75 76	3510-3140	PRM-b	AP 40% of TP. Oak and alder major components.		
78 79			Blue-grey organic clay	Marine incursion	
81	4034-3790		Grass pollen important. Wide range of wetland herbs. Chenopodiaceae herbs mid-zone.	Salt marsh Fen carr, reedswamp and brackish communities	
82.5 84 121 125	4234-3980	PRM-a	AP 50% of TP. Oak, elm, and alder major types. Less hazel and birch. Elm Decline marks the upper boundary. Grass pollen high with a wide range of coastal and freshwater herbs.	Salt marsh	Elm Decline Mixed woodland with oak and elm. Understorey of hazel. Alder at wetland margin.
			Blue-grey silty clay	Marine/estuarine silting	

In the north-east, Bidston Moss had been inundated for the first time and transformed into part of a wide estuary after *c*3690-3360 Cal BC (4740±70 BP; SRR-2924). This effectively ends the value of this site for identifying human influence on the vegetation in the vicinity. Boreholes from the coast to the north-west, however, show that in this area the subsequent drier conditions lasted longer, with woodland (the Upper Peat/Forest Bed) being found extensively behind the present beach, suggesting that it extended out beyond the present intertidal zone during this period. Samples of pine and oak from this peat bed at Wallasey Embankment near Leasowe are dated *c*2858-2049 Cal BC (3980±70 BP; Birm 1013; 3800±40 BP; SRR-1495). Several kilometres to the south, in Moreton, wood from the same horizon produced a date of 2139-2038 Cal BC (3695±10 BP; Q-620) (Kenna 1986).

Settlement evidence is slight for this period, spanning the late Neolithic/early Bronze Age, across the whole of North Wirral, although the incidence of previously recorded single findspots is stronger around the north-west part of the coast (Fig 72). A sherd of late Neolithic pottery comes from the beach at Meols (A 5), and on the low ridge to the south of the Newton Carr basin a single discoidal core was found in fieldwalking (Me 002). At Hilbre island (A 8) a number of probable oblique arrowheads is recorded. This pattern continues into the early Bronze Age, with barbed and tanged arrowheads from the peat beds at Dove Point (A 5), from the West Kirby ridge, and from Hilbre Island (A 8). These probably reflect the continuing exploitation of the woodland for hunting, marked by the Upper Peat/Forest Bed, that was temporarily more extensive across the modern intertidal zone. There is no pollen evidence available from any of the sites for this period to show how far the woodland was being affected by this activity. Indirect evidence, from early Bronze Age burials, suggests woodland may have been lighter by this time, as visibility is probably an important feature of burial distribution. A series of inurned cremations comes from the sandstone ridge, overlooking the Dee, to the south-west of Newton Carr (A 11). A bucket urn (A 9) has also come from Middle Eye, although this find is not demonstrably associated with a burial, and could therefore potentially mark a settlement location.

There is a similar lack of pollen data for the area to the north-east, which makes it difficult to determine to what extent the clearance and cultivation around Bidston Moss in the earlier Neolithic continued. The archaeological evidence, unfortunately, does not provide any good indication in the absence of palaeoenvironmental evidence. There are no sites or artefacts dated to the late Neolithic period, unless some of the stone axes date from this, rather than from the earlier part of the Neolithic (Smith 1979).

Evidence from the Bronze Age is hardly better represented. A barbed and tanged arrowhead was included in the flint scatter site at New Brighton (A 6), and a stone axe hammer (A 12) is recorded from the silt of the early Birket channel to the west of Bidston Moss, suggesting that some settlement and farming may have existed at this time in the locality. To the north-west of the moss a bone midden is also recorded (A 17), containing red deer and oxen, which is dated to *c*2861-2310 Cal BC (3980±70 BP; Birm-1013).

The use of woodlands seaward of the present coast must have become more restricted through the early Bronze Age, as after *c*2460-1774 Cal BC increasing wetness led to several centuries of alder fen carr and raised bog conditions. This may have been as a result of rising sea-level, although climatic conditions cannot be ruled out (Kenna 1978). This culminated in the death of the trees in the present zone *c*2012-1680 Cal BC (3490±60 BP; GU-1271) as local water-tables rose, although in general the rate of sea-level decreased after this time (Kenna 1986). To the east of Leasowe, landward clays and silts were being laid down in lagoonal type environments behind the sand barrier, approximately along the present line of the coast, during this time (Kenna 1986).

Late prehistoric period *c*1650 Cal BC – AD 43

The lack of a satisfactory palaeoenvironmental context for human activity through the third millennium Cal BC continues into the late prehistoric period. The peat profiles in both of the wetland embayments relating to this period are truncated, and the coastal peats seem to have suffered a high degree of erosion and disturbance. A peat horizon found in the Leasowe area has produced a pollen spectrum of open woodland and well represented ruderal herbs, although the date of 1030-810 Cal BC (2750±55 BP; GU-1270) is not completely certain due to possible disturbance (Kenna 1986). Elsewhere there appears to be a substantial break in the stratigraphic record at about this time. The fairly extensive peaty clay 'Soil Bed' , mostly found above the late Neolithic/early Bronze Age Upper Peat/ Forest Bed, seems to belong to the late medieval period. It may correspond to a phase of dune building noted on the Sefton coast (Tooley 1978b), which on the Wirral may have led to a series of dune slack environments across the coastal plain.

In keeping with most of North West England, there is little archaeological evidence for this period in North Wirral, and that which exists tends to be restricted mainly to later Bronze Age metalwork (Fig 72). In North Wirral even this is scarce. A socketed axe from Hilbre Island (A 8) is the only artefact recovered to date from the mossland margins, while a middle Bronze Age dirk (A 13) has come from the present beach, and there is a thin scatter of metalwork from across the sandstone outcrop upon which Wallasey has grown up, on the north edge of the moss (Cowell 1991a). However, the Dove Point promontory at Meols can be traced from early maps as extending some distance out into the present intertidal zone (Kenna 1986, fig 2), from where collectors during the later nineteenth century gathered one of the richest assemblages of artefacts from any site in the North West (Hume 1863). These were mostly of the historic period, but several first century BC coins were found (A 5), some of them of Mediterranean origin, and these may mark the beginnings of the rise to prominence of what became an extremely important coastal settlement in subsequent centuries. The site may have been located in order to take advantage of the natural attractions of Dove Point as a harbourage. In the Roman period it may well have acted as a transfer point for river boats to negotiate the Dee with material destined for the legionary town of Chester (Philpott 1991).

How well established the Dove Point settlement was in the late Iron Age is less certain. The evidence for the pattern of settlement in this period is just as sparse away from the wetland zone. A little evidence is available on the sandstone ridge overlooking the Dee estuary *c*4km to the south, where an oval cropmark near Irby potentially marks a farmstead of the late prehistoric period. A kilometre to the north-east, in Irby, current excavations on a Roman farmstead have also produced a little pottery that may reveal traces of an earlier, probable Iron Age settlement on the same site (Philpott 1993, 22). Although this evidence is extremely slight, this is the earliest evidence in the county for fixed farmstead sites. These are located on the light, sandy soils of the main sandstone ridges in the Wirral. If the pattern of Iron Age sites mirrors that of the Roman period, for which there is more evidence (Philpott 1993, 23), then proximity to wetland areas seems not to have been an important locational feature. Without pollen evidence from these wetland areas, however, it is difficult to know how far the pattern is biased to the sandstone areas, where fieldwalking and aerial photography are more productive of evidence than are the wetland areas.

Romano-British period AD 43–410

The Romano-British period is, with one exception, likewise poorly represented in the wetland zone of North Wirral, nor is the palaeoenvironmental evidence strong enough to provide a context for the little that is known. Roman coins are well represented in casual finds from within the present dune areas around Meols (R Philpott pers comm) and it may be suspected that this has some link with the development of the Dove Point, Meols site (A 5) in the Roman period. Here, over 80 coins were found along the beach during the nineteenth century, in addition to over 70 brooches, suggesting occupation of the site from the first to fourth centuries AD (summarised in Philpott 1991), and third century AD Roman pottery has also been excavated from Hilbre Island (A 8). The Dove Point settlement was an established port at the mouth of the Dee in the seventeenth century, acting as a transhipment point for journeys down the river to Chester. This commercial role, recognised by a little imported pottery and the range of coins, may have been the main reason for its Roman pre-eminence, although the earlier coins from the site suggest that this role may have its origins in the Iron Age. Evidence of rural Roman settlements is mainly known along the sandstone ridges of the peninsula, *c*5km or more from the wetland areas, the richest area, with coin and metalwork evidence, being around Irby/Thurstaston overlooking the Dee valley (Philpott 1991, 65). This may be a reflection of the influence that the Meols site had on the surrounding countryside with its suggested commercial, administrative, industrial, military, and agricultural role within the region (Philpott 1991, 68).

There is little evidence that Roman settlement directly exploited the edges of the mosslands anywhere in Merseyside, although this type of exploitation has been suggested by Jones for West Lancashire (1980, 79) and illustrated by work in West and North Lancashire (summarised in Philpott 1991). In a recent evaluation of a post-medieval farm site in Bidston village, a few unstratified sherds of Roman pottery were found (Cleary and Philpott 1993). These may only reflect arable activity around the sandstone slopes of Bidston Hill but, as at Irby, it might be suspected that associated farmsteads are more likely to have lain in the vicinity of the sandstone ridges around Bidston, rather than on the edges of the moss to the north. It is here, on the western sandstone slopes of the ridge, overlooking the River Fender, that the medieval village was located.

Early medieval period AD 410–1066

In the immediate post-Roman period there is little evidence to aid an understanding of the landscape and settlement pattern. There is some evidence that the North Wirral was a reasonably well populated area at this time, although how the wetland areas contributed to this is not clear. The importance of the area was probably built around Meols, which continued to be a thriving coastal port with middle and late Saxon, and Hiberno-Norse, metalwork and coins coming from the present beach in the nineteenth century (Bu'Lock 1960). Much of the former coastal wetland area was probably sand dunes by the tenth century, indeed the name Meols is from the Old Norse *melr*, a sandhill (Dodgson 1972). Several of the medieval villages to the south of the wetland belt seem to have had pre-Conquest origins, which might suggest that remaining wetland areas were marginalised for settlement locations, although still valuable as a resource. Wallasey village, lying on the summit of the ridge on the eastern side of Bidston Moss and the Pool, is of probable early foundation around the circular enclosure of St Hilary's church. The place name means 'Isle of the Welsh (or Britons)' from the fact that the tidal Pool would have isolated the area from the rest of the Wirral at high tide. St Bridget's at West Kirby, in the north-west coastal area, may also be an early

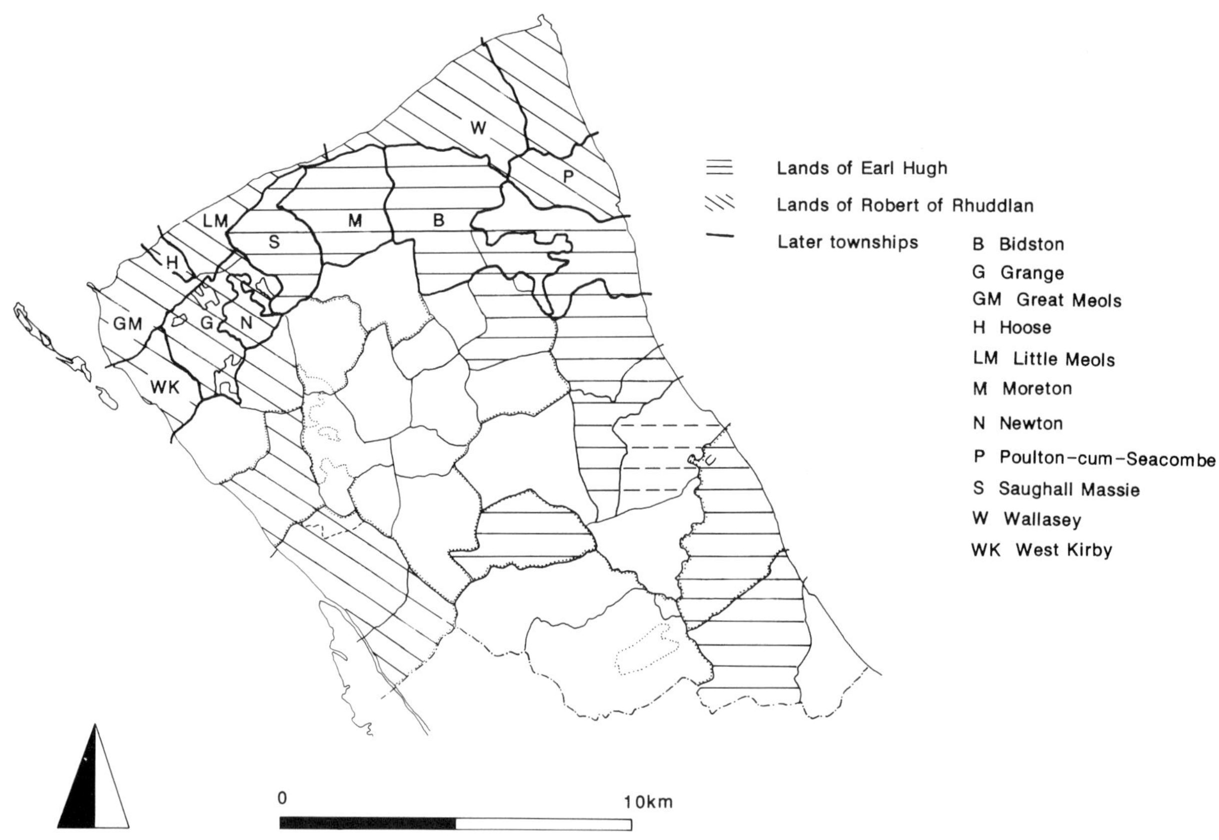

Fig 7 Wirral: Domesday manors and later townships (after Chitty 1978)

dedication, while a number of cross heads, shafts, and carved stones of tenth and eleventh century date also come from here, with a further one from Hilbre island (Chitty 1978, 8).

The settlement at Meols appears to have migrated south-westwards along the beach in the early medieval period, probably as a response to encroaching sand dunes. Most of the coastal strip was later used as common pasture land shared by a number of settlements, and it might be suspected that this was a continuation of an earlier pattern of landuse away from Meols. The economic importance of the mosses in the embayments in the pre-Conquest period is unknown, although it might be suspected that they played a role similar to the dune areas in providing grazing land as well as resources such as turf and wood.

Later medieval period AD 1066–1500

By the later medieval period the townships, some of which must have had their origins in the pre-Conquest period, can be traced across this part of the Wirral (Fig 7). Bidston Moss lay mostly within the township of Bidston-cum-Ford, and Newton Carr within the townships of Grange and Newton. The coastal belt was divided between Wallasey in the north-east, and Little Meols and Great Meols in the north-west, around present day Hoylake.

In Bidston, the moss was marginal to the main location of settlement, which was centred around the area where the present village now stands. The Kingston Estate map of 1665 by Thomas Taylor (copy in Merseyside SMR) preserves the layout and location of the medieval village around the parish church of St Oswald. The accompanying survey records seven farms and a number of cottages which exploited the well-drained arable land around the foot of Bidston Hill, together with the moss lying on the north side of the ridge. The moss, although marginal land for settlement, was still a vital resource for the inhabitants of the area. Documents suggest that, in addition to the peatlands, its landscape included woodland, probably scrub, and

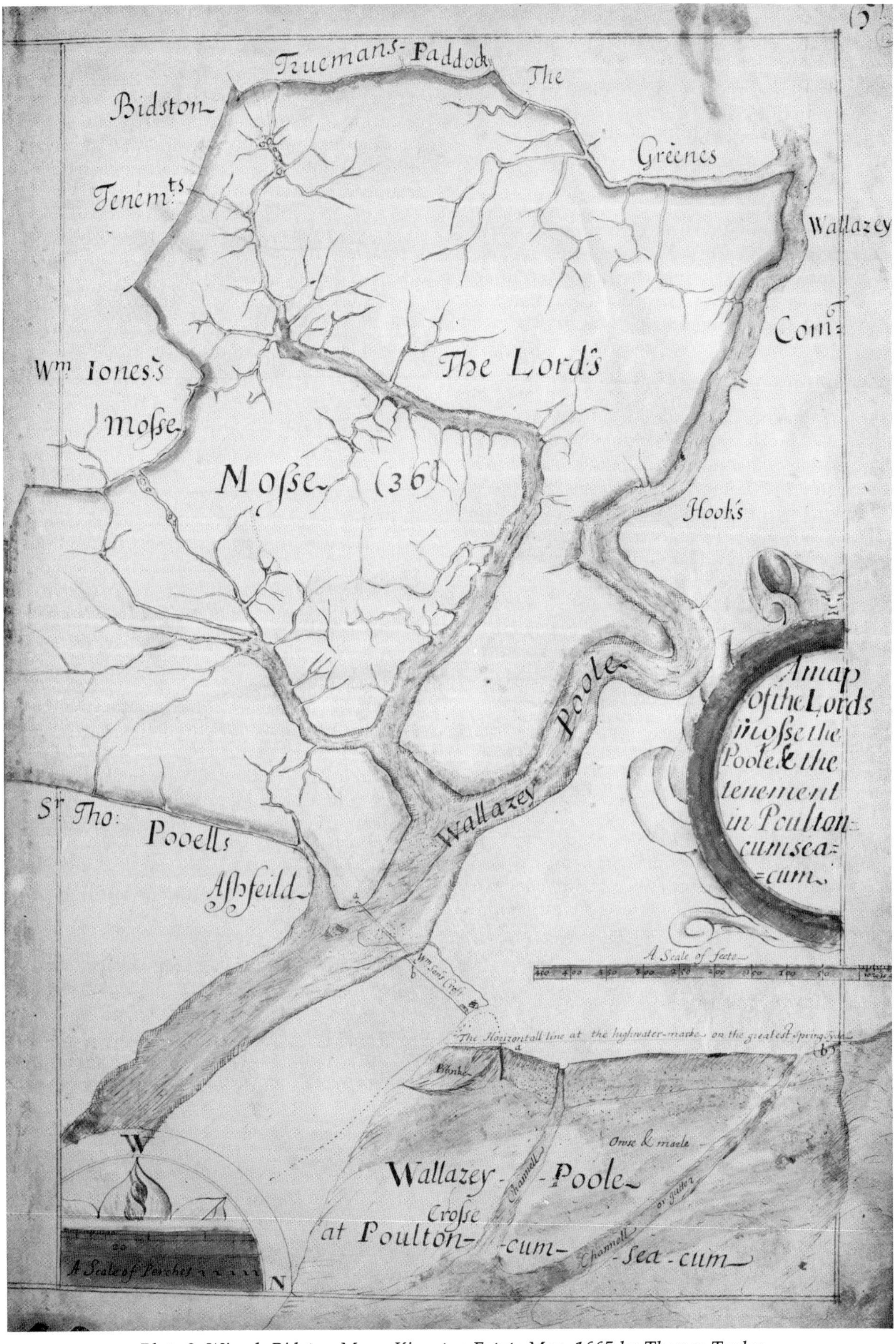

Plate 2 Wirral: Bidston Moss, Kingston Estate Map, 1665 by Thomas Taylor (reproduced with the permission of the Trustees of the Vyner Estate)

rough grassland at this time. The Forest Rolls provide evidence of exploitation for fuel, while the prior of Birkenhead claimed rights of turbary (cutting peat for fuel) and pasture on the moss in Bidston and Moreton in 1353 (Stewart-Brown 1925, 24). An Abbot Thomas was also allowed to take wagon loads of turf from the moss to his manors of Irby and Greasby (Brownbill 1936, 164).

There is less information from contemporary documents for the landuse in the other areas of wetland, but it seems very likely that the same kind of arrangements were being made for land that was largely held as common grazing.

Post-medieval period AD 1500–1850

In the early post-medieval period, Bidston Moss was still largely unreclaimed, being divided into a Town Moss, for the villagers' use, and the Lord' s Moss (Merseyside SMR). A series of enclosed strip fields ran down the northern slopes of the hill to the wetland edge, probably representing former strips in the medieval arable field of the township. The late post-medieval period saw the moss gradually being enclosed for farmland, although this was still not complete by the time of the tithe map of *c*1842 (LRO, EDT 47/2), when almost half the area still remained as unreclaimed mossland. In the late eighteenth century, the north side of the moss saw some industrial development exploiting the tidal currents of Wallasey pool, with mills and associated buildings. In the mid nineteenth century, when the docks were constructed, much of this was swept away.

With the growth in population along the North Wirral coast at about the same time as the construction of the docks, developments took place that opened the moss up to exploitation. Initially this came from the growth of the transport system to serve the new centres of population.

Palaeoecological survey

Bidston Moss (BM2) (SJ 285909)

The description of the gross stratigraphy of the Moss was provided by four cores from a 2m thick peat deposit resting on blue-grey clay and overlain in turn by a surface clay layer. Core BM-1 revealed a sequence similar to the others, with the exception of a thin clay layer in the lower part of the peat which may have been due to the effects of a stream to core BM-2. The altitude of the top of core BM-2 was levelled at +2.31m OD.

Pollen analysis

Samples were taken from core BM2 at 40mm intervals and the resulting pollen diagram is shown as Figures 8 and 9. Three samples were submitted for radiocarbon dating and the results are shown on the diagrams and in detail in Table 2. Five local pollen assemblage zones were recognised, and are described below.

BM2-a 2.00-1.98m The lowest pollen level constitutes a separate zone and is dominated by *Pinus*, with substantial frequencies of *Betula*, *Quercus*, and Coryloid pollen. Wetland taxa *Salix*, Gramineae, and Cyperaceae are present but very little other herb pollen occurs. Trees and shrubs account for over 80% of total land pollen. The base of the peat profile (1.96-2.0m) is dated to 6400-6080 Cal BC (7360±60 BP; SRR-2926).

BM2-b 1.98-1.74m Tree pollen values attain over 60% of total pollen, the main contributor being *Alnus*, but with significant frequencies for *Betula* and *Quercus* also. *Pinus* declines throughout the zone. Coryloid is low but consistent, while other shrubs *Hedera*, *Ilex*, and *Prunus* also occur. Herb values are low and confined mainly to aquatic types. *Pteridium* and *Filicales* fern spores achieve significant values.

BM2-c 1.74-0.86m Tree pollen values rise and contribute over 70% of total pollen, although they decline late in the zone. *Alnus* dominates the assemblage. All other tree types occur in low values and Coryloid pollen also is reduced in frequency. Sporadic grains of other shrub taxa occur. Herb pollen of both wetland and ruderal affinities are recorded, but only in low values. A range of marsh and fenland ferns is consistently present.

BM2-d 0.86-0.50m *Alnus* declines sharply in value and *Quercus* becomes the main tree pollen type. *Betula*, *Ulmus*, and *Fraxinus* increase in frequency, but tree pollen as a whole falls to around 40% total land pollen. *Calluna*, *Empetrum*, and Gramineae all increase in value and the range of shrub taxa recorded is reduced. Herb pollen increases, especially for dryland ruderal types including cereal-type grains. Spores of *Filicales* and *Pteridium* are prominent in the assemblage. *Sphagnum* is important for the first time.

BM2-e 0.50-0.32m *Ulmus* temporarily declines at the start of the zone and tree pollen frequencies as a whole are reduced, mainly due to a sharp rise in Gramineae pollen values. *Quercus* remains the most abundant tree pollen type. *Pinus* increases slightly and Coryloid pollen and *Betula* values also rise. *Calluna* remains important. A wide range of ruderal weeds occurs, including cereal-type pollen. *Filicales* and *Sphagnum* spores are still present in high values. The upper boundary of this zone (0.32-0.36m) is dated to 3690-3360 Cal BC (4740±70 BP; SRR-2924).

Ecological history

Although the basal grey clay was sampled for diatoms and pollen, no microfossils were recovered from the sediment and its environment of deposition must remain conjectural at present. It is certainly water lain but may be either estuarine or freshwater in origin. If the clay is marine or estuarine in origin, however, the later clay/peat contact would be a regressive overlap, indicating a fall in sea-level and the growth of peat upon the exposed marine clay surface. In this case it would be expected that the pollen of nearby saltmarsh plant communities would be present in the basal peat sample. No such indicator taxa are recorded and so it might be more likely that the lower grey clay is a freshwater deposit, formed under swamp or lake conditions. The altitude of the top of this clay, at +0.31m OD, would also be rather too high to correlate satisfactorily with the known height of sea-level in North Wirral about 6400-6080 Cal BC (7360±70 BP; SRR-2926) as deduced from its coastal history and other sea-level index points.

The onset of biogenic accumulation records the colonisation of the site by reedswamp communities dominated by *Phragmites*, the rhizomes of which are present in the lower peat and basal clay. The substantial grass pollen frequencies are probably attributable to this plant, and the identification of *Typha angustifolia* and *Nymphaea* pollen reflect this open water/reedswamp environment. Succession to fenwood communities took place, as *Alnus* pollen is abundant and sedge-fen taxa are prominent, including *Filipendula*, *Caltha*, and *Thelypteris*. The high early record of *Salix* and *Betula* may represent an early stage in the establishment of fen carr communities before water depth and quality had progressed far enough to permit the formation of high nutrient status alder swamp that the pollen implies.

A mosaic of open water, fen, and alder swamp seems to have persisted for much of the site's ecological history. The termination of the *Alnus* swamp phase occurs at the end of pollen zone BM2-c when the increase in *Quercus* values may signify the change to oak-dominated drier fenwood vegetation. Some acidification of the environment may also have occurred, if the rise in *Calluna* pollen is associated with mire development rather than changes on the nearby higher ground around the site. Until this point little change in the alder swamp landscape seems to have taken place and alder wood was recovered from this part of the profile. Organic accumulation was relatively slow for the peat is well humified and compact. The long persistence of alder swamp communities suggests that the site was subject to seasonal flooding for much of its history, providing a regular influx of nutrients essential for the continuation of this minerotrophic ecosystem. The low elevation of the basin relative to Ordnance Datum may have been an important factor here. Rosaceae, Ranunculaceae, *Potentilla*-type, *Polygonum*, *Succisa*, and Umbelliferae are all found in marsh-fenwood habitats while *Anemone*, *Malva*, and *Sanguisorba* reflect rather drier but still damp locations. Ferns were abundant within the rich damp shady habitats. *Osmunda* and *Polypodium* are prominent, the latter probably in epiphytic situations and the former colonising the drier hummocks within the fenwood, within which *Rhamnus*, *Euonymus*, and *Viburnum* all found suitable habitats.

The upper two pollen zones at the site (BM2-d/e) reflect a much drier, perhaps more acid environment probably due to terrestrialisation through vegetation succession and reduced groundwater influence. A marked reduction in water depth and quality is shown by the appearance of the bog moss *Sphagnum*, the dwarf shrubs *Calluna* and *Empetrum*, and a rise in Cyperaceae pollen. *Eriophorum* was noted macroscopically and a transition to the early stages of an ombrotrophic system seems to have been under way at this time. This trend continued and the site developed into a raised mire, eventually being drained and reclaimed in historic times. The thin organic clay layer which forms the surface deposit is much affected by oxidation and other disturbance. It does appear to be a discrete clay horizon similar to that described by Davies and McMillan (1956), although there are no contained micro or macrofossils to suggest a freshwater or estuarine origin. The relatively low altitude of the layer at +2m OD at *c*3690-3360 Cal BC (4740±70 BP; SRR-2924) would suggest that it may well be an estuarine facies.

Landscape history and forest disturbance

Several phases are recognised throughout the diagram in which the presence of open habitat dryland weed pollen grains suggests the creation of breaks in the forest cover through vegetation disturbance. These often coincide with fluctuations in tree and shrub pollen curves which support some opening of the tree canopy. The history of woodland disturbance at the site is described on Figures 8 and 9, upon which the following sixteen phases are based.

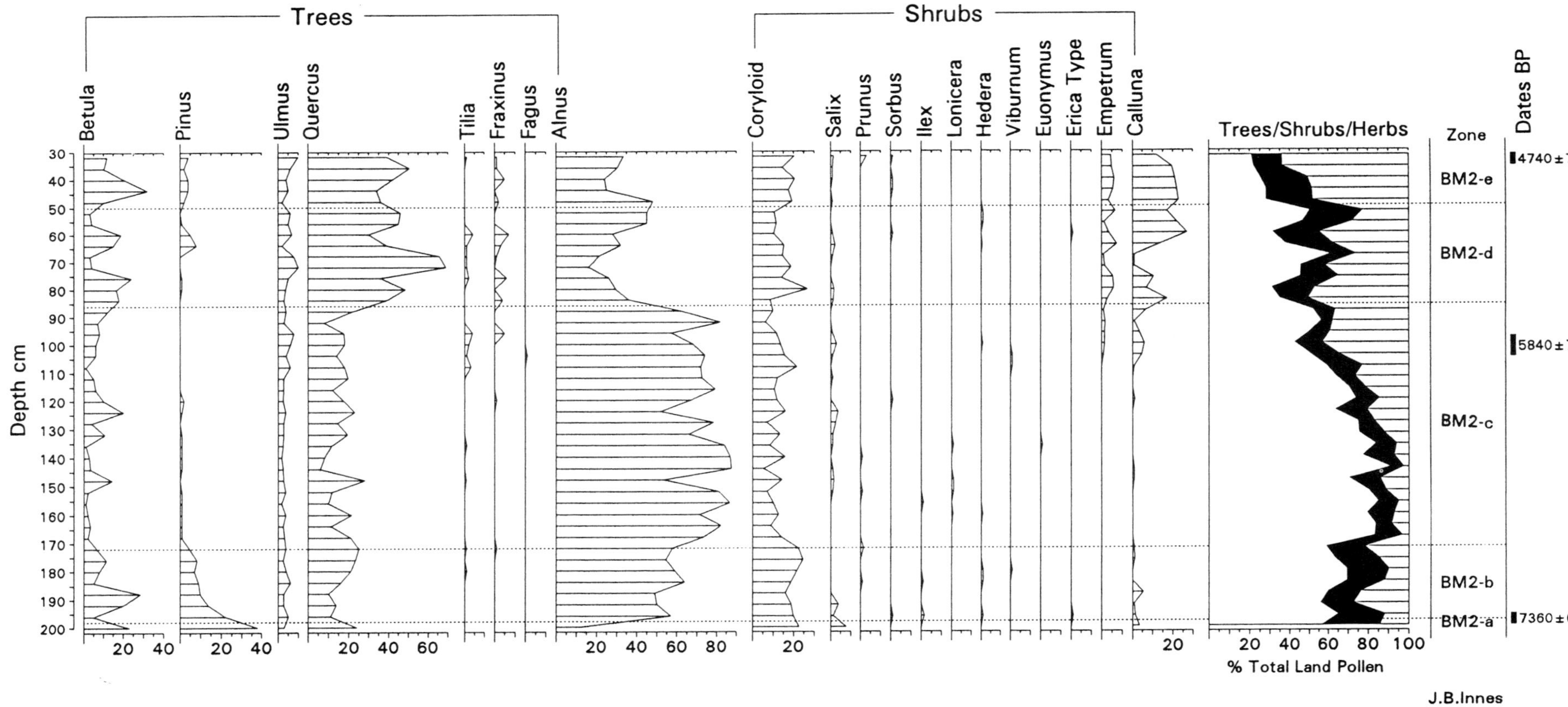

Fig 8 Wirral: Bidston Moss 2, percentage arboreal pollen diagram (trees plus group)

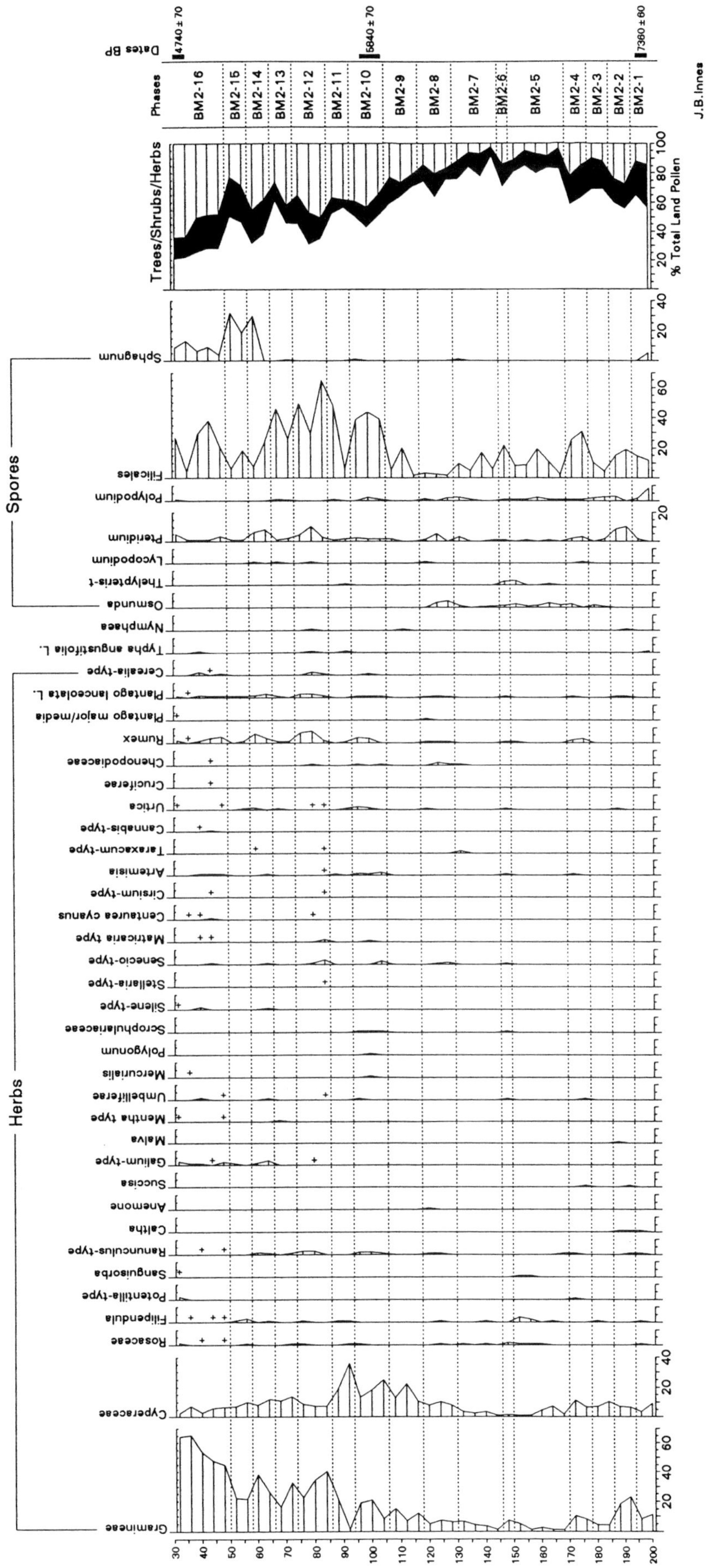

Fig 9 Wirral: Bidston Moss 2, percentage herb pollen diagram (trees plus group)

Pollen phases

BM2-(1) 2.00-1.94m The basal phase records a landscape of mixed woodland upon the drier areas around the site itself. *Pinus*, *Quercus*, and *Betula* were the main constituents, the low *Alnus* values supporting the radiocarbon date of 6400-6080 Cal BC (7360±60 BP; SRR-2926) for peat inception just prior to the Flandrian I/II boundary. There is no pollen evidence of disturbance to woodland at this stage.

BM2-(2) 1.94-1.86m There is limited evidence for the creation of open ground during this phase, with *Ulmus*, *Quercus*, and *Alnus* all slightly reduced in value. The steady decline of pine may also reflect this change, but it is also to be expected in early Flandrian II as part of a natural process recorded widely around this time, when it was being replaced in the woodland cover by other species, particularly *Alnus* (Godwin 1975). The appearance of *Plantago lanceolata* is supported by increased frequencies of Gramineae and *Pteridium* and small peaks in *Betula* and *Calluna*. These changes suggest the localised creation of grassy clearings and subsequent regeneration to woodland.

BM2-(3) 1.86-1.78m The minor forest opening of Phase 2 apparently was only temporary, for dense woodland, dominated by *Quercus*, *Alnus*, and *Ulmus* became established in phase 3. There are no indications of continued disturbance of the woodland cover, herb values being very low and confined to probable wetland types represented by the curves for grasses and sedges. The few grains recorded of dryland shrubs, such as *Ilex* and *Prunus*, are not evidence of a broken tree canopy. *Alnus* remains dominant probably due to its local abundance.

BM2-(4) 1.78-1.70m Renewed woodland disturbance, although of a limited scale, occurs in this phase. Tree pollen values are depressed, with *Alnus* most affected, and herb taxa representative of open ground reappear in the assemblage. The presence of *Plantago lanceolata*, *Rumex*, and *Artemisia* indicates some clearings within the forest, and Gramineae, *Calluna*, Coryloid, and *Pteridium* all increase in frequency and suggest a more open vegetation.

BM2-(5) 1.70-1.50m This phase represents a period of woodland regeneration and stability during which disturbance of the vegetation seems to have been absent. Tree pollen accounts for over 70% of total pollen, and although much of this is due to local *Alnus* abundance, the frequencies of dryland trees, *Quercus* and *Ulmus*, are also high. Herb pollen counts are confined to wetland varieties and heliophyte taxa are very poorly represented. Even the curve for Coryloid pollen is low. If any forest disturbance was taking place in the area, the screening effect of the dense local stands of alder may have prevented it being recorded in the pollen assemblage.

BM2-(6) 1.50-1.46m At the single level described by this phase a brief but significant episode of woodland disturbance takes place. *Alnus* frequencies fall sharply while those of *Betula* and *Quercus* increase. This is probably due to improved transport of pollen to the site after opening of the dense local stands of *Alnus*. *Betula* and *Quercus*, together with *Ulmus*, therefore, probably formed the woodland on the drier ground away from the mire itself. Among the indicators of more open conditions which increase in frequency are Gramineae, *Salix*, Coryloid, and Rosaceae. *Plantago lanceolata*, *Rumex*, *Artemisia*, and *Senecio*-type also occur.

BM2-(7) 1.46-1.30m This phase represents a period of regeneration of closed woodland, once again dominated by local alder but with mixed deciduous forest forming the regional vegetation. There is no evidence of any disturbance of these woodland communities. Non-tree pollen is very low, and confined to a few wetland herb types of very local origin.

BM2-(8) 1.30-1.18m Ruderal herb taxa return to the pollen curves which suggest a period of significant woodland recession. *Plantago lanceolata*, *Plantago major*, *Rumex*, Chenopodiaceae, and *Senecio*-type occur, and *Pteridium* values also rise. Local *Alnus* populations seem to have been most affected by this disturbance. Once again *Quercus* and *Betula* are the trees displaying increased values as *Alnus* frequencies fall, but a small peak of *Pinus* pollen also occurs. Slight rises in *Salix* and *Calluna* values take place. The pollen evidence suggests a small-scale removal of the alder carr near to the sampling site and its temporary replacement by open ground and regeneration communities.

BM2-(9) 1.18-1.06m Regeneration of woodland seems to have been complete after the limited disturbance of the previous phase. *Alnus*, with *Quercus* and *Ulmus*, once again dominates the tree flora with Coryloid the only significant understorey shrub. Cyperaceae and Gramineae values increase but there are no indications of continued woodland disturbance.

BM2-(10) 1.06-0.94m This phase records a major episode of forest disturbance with a sharp increase in the non-tree pollen count and a fall in *Alnus* frequencies in particular. *Tilia*, *Fraxinus*, and *Pinus* are all increased and *Fagus* also occurs, suggesting some change in local dryland woodland composition at this time. The shrubs *Calluna, Salix,* and Coryloid increase in value and may reflect more open conditions. Total herb pollen values are high and, while much of the rise is due to increased Gramineae and Cyperaceae counts, peaks also occur in ruderal herb indicators of forest clearance and open ground. Among these are *Plantago lanceolata, Rumex, Artemisia, Senecio*-type, Umbelliferae, *Matricaria*-type, Chenopodiaceae, and *Polygonum. Pteridium* values are also significant. Most importantly, Cerealia-type pollen, of *Triticum*-type, is recorded in association with this range of disturbance evidence. The centre of this phase, between 0.97-1.03m, has been radiocarbon dated to 4900-4530 Cal BC (5840±70 BP; SRR-2925).

BM2-(11) 0.94-0.86m A short period without disturbance occurs in this phase, during which tree pollen values recover most of their former dominance within the assemblage. Most of the total non-tree pollen may be attributed to local mire sources such as Cyperaceae. *Alnus* again dominates the pollen record, with *Quercus* and *Ulmus* the major dryland forest trees. Sporadic records of weed pollen are not sufficient to indicate continued open ground.

BM-2-(12) 0.86-0.74m Extensive forest clearance is recorded in this phase, with tree pollen reduced to 30% of total pollen, and a wide range of indicators of open and disturbed ground are recognised. *Alnus* values fall sharply and are replaced by *Betula* and *Quercus* as the main tree types. A major removal of the local dense alder carr seems to have occurred. High values for Coryloid, *Calluna, Betula,* and *Salix* indicate a more open vegetation. Among grassland herbs *Rumex* reaches almost 20% of tree pollen and *Plantago lanceolata,* Gramineae, and *Pteridium* are present in high frequency. *Taraxacum*-type, *Urtica, Ranunculus*-type, and *Senecio*-type are also prominent. Possible weeds associated with agriculture include *Matricaria*-type, *Centaurea* cyanus, Chenopodiaceae, *Stellaria*-type, and *Cirsium*-type. Several pollen grains of Cerealia-type are also recorded.

BM2-(13) 0.74-0.66m This phase forms a brief interlude of forest regeneration as ruderal pollen frequencies are much reduced and tree pollen values rise. With the removal of local *Alnus* dominance, *Quercus* is clearly the major woodland tree and contributes up to 70% of the tree pollen count. *Ulmus* is also common in the woodland. Open habitat taxa are almost absent and *Calluna*, Coryloid, and *Betula* are very low, which suggests that regeneration in this phase was to mature deciduous woodland rather than secondary or scrub communities. Replacement of alder carr by oak fenwood could also account for some of the enhanced *Quercus* frequencies. There is no evidence of human activity at this level.

BM2-(14) 0.66-0.58m This is a further phase of forest disturbance during which *Quercus* values are greatly reduced and *Betula, Pinus, Tilia,* and *Fraxinus* all increase in frequency as a result. There is a small decline in *Ulmus* pollen values. No cerealia-type pollen is recorded, however, and a reduced range of weed pollen occurs which is characterised by grassland rather than cultivation related taxa. *Plantago lanceolata* and *Rumex* are again important, and are joined by *Urtica* and *Galium*-type, although some species of the latter could be natural components of fen vegetation. Only single grains of *Taraxacum*-type, *Artemisia,* and *Senecio*-type occur. Peaks of *Pteridium* and Gramineae occur. *Calluna* and *Betula* increase markedly.

BM2-(15) 0.58-0.50m This phase is one of woodland regeneration with tree pollen values increased to about 50% of total pollen. *Quercus, Ulmus,* and *Alnus* are the main tree taxa which recover and presumably recolonise previously cleared areas. *Calluna* remains high and some areas may well have failed to regenerate tree cover. No active disturbance was taking place, however, for ruderal pollen counts are very low and most herb grains are of wetland type.

BM2-(16) 0.50-0.32m The final phase records a major recession of woodland cover and its replacement by a landscape dominated by grassland and areas of probable cultivation. Tree pollen declines to only 20% of total pollen. *Ulmus* falls at the start of the phase before recovering. Other tree and shrub types do not fluctuate greatly and it seems the character of the remaining woodland was not much changed, although *Betula* in particular, and *Pinus* and *Fraxinus* increase their values slightly. The big rise in non-tree pollen is due mainly to greatly increased Gramineae and a real spread of grassland seems to have occurred, although mire grasses will have contributed to the grass pollen curve. *Plantago lanceolata, Rumex, Galium*-type, and *Pteridium* all reflect the increased open conditions, as does the maintained *Calluna* curve. The earlier part of the phase, however, is characterised by indicators of cultivation. Cerealia-type pollen is consistently present with *Plantago major/media, Artemisia, Matricaria*-type, Cruciferae, Chenopodiaceae, *Centaurea cyanus,* and Caryophyllaceae, all of which may represent arable weeds. The upper boundary of this phase (0.32-0.36m) was radiocarbon dated to 3690-3360 Cal BC (4740±70 BP; SRR-2924).

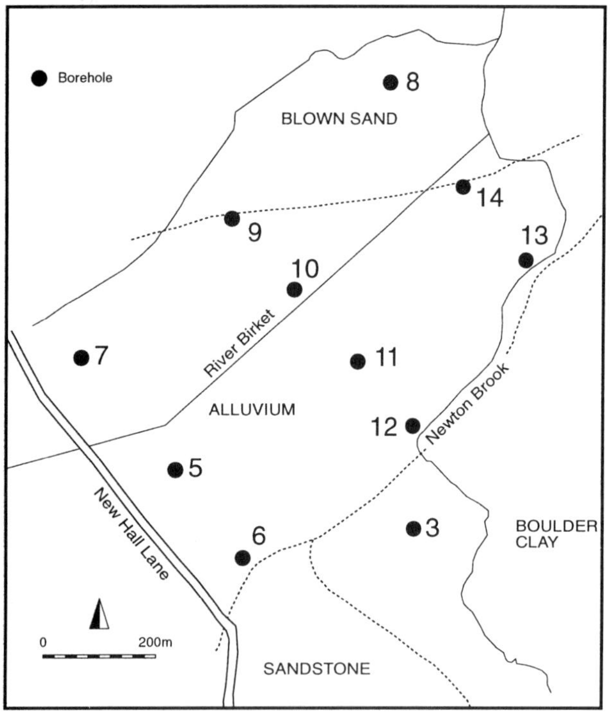

Fig 10 Wirral: Newton Carr borehole locations (Innes et al 1990)

Newton Carr (SJ 225885)

The surface sediment distribution and the location of the previously existing boreholes in the Newton Carr depression are shown on Figure 10. Based upon this expanded stratigraphic data (Fig 11), the deepest Holocene sediment is a thin, highly compact peat with wood fragments which rests on sand over boulder clay at an altitude of -5.3m to -5.1 m OD in the most westerly borehole (7). It is overlain by a cohesive blue-grey silt which gives way to a second thin peat between -2.45m and -2.25m OD. Peat is also present at this level in boreholes 8, 9, 10, and 14. In boreholes 8 and 9 this peat overlies silts and sands, but in boreholes 10 and 14 the peat rests upon the pebbly clay surface of the pre-Holocene drift. As with the lower peat in core 7, these latter two strata are basal peats which allow the recognition of the rising drift surface towards the southern rim of the basin. This mid-altitude peat layer is overlain in all cases by clayey silts or silty sands, the sandier facies predominating in a seaward direction. The clayey silt facies of this unit is also recognisable in the next, more southerly, transect of cores, where it lies directly on the boulder clay surface in cores 5, 11, and 13. There appears to be

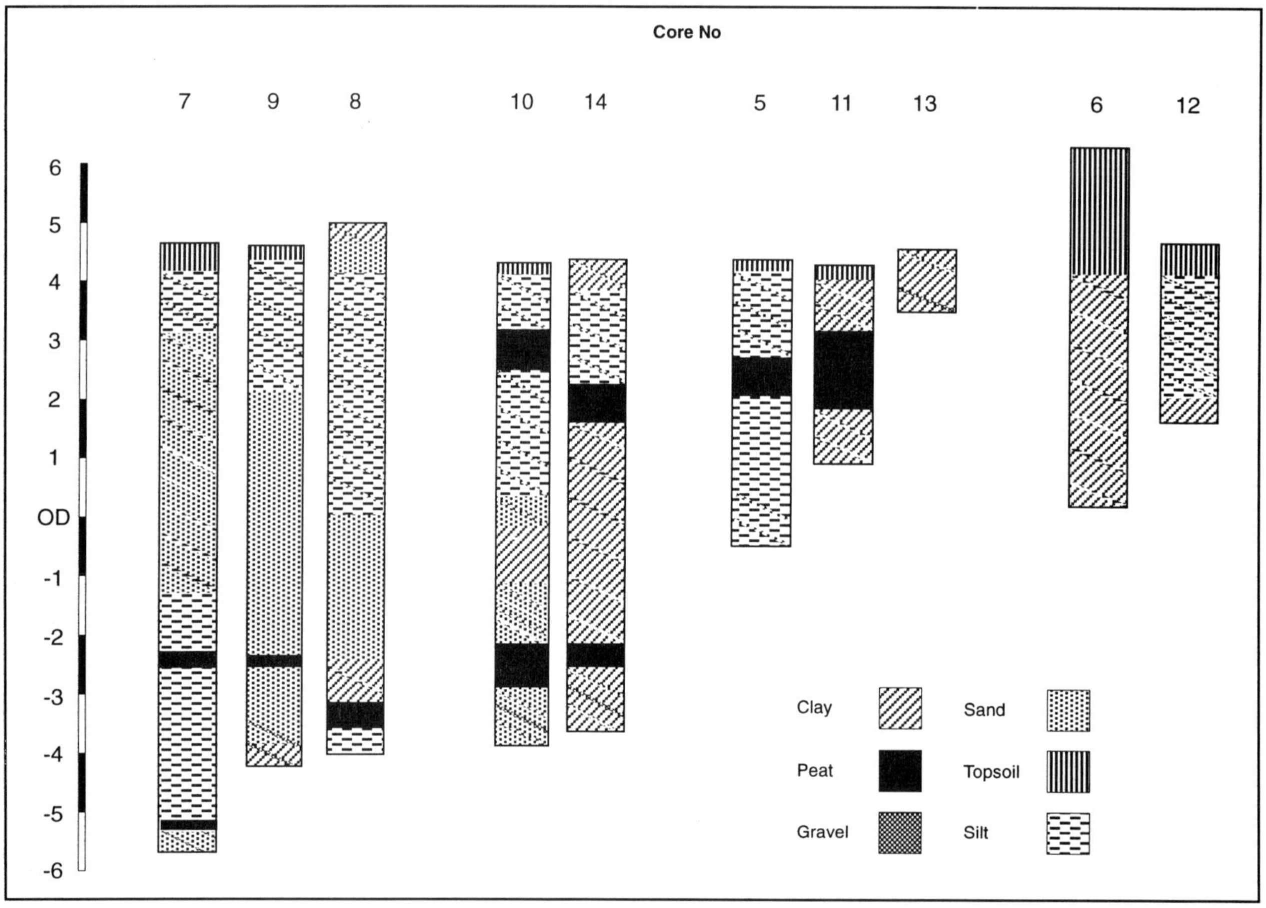

Fig 11 Wirral: Newton Carr stratigraphic diagram (Innes et al 1990)

a major break of slope between transect 10 to 14 and transect 5 to 13, the drift surface rising from about -3m to about +1m OD over a relatively short distance. The edge of the true depression feature itself may lie between these two transects, as the landward gradient from this point is shallow. No basal peats are recorded in, or upslope from, transect 5 to 13.

Resting upon this clayey silt facies is an upper peat horizon which appears only in the central part of the basin, being present in cores 5, 10, 11, and 14. To the north silts and sands continue to the surface, while to the south clays or the pre-Holocene land surface are present at this altitude. This upper peat layer, lying between about +2 and +3m OD, may be regarded as analogous with the Upper Peat/Forest Bed of earlier workers (de Rance 1872a; Morton 1897). Kenna (1986) has suggested that in places in North Wirral this upper organic succession is divided by a thin silty clay. At Newton Carr such a thin clay was proven in a number of cores between boreholes 10 and 14, and has also been recorded at other sites on this coast, as at Park Road, Meols (*see below*), apparently confirming Kenna's observations. The age of this thin clay intrusion will be discussed further below.

The upper peat bed at Newton Carr is covered by a bed of alluvial silty clay which extends across the whole site, forming the upper unit of the minerogenic succession in cores 7, 8, and 9, and laps over the pre-Holocene surface in cores 6, 12, and 13, where it forms the most landward extension, at the highest altitude, of the Holocene sediment sequence which fills this major basin. To the south and east of these points, the boulder clay outcrops at the surface at about +5m OD. The silty clays at core 8, elsewhere exposed at the surface, are covered by a thin sheet of blown sand, the landward limit of the late Holocene dune system which occupies the North Wirral fringes from West Kirby on the Dee to New Brighton on the Mersey.

Pollen analysis

Pollen analysis was undertaken from three sites at Newton Carr, where three boreholes were selected for more detailed examination, and their detailed stratigraphy is described below.

Newton Carr 11 (NC11) (SJ 227882)

Here a sampling interval of 100mm was used, closing to 50mm at the upper contact to ensure that conditions at the transition from peat to clay were examined. Pollen preservation was generally good. The results are shown in Figures 12 and 13. The following four local pollen assemblage zones were recognised.

NC11-a 2.90-2.45m The lowest pollen zone is characterised by *Quercus*, *Ulmus*, and *Alnus*, with a little *Betula* and Coryloid. Several herb pollen types are recorded, the most significant of which are Chenopodiaceae, *Artemisia*, and Gramineae. Tree and shrub pollen represents up to 80% of total pollen. The upper boundary of the zone is defined by a fall in *Ulmus* pollen values.

NC11-b 2.45-1.85m This is characterised by *Quercus* and *Alnus*, again with *Betula* and Coryloid. *Salix* and *Tilia* percentages rise. Tree pollen totals fall to 50% of total pollen. Gramineae frequencies increase and the range of herb types present increases; *Aster*-type pollen is prominent. Aquatic pollen taxa are strongly represented during this zone. Chenopodiaceae and *Artemisia* maintain consistent curves throughout, and some grains of ruderal pollen types like *Plantago lanceolata* are also present. The upper boundary of the zone is defined by a fall in *Quercus* pollen and a rise in *Betula*.

NC11-c 1.85-1.45m *Betula* dominates the assemblage, with lower values of *Alnus*, Coryloid, and *Quercus*. Tree pollen achieves 80% of total land pollen. All herbaceous pollen types, including Gramineae, fall to very low frequencies indeed. There is a slight peak in *Sphagnum* values. The upper boundary of the zone is defined by a fall in *Betula* frequencies and a rise in *Alnus*.

NC11-d 1.45-1.35m The zone is characterised by *Alnus*, with lower frequencies of *Betula*, *Quercus*, and Coryloid, the latter rising sharply at the end of the zone. The range of shrub and herb types recorded rises and isolated peaks of *Ilex* and *Ulmus* occur. Tree pollen totals fall throughout the zone. Many types of coastal and ruderal herbs are recorded at the upper boundary of the zone, with Chenopodiaceae prominent.

Newton Carr 7 (NC7) (SJ 221883)

Pollen analysis was also undertaken on the peat layers in the Newton Carr 7 profile, and the following three pollen assemblage zones were recognised (Fig 14).

NC7-a 9.98-9.92m The lowest zone is characterised by *Quercus*, *Betula*, and *Pinus*, with high frequencies also for *Ulmus* and Coryloid. Very high Gramineae values occur, but other herb pollen frequencies are low. The upper boundary of the zone is characterised by a sharp rise in *Alnus* pollen values and falls in *Quercus* and Gramineae.

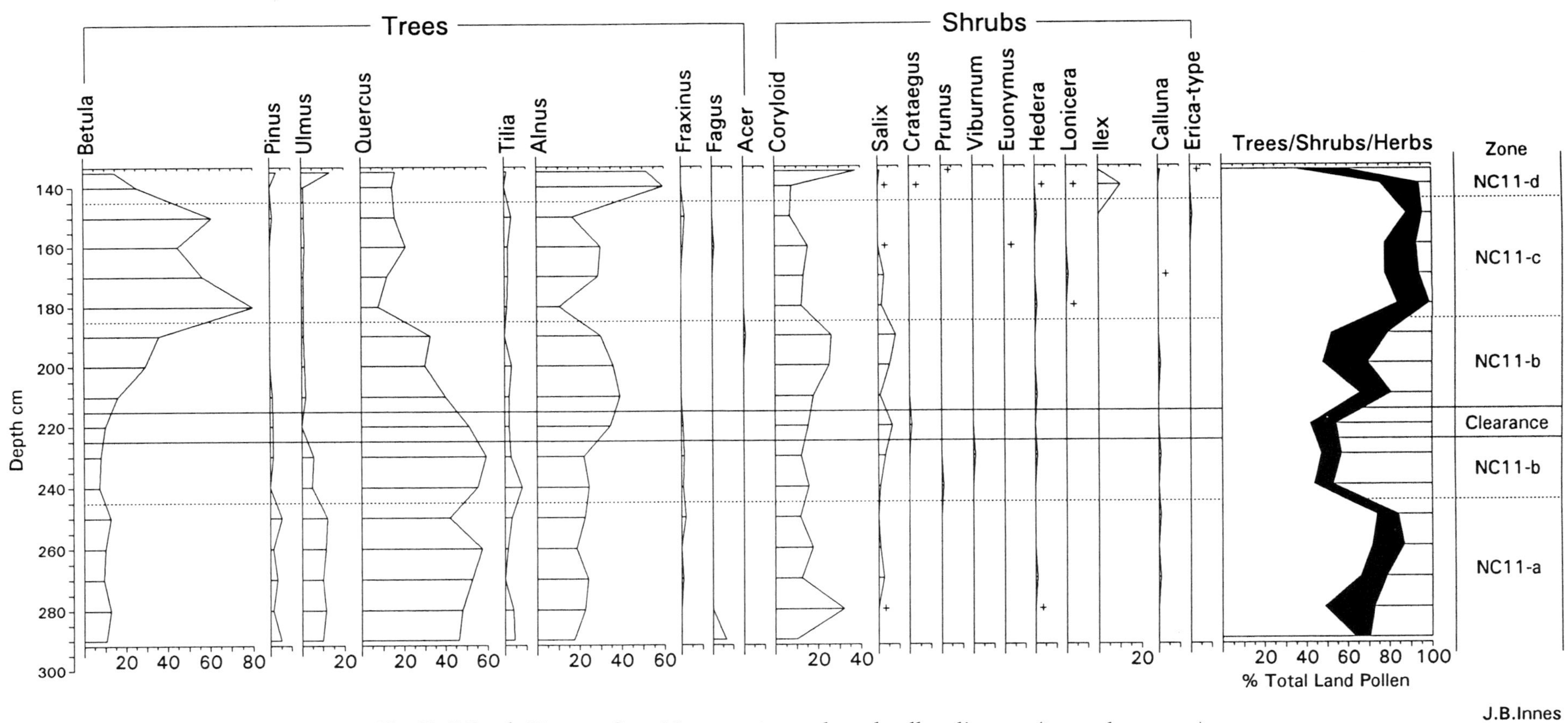

Fig 12 Wirral: Newton Carr 11, percentage arboreal pollen diagram (trees plus group)

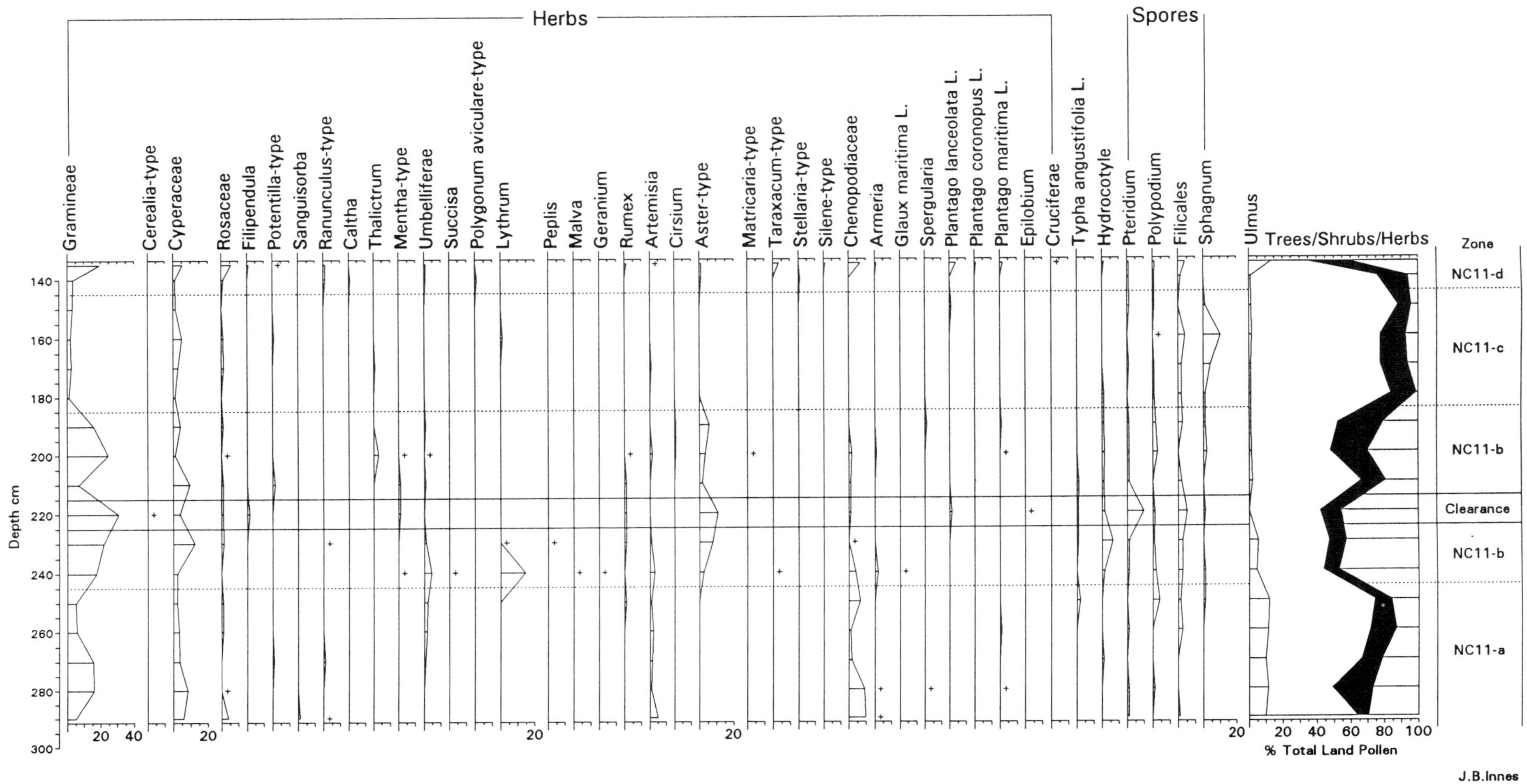

Fig 13 Wirral: Newton Carr 11, percentage herb pollen diagram (trees plus group)

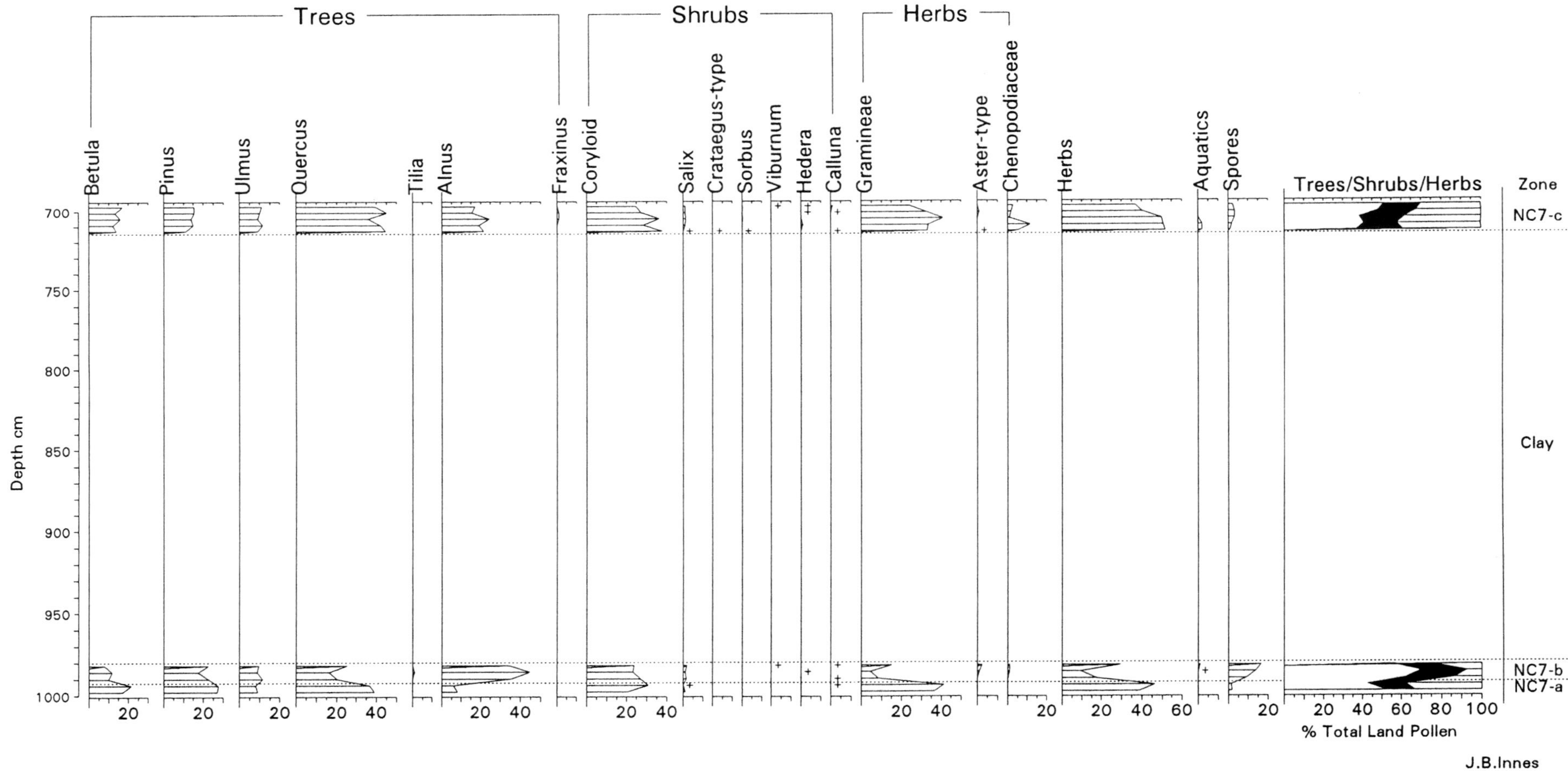

Fig 14 Wirral: Newton Carr 7, percentage arboreal and herb pollen diagram (trees plus group)

Fig 15 Wirral: Newton Carr 8, percentage arboreal and herb pollen diagram (trees plus group)

NC7-b 9.92-9.84m The zone above is characterised by *Alnus*, although all other major tree pollen types are present in high frequencies. Herb pollen is poorly represented, but coastal herb frequencies increase towards the top of the zone, including Chenopodiaceae and Compositae Tubuliflorae (*Aster*-type). The upper boundary of the zone is placed at the transition from peat to clay sediments.

NC7-c 7.14-6.93m The uppermost zone is characterised by *Quercus*, *Alnus*, and Coryloid. *Betula*, *Pinus* and *Ulmus* are also important. Gramineae is the main herb type but a wide range of other herb types occur, many of which are of coastal habitats, particularly Chenopodiaceae.

Newton Carr 8 (NC8) (SJ 228888)

At the third site pollen analysis was undertaken on the thin peat bed recorded in mid-profile and the following pollen assemblage zone was identified (Fig 15).

NC8-a 8.57-8.22m This is characterised by *Alnus*, *Quercus*, and Coryloid pollen with lesser amounts of *Betula*, *Pinus*, and *Ulmus*. Gramineae is the only herb type present in quantity but a range of other herbs is present in small values. Coastal herb pollen is recorded at the top and bottom of the zone adjacent to the transition from peat to clay sediments.

m OD	Lithology	Pollen Analysis	Approx age (Cal BC)	Chronozone
	Upper Marine Facies			F III
+3.2	Upper Peat	NC11-d NC11-c NC11-b *Quercus-Alnus*	3780	
		NC11-a *Quercus-Alnus* *Alnus*		F II
-1.5	Middle Marine Facies			
-3.0	Middle Peat	NC7-c *Quercus-Ulmus* NC8-a *Alnus*		
-3.5	Lower Marine Facies			
-5.1	Lower peat	NC7-b *Quercus-Ulmus* *Alnus*	6050	
-5.3		NC7-a *Quercus-Pinus*		F I

Table 6 Wirral: Newton Carr environmental record summary (after Innes et al 1990)

Ecological history

By analogy with the extensively radiocarbon dated regional palynological type site for the Flandrian from Red Moss (Hibbert *et al* 1971) Zone NC7-a may be assigned to Flandrian chronozone I due to its low *Alnus* and high *Pinus* values. The boundary between zone NC7-a and NC7-b may be relatively dated, therefore, to a little before 6000 Cal BC. High *Ulmus* and *Alnus* pollen frequencies in zones NC7-b, NC7-c, NC8-a, and NC11-a allow these zones to be referred to Flandrian chronozone II, between *c*6180-5720 Cal BC and 3990-3640 Cal BC (7107±120 to 5010±80 BP; Q-915/Q-912). Reduced *Ulmus* pollen values in zones NC11-b, NC11-c, and NC11-d suggest that these zones are of post-elm decline date and thus later than *c*3990-3640 Cal BC in chronozone Flandrian III.

The oldest deposit, in Zone NC7-a, is a basal peat and reflects conditions around Newton Carr during the period of peat inception, probably due to rising water-tables locally. This in turn may well have been due to the effects of sea-level rise at this time. The area appears to have been heavily wooded, but with an increasing degree of wetland plant communities coming into existence as the mire expanded. Mixed pine, birch, and hazel woodland was the dominant local vegetation. The high grass count probably results from the creation of local reedswamp conditions with the rising water-table.

Ecological conditions during Flandrian II are dominated by the effects of repeated marine inundation of the site, most clearly seen in the stratigraphy by the presence of layers of marine clay, but also causing the change from freshwater environments to estuarine environments via saltmarsh communities. The marine origin of the blue-grey clays is demonstrated by the presence of saltmarsh taxa at the peat-clay boundaries, as well as by the presence of marine and brackish diatoms within the clay itself. The middle phase of peat accumulation of Flandrian II age (Fig 11; Table 6) represents a period of relative fall in sea-level and the expansion of terrestrial and freshwater ecological conditions into the Newton Carr depression. These wetland areas were dominated by fen carr and reedswamp, with alder very common and with oak and birch also present in the carr community. The more regional dryland vegetation comprised mixed deciduous forest in which elm, oak, and hazel were most abundant. Pine was still able to find suitable habitats in the area, probably upon thinner sandy soils within this coastal zone.

Near the end of Flandrian II, before 3990-3640 Cal BC, a greater extension of drier conditions and withdrawal of marine influence occurred, and the pre-elm decline vegetation around Newton Carr 11 was dominated by woodland in which oak was more abundant than alder. Other trees such as pine and birch had by now declined in abundance. Wetland habitats were still characteristic of the local vegetation, however, and soon after the elm decline, with the regional vegetation maintaining a dense oak/alder woodland, there are signs of increased wetness within the peat profile. Aquatic pollen types increase sharply and some coastal types occur. It is probable that a rise in sea-level at that time caused a local raising of the water-table, although Newton Carr 11 itself was not directly transgressed by the sea. Following this local flooding phase drier conditions occur in zone NC11-c, and while oak and alder woodland probably remained dominant regionally, birch carr communities occupied the immediate environs of the site. These drier and more acidic conditions prevailed until a further rise in sea-level again caused flooding of the peat at the site, allowing alder carr to replace birch and finally culminating in the renewed inundation of the peat surface by estuarine conditions after passing through a phase of saltmarsh communities.

The relationship of the pollen analytical work to the lithology is summarised in Table 6.

Landscape history and forest disturbance

The pollen records from Newton Carr 7 and 8 do not contain characteristics that might suggest human impact upon the environment during the earlier Flandrian II period. In the longer pollen record from Newton Carr 11, however, there are indications in the middle of zone NC11-b of some limited forest clearance. A single grain of cereal-type pollen occurs at 2.20m depth at a point not long after the elm decline and so in a possible Neolithic context. Some caution must be exercised in definitely attributing such a grain to cereal cultivation for, as discussed elsewhere, there are a number of other grass pollen types of similar size and character which, like *Agropyron*, are found in coastal situations. This large grass grain is, however, of the correct size and morphology for *Triticum*-type pollen and occurs with a number of other indicators of forest clearance such as a peak of *Plantago lanceolata* and *Pteridium*. While herbs of disturbed or open ground may occur naturally at the coast, a phase of Neolithic cultivation in the coastal zone during the period of lowered sea-level in early Flandrian III could be one explanation for its presence.

Park Road, Meols (PRM) (SJ 239901)

The site chosen for detailed analysis contained a complex sequence of intercalated organic and clastic sediments. A lower blue-grey clay was overlain by a thick (*ie* 0.65m) peat bed which itself rested beneath a grey clay. This upper clay graded into a disturbed and oxidised surface deposit. Within the upper part of the peat bed a thin (*ie* 10mm) grey clay layer occurred which seemed to represent the extreme landward limit of a clay wedge which became thicker to seaward and formed a significant stratigraphic layer closer to the modern coast. Samples through this sequence were collected in monolith tins by R Kenna and P Tomlinson and returned to the laboratory for further analyses. The surface altitude of this section from Park Road was levelled to +3.52m OD.

Pollen analysis

Pollen analysis was conducted on the organic strata in the Park Road succession at intervals of no more than 30mm in the centre of the peat layers, and closing to 10mm near the peat/clay boundaries. Pollen preservation was generally good, although some corrosion of grains was noted in the upper levels of the upper peat. The results are shown in Figures 16 and 17. In addition to the radiocarbon date previously obtained by Kenna for the boundary between strata 4 and 5, three samples were submitted for radiocarbon dating and the results are shown on the diagrams and in detail in Table 5.

PRM-a 1.25-0.825m The lowest zone is characterised by *Quercus*, *Ulmus*, and *Alnus* with lesser frequencies of Coryloid and *Betula*. Tree pollen was consistent at around 50% of total land pollen. Although Gramineae alone reaches high values, a wide range of herbaceous pollen types is recorded, including taxa both of coastal and freshwater wetland affinities. Chenopodiaceae pollen is common at the base of the zone, with types like *Artemisia* and *Plantago maritima*. The base of the peat is dated to 4234-3980 Cal BC (5250±50 BP; SRR-2694). The upper boundary of the zone is defined by a fall in *Ulmus* pollen frequencies.

PRM-b 0.825-0.75m This zone is characterised by *Quercus* and *Alnus*, with lesser values for *Ulmus*, Coryloid, and *Betula*. Tree pollen is reduced to about 40% of total land pollen, but little other significant change occurs in the assemblage. Gramineae remains important and a wide range of wetland taxa occurs. A peak of Chenopodiaceae pollen occurs in mid-zone. The base of the zone is dated to 4034-3790 Cal BC (5120±50 BP; SRR-2929), while the clay layer which occurs in the upper part of the zone is dated to 3510-3140 Cal BC (4620±50 BP; SRR-2693). The upper boundary of the zone is defined by a fall in *Quercus* and a rise in *Alnus* pollen.

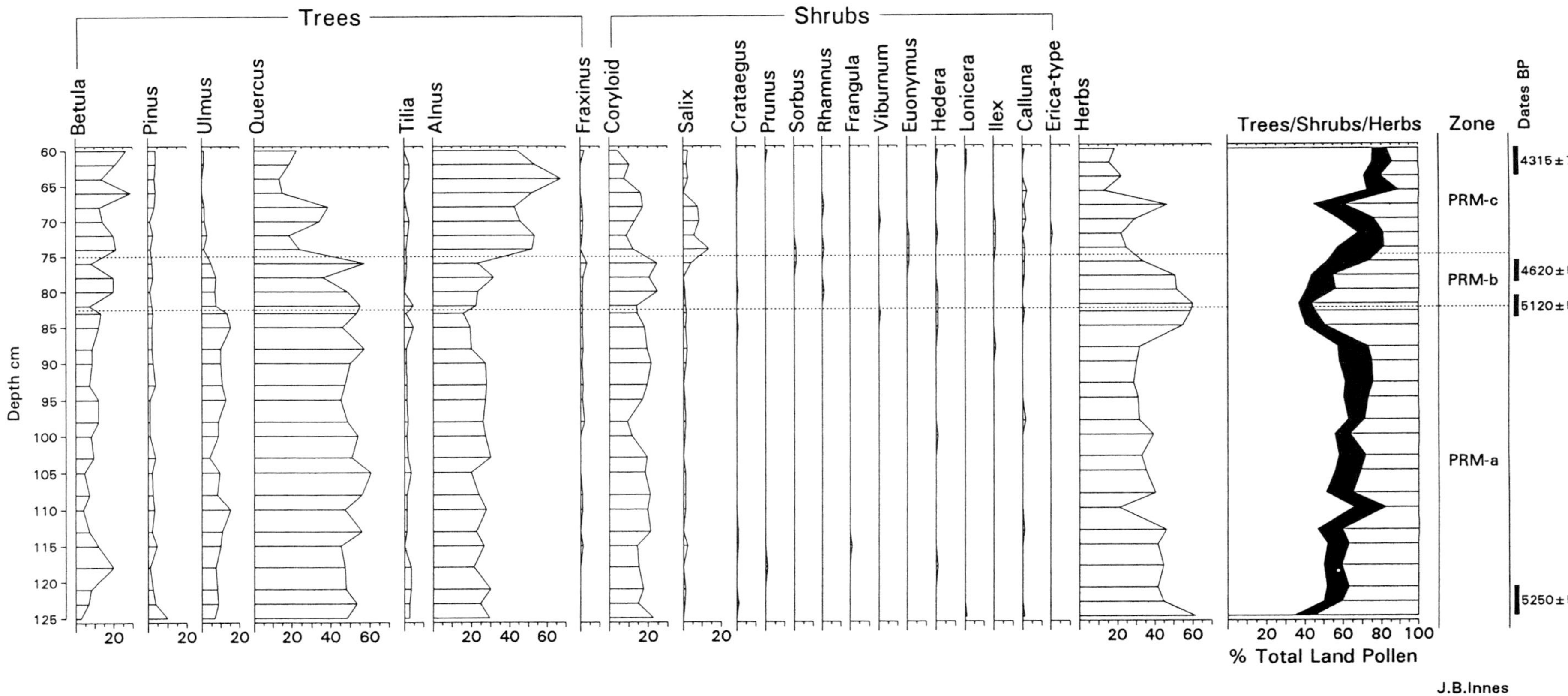

Fig 16 Wirral: Park Road, Meols, percentage arboreal pollen diagram (trees plus group)

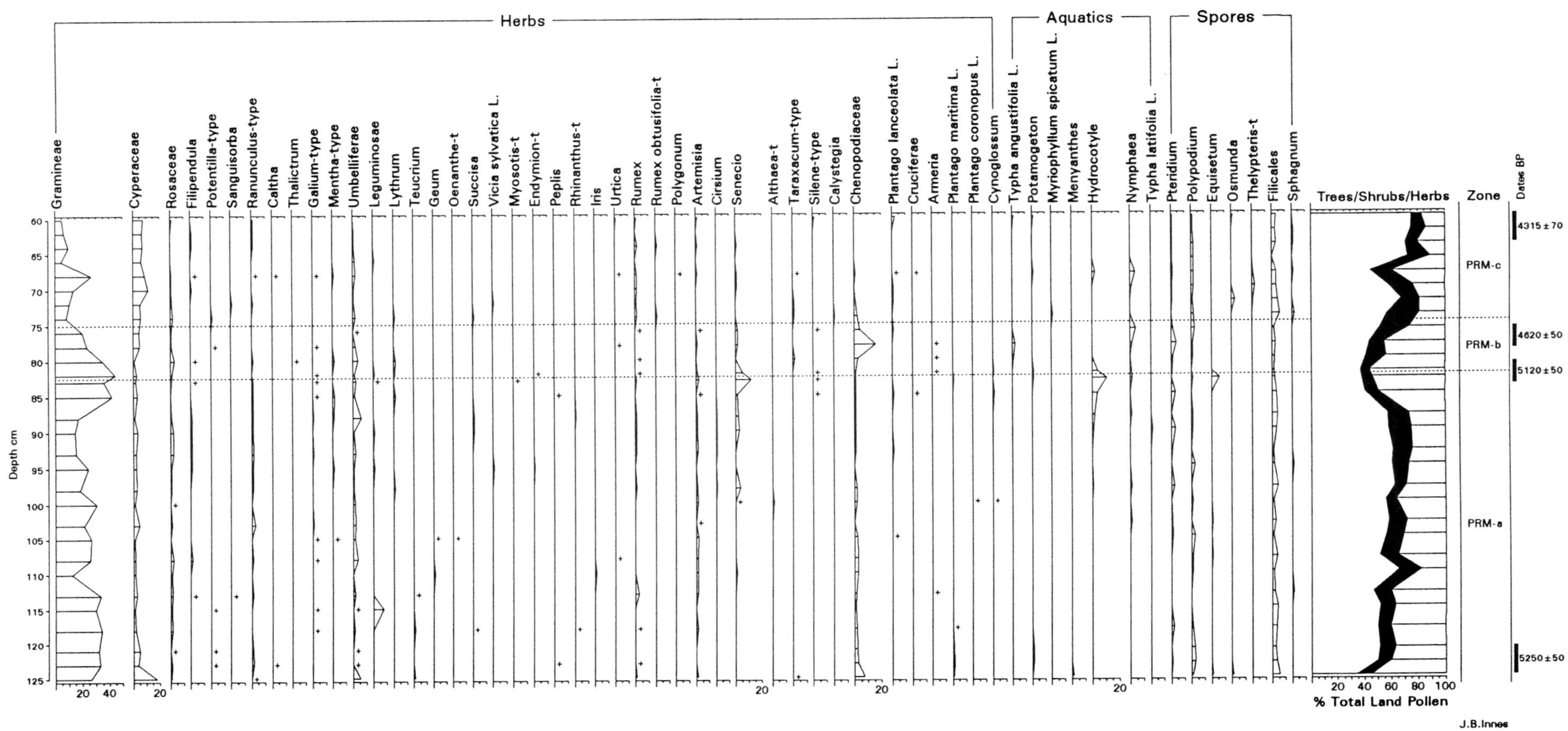

Fig 17 Wirral: Park Road, Meols, percentage herb pollen diagram (trees plus group)

PRM-c 0.75-0.60m The uppermost zone is characterised by *Alnus*, with lesser values for *Quercus*, *Betula*, and Coryloid. Tree pollen increases to 70% of total land pollen at the top of the zone. A peak of *Salix* pollen occurs at the start of the zone, and wetland shrub pollen types generally increase in importance. Gramineae values are on average lower than previously, but the range of wetland herbs recorded is not diminished. Peat from the upper boundary of this zone, where it is overlain by silty clay, has previously been dated to 3100-2704 Cal BC (4315±70 BP; GU-1312) (Kenna 1986).

Ecological history

The pollen and radiocarbon data agree and provide a chronological context for the ecological changes which occur at this site. The date of 4034-3790 Cal BC for the Zone PRM-a/b boundary shows that the fall in elm pollen frequencies at the end of PRM-a represents the major elm decline which is a regional feature in North West England and which forms the end of the Flandrian chronozone II (Hibbert *et al* 1971). This date may therefore function as a standard date for the elm decline in North Wirral, which has hitherto been unavailable, with which the same feature from the other diagrams may be compared. Zone PRM-a therefore falls within Flandrian II, and the later zones relate to Flandrian III.

The earliest indications of ecological conditions at this site come from the basal silty grey clay. As sediment samples were recovered by monolith tins from exposed sections, the bottom of this clay has not been probed and it is likely that further, perhaps basal, peats may exist at depth. More fieldwork is required to ascertain this. The ecological record at Park Road as presently known, however, starts towards the end of the period of deposition of this blue-grey silty clay. The relatively low altitude (+2.27m OD) and late Flandrian II age (4234-3980 Cal BC) of the clay/peat boundary suggest that this basal clay should be of marine or estuarine origin in this area. Preliminary diatom analyses by D Bedlington have confirmed this, the clay containing a diverse marine and brackish water flora which includes *Paralia sulcata* and *Diploneis interrupta*. Saltmarsh pollen taxa in the basal peat layers, such as *Plantago maritima*, Chenopodiaceae, and *Artemisia*, confirm the marine origin of the clay. The stratigraphic change from clay to peat at the top of the basal deposits therefore forms a regressive contact which may represent a relative fall in sea-level on this coast. The transition from a marine and brackish diatom flora to a *Phragmites* dominated reedswamp via saltmarsh plant communities is a classic example of the gradual ecological response to the change from saltwater to freshwater-dominated environmental conditions. Coastal plant communities still existed nearby during almost all of Zone PRM-a, however, as saltmarsh pollen taxa continue to be recorded in low frequencies. Extremely wet fen/reedswamp environments existed at the site, and must have been of high nutrient status to support the extremely rich aquatic and semi-aquatic herb flora recorded in the pollen diagram. Taxa such as *Lythrum*, *Teucrium*, *Caltha* and *Ranunculus* are typical of the rich diversity of the herb flora at this stage. True aquatic types like *Nymphaea*, *Hydrocotyle*, and *Potamogeton* are present and increase towards the end of the zone, showing that areas of deeper water occurred within the wetland landscape. There are no real indications of anything other than eutrophic communities.

Alder percentages are relatively low in comparison to those of oak, and *Salix* is not heavily represented, and so it is unlikely that carr vegetation containing woody taxa was important around the site. The higher oak representation probably reflects the more regional terrestrial vegetation which comprised mixed oak and elm forest with a hazel understorey on the drier ground, probably with an important alder component in damper locations at the wetland edge. *Tilia* is also consistently present and may have been an important forest tree locally, being poorly represented on the pollen diagrams due to its low pollen productivity. The sedimentation rate during zone PRM-a was high, about 400mm of reedswamp peat accumulating in about 130 radiocarbon years, although after calibration this period could be rather longer in calendar years. Water-tables remained high, however, and even seem to rise further near to the end of the zone, with aquatic types more abundant and forming a distinct pollen peak (mainly comprising *Hydrocotyle*) at the end of the zone. Since saltmarsh taxa are present throughout, it would seem that freshwater sedimentation may have been doing no more than keeping pace with rising sea-level, and terrestrialisation of the environment did not occur. Indeed, diatom analyses of the organic sediments of zone PRM-a reveal a mixed assemblage in which marine and brackish taxa are still important, but have been joined by a freshwater diatom flora with high frequencies of, for example, *Pinnularia gentilis*. Approaching marine conditions may well have caused the ponding of freshwater and pool formation at the end of PRM-a, for these flooding conditions presage a short-lived marine inundation of the site early in zone PRM-b during Flandrian III, around 3510-3140 Cal BC, which deposited a thin marine clay. The site must have been very near the landward limit of this inundation, for the clay layer here is only 10mm thick. The diatom flora from this clay layer shows a switch to marine/brackish domination and the pollen assemblage also shows a peak of saltmarsh types, for example Chenopodiaceae and *Armeria*. Fen and reedswamp pollen are still prominent and little change occurs in the overall vegetation. The site must have moved from within the margin of the perimarine zone to within the margin of the intertidal zone.

This phase of marine sedimentation did not last long and the combination of continued sedimentation and perhaps a fall, or at least a reduced rate of rise, of sea-level caused a renewed removal of marine influence. From 3510-3140 Cal BC until the end of peat formation at the end of Zone PRM-c, at 3100-2704 Cal BC, there is a progressive reduction in evidence for marine conditions and a gradual succession towards more terrestrial freshwater environments. Diatom assemblages become dominated by freshwater types and saltmarsh pollen taxa fade from the diagram. In Zone PRM-c both grass and oak pollen frequencies fall. The former shows the direct replacement of reedswamp vegetation by hydroseral succession and the latter shows the replacement of the regional oak pollen component by one derived from the denser local carr vegetation which screens out more of the regional pollen rain. The marked rise in *Alnus* and *Salix* frequencies and the increase in wetland shrub types recorded, for example *Rhamnus*, is caused by the occupation of the local area by carr and fenwood vegetation.

The final ecological change recorded at the site is a quite abrupt return to domination by marine sedimentation, laying down silty clay with fragments of brackish water diatoms within it, of which very little warning is given in the preceding pollen or diatom evidence. Diatoms in the latter stages of Zone PRM-c are very poorly preserved, and entirely fresh, while unequivocal indicators of saltmarsh conditions no longer appear in the pollen record. Alder frequencies rise to very high levels and are accompanied by some aquatic herbs. If this is evidence of raised water tables, it may reflect some ponding of freshwater by approaching marine conditions. The lack of microfossil evidence of gradual marine transgression of the site may suggest that this transgressive contact is eroded and the radiocarbon date may thus be a little earlier than the true age of the reversion to marine sedimentation. The alternating record of marine and freshwater sedimentation may be correlated to the other North Wirral sites quite well. The thin marine clay dated to 3510-3140 Cal BC is of comparable age to the upper clay at Bidston Moss (3690-3360 Cal BC) and is 0.2m higher in altitude. This confirms that the upper Bidston Moss clay logically must also be marine and, at the slightly lower altitude, remained within the direct influence of marine sedimentation, while the Park Road site was able to revert to organic sedimentation when a slight relative fall in sea-level occurred. Evidently in the low gradient coastal zone of North Wirral only slight variations in altitude were sufficient to make considerable differences in the ecological histories of individual sites. The palaeogeography of the area, and the distribution of settlement and resource potential for human groups, must have been very sensitive to quite small alterations in sea-levels and water-tables. The brief marine interlude at Park Road has also been recognised at an analogous date at Newton Carr, as a phase of freshwater flooding and aquatic vegetation and an increase in saltmarsh pollen at Newton Carr 11, and as a brief period of marine clay deposition at Newton Carr 7 (Bedlington in prep). This early Flandrian III marine episode clearly occurred throughout North Wirral at altitudes up to about +2.75m OD, its expression and distribution depending upon local topographic factors. It has major implications for Neolithic settlement and activity in this area.

Landscape history and forest disturbance

There is little direct evidence for human impact upon the environment in the pollen data from Park Road, Meols. Few herbaceous indicators of forest opening occur, mainly restricted to *Plantago lanceolata* and *Pteridium*, and in such low values as to be insignificant. These taxa do increase after the elm decline but are still very low in frequency. The location of the site within an extremely wet area at the seaward margin of the fen/reedswamp zone must have restricted human access to it and interest in it. It would have lain within a large area of homogeneous swamp and fen wetland, relatively far from the edge of drier land which would have been attractive to human activity. The pollen source areas for any human impact on the landscape were therefore too far away for it to become recorded in the Park Road pollen rain, unlike the Newton Carr and Bidston Moss examples where the wetlands lay adjacent to higher land and could record, and also attract, human activity.

Conclusions

The main points of archaeological and palaeoecological significance from the survey in North Wirral that have particular relevance to research questions in the region or beyond are illustrated here, while the wider and more detailed discussion of these issues is left to Chapter 7, where they are seen in the context of patterns within the county. The main locations that may have the potential to develop this research are described here, as well as the threats that might destroy the information. This evidence can be used to set the criteria for preservation of wetland areas through management strategies, or for developing its research potential, through rescue archaeology if necessary.

The problems caused by the nature of landuse in North Wirral meant that the survey was unlikely to make substantial increases in the archaeological knowledge of the area. A treatment of the human influence on North Wirral still has to rely on the archaeological evidence provided by nineteenth century antiquaries, and although they recorded several locations which were relatively productive of evidence, the resulting pattern of settlement is still patchy. The interpretation of this evidence, however, has been greatly enhanced by recent palaeoecological research, much of it published here for the first time. This allows a reasonably detailed interpretation of landuse for the earlier prehistoric period in North Wirral, which has some potentially important implications for developing an understanding of the processes by which the first farming communities in the area became established.

Prior to the NWWS, the relatively high incidence of chance finds and sites recorded from North Wirral was greater than in any other part of the lowlands of the region, apart from North Lancashire. Unless this feature is attributable to bias in collecting patterns, it suggests that the area lay at the heart of one of the more favoured areas for settlement in the prehistoric and Roman periods. This suggestion is enhanced by the large range of material recovered from more intensive field survey in the coastal area of Sefton, to the east of the Mersey estuary (*see Chapter 4*). This evidence contrasts strongly with the density of material from inland areas of the county, pointing to the conclusion that coastal settlement was a dominant feature of the prehistoric period in Merseyside (*see Chapter* 7).

When attempting to understand the nature of the evidence for settlement and landuse in North Wirral, the palaeoecological evidence is of prime importance in providing a context for the interpretation of the archaeology. Unfortunately, the pollen record itself only provides a detailed record for those periods which predate the mid fourth millennium Cal BC, as marine clays tend to interrupt peat growth in the basin areas after this date. This is partly offset by Kenna' s (1978, 1979, 1986) work on the coastal stratigraphy, which provides additional information to the end of the early Bronze Age (*c*1650 Cal BC), and intermittently thereafter. The palaeoecological background largely covers the period which is represented by the bulk of the archaeological evidence. After the early Bronze Age, both sources of evidence are too sparse to allow a coherent pattern to emerge.

The palynological evidence from Bidston Moss suggests that this wetland was an important resource for human exploitation between the sixth millennium Cal BC and the mid fourth millennium Cal BC. Elsewhere, in general, a clear link between earlier prehistoric settlement and associated landuse, and the areas of freshwater wetland, is more difficult to illustrate. However, although the archaeological evidence is very sketchy, overall it appears to be associated more with former river channels than with lakes or mosses.

The first good evidence for settled farmsteads in North Wirral does not appear until the Romano-British period. By this time the earlier close relationship that may have existed between settlement and the channels and mosslands of the coastal wetland seems to have been largely broken. With the exception of the coastal site of Meols, well drained, light soils, adjacent to a variety of contrasting soil types, were now the favoured locations. These are found several kilometres to the south of the coastal wetland belt, although proximity to rivers would still have been of paramount importance for water. They could also potentially contain valley mires, although none of the latter have been identified close to Roman sites from the work so far carried out in North Wirral. There are hints, in the form of vestigial stuctural and finds evidence from Romano-British sites, that the preceding Iron Age pattern may have been similar.

The historic period sees the wetland areas being incorporated into territories associated, on the whole, with nucleated settlements, although the process by which the change came about from the preceding pattern based on single farmsteads is unclear. Various medieval documentary sources suggest that it is likely that the mossland resources were considered valuable, but were less places to live on or near than to exploit from a distance, as they were located on the margins of the townships. The finds collected in the nineteenth century indicate that the settlement at Dove Point continued to flourish until about the fourteenth century when, at a period of dune formation in the North West (Tooley 1978b), the settlement had ceased to exist.

Significance of the North Wirral wetland evidence

In the earlier prehistoric period, the area around Bidston Moss provides tantalising evidence of landuse around the wetlands. The potentially most significant evidence is mainly palaeoecological. Possible evidence for repeated small clearings around Bidston Moss is recognised during the sixth millennium Cal BC, suggesting that the fen carr was being utilised for hunting, fishing, and gathering. This evidence continues into the fifth millennium Cal BC, perhaps the least understood part of the prehistoric period, during which the change from hunting and gathering to farming may have taken place in many areas of Britain (Zvelebil 1986; Zvelebil and Rowley-Conwy 1986). Although there are difficulties of interpretation of the evidence, which are dealt with in detail in Chapter 7, the pollen record implies that some activity normally associated with Neolithic culture (*ie* arable cultivation) may have been taking place at the same time, 4900-4530 Cal BC, as late Mesolithic groups were active around Bidston Moss. This phenomenon may have lasted for approximately 500-800 years until, in the chronological Neolithic (*ie* after *c*4000 Cal BC), relatively more intensive woodland clearance and more securely interpreted arable farming occurred around the moss. Although local dated Neolithic assemblages do not exist, this later activity might be expected to have been undertaken by people who were using Neolithic cultural artefacts, *eg* stone axes and leaf arrowheads, a number of which are found on the sandstone slopes around the moss.

Apart from problems of interpretation of the validity of the basic data, including misidentification or contamination of the pollen (Kinnes 1988), a number of alternative explanations are possible for the processes involved during this long period when the basic method of subsistence was changing (Zvelebil 1986). The evidence at Bidston only hints at possible explanations, but it would be consistent with a view that is accepted as being eminently feasible on a theoretical basis, but for which there is no real archaeological evidence (Kinnes 1988; Bradley 1984b, 8; Zvelebil and Rowley-Conwy 1986; Simmons and Innes 1987). This suggests that the initial adoption of farming resulted from a gradual and selective adoption of some aspects of agriculture within a Mesolithic framework. Several other occurrences of this phenomenon are found in Britain, mainly, but not exclusively, along the western seaboard (Edwards and Hirons 1984). This suggests that the Wirral coast (and coincidentally the Sefton coast, where the same phenomenon is found (*see Chapters 4 and 7*), was well placed to receive influences from the continent where agriculture was established earlier than in Britain. The potential that the Bidston area may have for helping to clarify some of these problems is dealt with below.

The palaeoecological evidence for the Wirral is also significant for the period after the elm decline. At Bidston, the difference in the scale of vegetational changes, compared to other sites in North Wirral and in the rest of the county, implies that a long period of potential adaptation to a new economic base culminated in an economy more dependent on clearance and farming. This pollen evidence suggests that this part of the Wirral is unique in the county, and possibly the region.

The Newton Carr area provides a contrast to the above pattern. There is little evidence of disturbance to the woodland in the sixth millennium Cal BC, although not all hunting, or even grazing of domesticates for that matter, need have resulted in vegetational changes that would show up in pollen diagrams (Smith 1984). Nor is there any evidence for the same kind of intensifying clearance up to and after the elm decline, as is found at Bidston. The only recognisable potential human interference is a small-scale disturbance immediately after the elm decline, which quickly regenerated. This implies that either the nature of post-elm decline human activity was different in this area, or that farming was taking place, but too far from the sampling site to be represented in the pollen record, although a possible cereal-type pollen grain was found in the clearance horizon.

There are two implications for research arising from the palaeoecological evidence. Firstly it points to the Bidston area being important during the late Mesolithic and the early Neolithic. The latter is a period for which the settlement evidence is very sparse nationally (Healy 1984). The nature of settlement is also generally thought to have been more sedentary than that in the Mesolithic in some areas, although still involving regular movement of cleared land as nutrient exhaustion and weed regeneration took place (Smith 1984; Thomas 1991). The occurrence of Mesolithic sites in the same locations as Neolithic sites is also commonly commented on, while Bonsall (1981) argues for the sedentary nature of coastal Mesolithic groups facilitating the adoption of sedentism in an agricultural context. The evidence in North Wirral suggests that evidence relating to these processes may have been present.

Secondly, the difference in palynological evidence between north-east Wirral and the rest of the peninsula, and the county to the east of the Mersey, may suggest a differential rate of agricultural adoption across the area. The identification of settlement trends in the sixth and fifth millennia Cal BC across contrasting geographical areas in the county would be of great value in charting the progress of the development of sedentism and exploring the links with the adoption of agriculture.

The palaeoecological evidence for North Wirral becomes sparse during the later prehistoric period, confined to only the occasional section recorded by Kenna (1986) which, when combined with the lack of archaeological evidence, makes it difficult to identify factors of significance after the late Neolithic/early Bronze Age. Once the dune belt had become established, by the early Bronze Age, marine influence will have played less of a role in determining site location in the coastal belt. The archaeological evidence for this period is very slight and allows little to be said about its potential. By analogy with the evidence from Sefton (*see Chapter 4*) and other western coastal locations, the dune areas might be expected to have been important for grazing and possibly cereal farming during the Bronze Age.

Only the occasional chance find such as the bronze dirk (A 13) or axe-hammer (A 12) from Leasowe, or the early Bronze Age bone midden from the coast a little to the north-east (A 17), might bear on this problem. The context of these finds, with the exception of the dirk, coming from silty clay layers in proximity to peat deposits, suggests that there is some potential for developing an understanding of this period.

The later prehistoric pattern is even more difficult to identify around the wetland zone, as there is a gap in the palaeoecological record everywhere in North Wirral after *c*1650 Cal BC. The only archaeological evidence in a wetland context for this and the subsequent Romano-British period lies at Dove Point, at Meols. This is one of the most important sites in the North West, but unfortunately now appears to be lost to the sea and hence to normal techniques of archaeological investigation. Nonetheless, some chance Roman coin evidence from the dunes behind the sea protection walls at Meols implies that not all the activity was necessarily seaward of the present coast. In the pre-Conquest and medieval periods, the dune areas behind Meols are likely to have been exploited from the Dove Point settlement and other villages that grew up on the landward side of the coastal belt. It seems likely that the moss embayments were also used primarily for grazing, fuel, and other resources during this period, and so may have less archaeological potential.

Archaeological and palaeoecological potential

None of the archaeological sites from North Wirral can be said to be true wetland sites, in that they are in a direct relationship with waterlogged deposits. They do, however, represent sites from which their inhabitants exploited the mires, an activity which may be represented in adjacent waterlogged deposits.

The strongest archaeological evidence, based on early twentieth century records of sites discovered through marine erosion, lies in the north-west of the peninsula. It spans the Mesolithic to the Bronze Age, with Mesolithic and Neolithic material being recorded from the same sites, including Hoylake (A 10). There is one example away from this north-west corner of Wirral, at New Brighton (A 6), although there is no stratigraphic context for this material. These two sites, at Hoylake and New Brighton, lie in open ground, where sandstone knolls outcrop on the coastal strip. Only their approximate position is known from existing records, and it is not clear if their full extent was destroyed at the beginning of the century. They do at least suggest that other sites may still survive in the vicinity, where the relationship between the two cultures might be studied. The New Brighton site is protected from the sea now by the defence wall, in contrast to the West Kirby coast.

Other sites adjacent to the major channel between the mainland at West Kirby/Hoylake and the small islands in the Dee estuary, including Middle Eye and Hilbre (A 8, A 9) may also contain potential evidence for clarifying this relationship. There is more potential for recovering further material from the two larger, higher islands, than from Little Eye, since most of the island now consists of a series of sea-smoothed rocks, and only an area of *c*50m^2 still survives with clay and topsoil. However, the black soil band beneath the topsoil from which most of the finds are reported also survives, and it could still retain some evidence, although no finds have been reported from the site in the last ten years.

In the north-east, where the palaeoecological evidence is strongest, the archaeological potential appears limited. The archaeological evidence around Bidston Moss is limited to only single findspots of Neolithic axes, which have come from the built-up areas that dominate the surrounding landscape. This urban development has probably removed much potential dryland evidence of settlement activity. There are still a few undeveloped areas around the lower slopes and on the margins of the moss, particularly in the north, which may have potential, especially as a major buried channel runs through this area (Fig 72).

At all these sites adjacent organic deposits are likely to be deeply buried. To the south of the high water mark, at West Kirby/Hoylake and New Brighton, the channel deposits are buried by blown sand and marine clays. The Newton Carr evidence suggests that these deposits could be as much as 5m deep, and the deeply buried intertidal channel deposits are even more inaccessible in these locations. However, in the embayments to the south of the coastal strip the problems of sea erosion are not as great, although in Newton Carr the extra depth of the infilled depression makes location of prehistoric settlement areas difficult to identify. The preliminary work of Bedlington is starting to define certain buried topographic areas that may have potential, but much more work is needed yet. This work is, however, beginning to show that the moss is particularly important for understanding the likely impact of the relationship between sea-level changes in the North West as a whole. The evidence points to the coast and the major channels flowing through it as being the important locational factors for settlement in this area, although the scale of this settlement does not show in the palynological history of Newton Carr.

Potential archaeological evidence on the shallower buried northern slopes of the basin, where the pre-Holocene land surface rises slightly towards the southern edges of West Kirby/Hoylake, is perhaps more accessible than in the core of the moss. It is, however, difficult to identify because of urban development here. The break in slope a little above the junction between the wet deposits and the boulder clay would always have been relatively dry, and it might be expected that it would have formed an attractive location for settlement. The fieldwalking results have shown this to be an unverifiable theory, although the generally poor fieldwalking conditions combined with the typical nature of prehistoric settlement in much of the county must be borne in mind, and the two sites located on the boulder clay ridge (Me 001, Me 002) may be more important than seems likely at the moment.

The intervening areas of the coast, between Meols and Leasowe, are generally lower and now filled with Flandrian clays, silts, and peats, which mask the prehistoric surfaces. Possible locations for settlement here during periods of direct marine influence may have been restricted, although slight eminences in the buried boulder clay, such as near Leasowe Castle, may have potential for location of material.

Threats

The most obvious threat to much of the coastal wetland belt of North Wirral is from the action of the sea, although this has largely been curtailed with the construction of a sea wall in the 1970s. This, incidentally, provided the opportunity for Kenna (1986) to monitor the stratigraphy at several points along the coast. Only the north-west corner, around Hoylake/West Kirby and the adjoining islands, is directly open to the sea, although the known sites are on small sandstone knolls surrounded by tidal flats. The low sandstone 'cliffs' have a very thin sandy soil on them. The island of Little Eye, with a boulder clay capping to the sandstone, is most at risk, although the small Mesolithic site (A 7) appeared to have largely eroded away by the early 1980s. The late Mesolithic submarine forest a little to the north-east is now reduced to an impersistent band of silty, woody peat *c*0.2m thick, overlying the boulder clay. In the embayments to the south of the coastal strip the problems of sea erosion are not an issue, as dune building and modern development behind the sea wall have created a barrier to the sea.

The most threatened areas of wetland are in the north-east, particularly around Bidston Moss. The moss itself has already suffered a large degree of development, although some of the industrial units are unlikely to have destroyed the peat or its boundary with the sand or clay beneath it. Monitoring of development in these areas should therefore be worthwhile, particularly around the southern fringes of the moss. Other areas, under landfill and spoil tips from the construction of the motorway, are completely lost to monitoring.

The depth of the sediments in the other embayment of Newton Carr preserves it from immediate destruction, but there are undetermined plans to use the Carr for waste management at some time in the future. This would seal off large areas to further research, and may also have a deleterious effect on the nature and hence preservation of the deposits in the long term.

The area along the coast, immediately behind the sea wall, consists of several metres of recent sand dune deposits *c*200-500m wide, ranging from an altitude of *c*5-8m OD. It is mainly given over to recreation purposes and appears to be safe from imminent danger of development. Peat is sealed at several metres' depth from the surface all along this coastal strip, and future developments may allow a degree of preservation of deposits. The increasing need for leisure facilities in the area, with the establishment of golf courses and public open spaces, may help delay the erosion of important deposits, but there is little information about the long term effects of such developments and the surfaces are in effect sealed, so limiting research.

The area to the south of the coast road flanking the recreational belt of land is, however, designated a development area where housing and industrial units will be allowed to infill most of the land between current developments. Two developments that were monitored in this area showed that in some cases the peat may be totally excavated before construction takes place, and in other areas buildings are placed on stilts drilled through the peat. The latter type of development, although apparently less harmful, will affect the hydrology and cause peat decay, and opportunities for recovery of evidence during construction are necessarily limited. The former type of development is more harmful to the waterlogged deposits, but provides a more complete opportunity to retrieve evidence.

Future research

Unfortunately the whole wetland area of North Wirral has suffered so badly from development of one kind or another, particularly over the last 30 years, that its true potential will probably never be realised. In an area with the particular characteristics of North Wirral, possibilities for research will largely arise only when development circumstances allow. Present landuse in the North Wirral wetland areas suggests that the only way evidence of value will be discovered in the north-east of the peninsula and along the coastal belt is through close monitoring of development through liaison with the planning system. This type of monitoring might, at the least, be expected to provide micro and macrofossil evidence to aid the understanding of the development of the landscape and vegetation. This will be of value in extending the palaeoecological record for the area, which is very patchy after the beginning of the Bronze Age and almost non-existent for the periods after the late Bronze Age. The two monitoring projects mentioned above, which took place during the survey, at the Leasowe Hospital site on the coastal strip and on the southern edge of Bidston Moss, did not produce any archaeological evidence, but stratigraphical evidence showing bands of peat and intervening marine layers was forthcoming (Simmons and Cowell 1993). A more detailed understanding of small localities would also allow predictions to be made of potentially attractive environments for settlement, away from the more obvious dryland slopes. The evidence from Ince Blundell (*see Chapter 4*) shows that where the early channels can be predicted then the potential may be especially high.

The only area where archaeological research can be targeted using archaeological criteria and not development pressures is around Newton Carr. The current palaeoecological evidence provides one of the best sequences in the area for understanding sea-level movements in North West England. This is somewhat diminished, however, by the lack of radiocarbon dates for any of the three peat cores, which prevents correlation with other sites across North Wirral. It would therefore be beneficial to acquire radiocarbon dates for these sites. The slopes to the south of the Carr also provide the only evidence for sites in the coastal zone that are not either difficult to locate or interpret, or have been destroyed. The sites are small and few in number, with only two or possibly three known, but landuse has only allowed a small area to be fieldwalked. The pasture fields in the area come under a rotation to be reseeded over many years, and one aim should be to take the opportunity to extend the current fieldwalking coverage. The few known sites, although not true wetland sites, are close enough to the mire to suggest that correlations could be made between the settlement and the various stages of marine and biogenic deposition, through excavation and limited palaeoecological sampling close to the sites. This might serve to provide the first step in linking a complex sequence of natural events to the human presence in North Wirral, which is relatively prolific in terms of some areas of the county, but is almost totally lacking any kind of context.

4

THE SEFTON MOSSLANDS

The mosslands treated in this chapter lie in the north-western part of the county, within the Metropolitan district of Sefton (Fig 5). They are on the western edge of a large mossland complex, centred on West Lancashire, whose boundary with the Merseyside peat is defined by the River Alt. The mosslands stretch from the Alt eastwards to Liverpool Bay, north of the estuary of the River Mersey. Topographically the Sefton mosslands are represented by a narrow belt of coastal peat, which is largely overlain by blown sand, and an inland belt associated with the channel of the River Alt.

The coastal mosslands lie on the western edges of the historic townships of Little Crosby, Ince Blundell, and to a lesser extent Great Crosby (Fig 18). The modern extent of the town of Great Crosby marks the southern edge of the peat, which amounts to c280ha in this belt. The only other urban growth in this coastal belt is at Hightown, close to the Alt estuary, and north of the Altmouth, at Formby. The area of coastal wetland north of the Alt, stretching as far as Southport and amounting to *c*200ha, constitutes a physical part of the West Lancashire mossland complex, and will be surveyed as part of the NWWS fieldwork in Lancashire (Middleton *et al* forthcoming).

The main area of the inland belt of peat, *c*190ha, lies in Ince Blundell, largely within the valley of the Alt. This peat deposit also extends along the valley into the northern parts of the historic townships of Thornton and Lunt, all three townships sharing the River Alt as their common north-eastern boundary. Each of these townships contains a small hamlet that gives its name to the township. These settlements lie a little to the west of the brow of the sandy ridge that overlooks the Alt floodplain to the south-east. Ince Blundell is the only historic settlement that has been affected by modern growth, albeit in a limited form.

The southern limit of the survey in the Alt Valley is marked by the village of Sefton, whose medieval parish church marks it out as one of the main historic settlements in the area. Here modern tipping and industrial growth from the expansion of Maghull obscure the valley deposits. On the north side of the Alt lie the townships of Maghull and Lydiate in Merseyside and Altcar in Lancashire. The peats to the north of the river, lying in the latter two townships, extend northwards to join the large expanse of Downholland Moss. This area too is being treated as part of the NWWS fieldwork in Lancashire.

Landuse

Most of the peat surveyed in these mosslands is interspersed with a series of earlier coastal deposits reflecting environments, such as mudflats, saltmarshes, and dune systems, which have been controlled by changes in sea-level over many millennia. The effects of this, and particularly the widespread distribution of blown sand along the coastal mossland belt, has conditioned the landuse of much of this area today. In the early medieval period English settlement appears to have avoided much of the coastal area, being concentrated further to the south in more varied topography. This pattern changed little in the later medieval period when much of the wetland areas were used for grazing. The settlement pattern that grew up was essentially one of dispersed farms and hamlets around these wet areas.

In contrast to the topographically similar North Wirral, this scattered historical pattern changed little during the late post-medieval period. The main expansion of settlement here did not come until the late nineteenth and twentieth centuries, with the growth of modern leisure and housing needs. Modern housing development in Sefton is largely confined to the south and east, away from the wetland areas, and only at Formby, and at the northern end of the county at Southport, has development substantially extended from the historic settlement cores onto the sandy wetland margins.

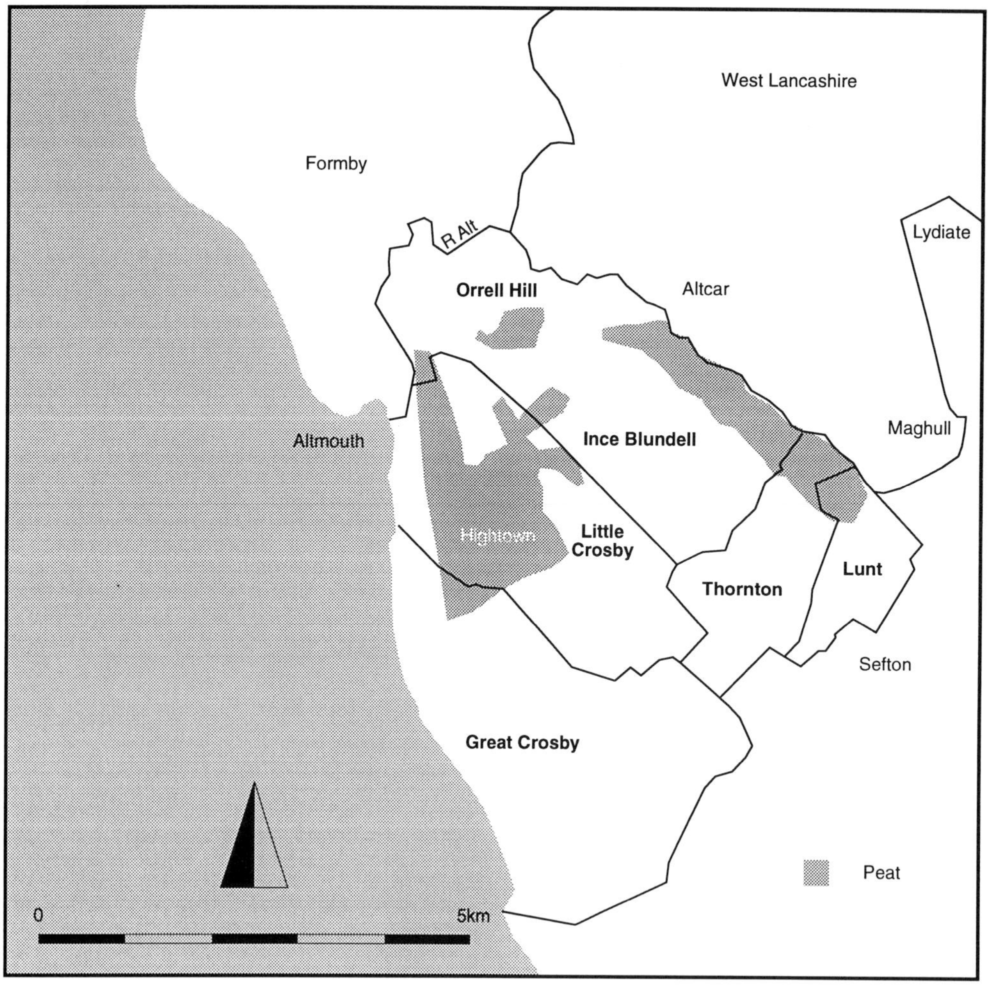

Fig 18 Sefton: Location of survey area

The landscape of the survey area has therefore remained undeveloped and retains many elements of its earlier agricultural nature, with activity mainly divided between agriculture and amenity. Most of the coastal zone between Southport and the Altmouth consists of one of the largest areas of sandflats and dunes in the country, which is managed almost entirely for conservation, recreation, and amenity purposes. It contains a series of National Nature Reserves and a string of SSSIs. The tidal belt from the Mersey estuary to the Ribble estuary is also one of the most important areas in Europe for wintering birds (Wood 1993). A Coast Management Scheme, administered by the local authority, aims to handle the great pressure imposed by visitors to the area. South of the Altmouth the recreational aspect of the landscape continues, with the Hightown dune system being largely developed as a golf course.

To the east of the dune system the land is given over mostly to agriculture, with a pattern of small hamlets and many scattered farmsteads. Most of the farmland is dominated by arable, although close to the eastern edge of the dunes the extreme sandy nature of the fields means that poor grassland or playing fields are almost as common. Most of the area away from the dune area is agricultural land with a heavy preponderance of arable, growing root crops, and brassicas in the sandy and peaty fertile soils. There is only one area of significant pasture in the survey area, amounting to 40ha, which is found in the central part of the mossland, around Orrell Hill in Ince Blundell, with sheep and cattle husbandry being a feature on several mixed farms. The nature of the landscape in the Sefton wetland areas has therefore allowed more peatland to survive the development that has occurred in other parts of the county, consequently improving the potential recovery of surface archaeology.

Previous research

The peat areas of Sefton have had little archaeological interest shown in them until the present work. A landscape survey of the district, undertaken in the early 1980s as part of the creation of the Merseyside SMR, provided some context for the historic settlement and landscape development of the district, although little fieldwalking survey was undertaken (Lewis 1982). A few chance finds, some of possible prehistoric date, are recorded in the SMR from the Altmouth area, but elsewhere only a few casual finds have been brought to the attention of archaeologists. The only significant piece from the wetland zone is a stone axehead from north of Little Crosby (A 18), close to the peat deposit at Flea Moss Wood.

The recognition of this area as a nationally important natural habitat (Atkinson and Houston 1993) has, however, led to a great deal of research on the natural development of the coastal and inland wetlands of the area. This part of south-west Lancashire has been the subject of interest since at least the middle of the nineteenth century in regard to its coastal deposits (Reade 1872b; Gresswell 1953, 1958). More recently Tooley (1978a, 1980, 1982, 1985) has provided a context for the sequence of sea-level changes and coastal palaeoenvironments in this part of the region. His work at the well known site of Downholland Moss, to the east of Formby, forms the essential background for the understanding of human impact in the area.

Some palaeoecological work was done specifically on these mosslands in the early 1980s by one of the authors (J I), as part of the creation of the county SMR, and the pollen diagrams for two sites in the coastal belt of peat, at Flea Moss Wood and Sniggery Wood, are presented in the palaeoecological survey section of this chapter. Monitoring of developments in urban areas along the coastal strip by Liverpool Museum also took place in the late 1970s. This located and recorded a number of peat horizons, most notably in Waterloo overlooking the Mersey estuary on the northern edge of Liverpool. These sites include Murat Street (Innes and Tooley 1993), Mount Pleasant, and Church Street. An assessment was also carried out recently on behalf of Sefton Council on the archaeological potential of the Rimrose valley, a little to the north of Waterloo. This contains a major stream flowing into the Mersey estuary south of the Alt, travelling mainly through the built-up areas north of Bootle. It was known to contain deep deposits of peat in its less developed lower part (Smith 1866b), which will be the subject of future development plans (Cowell and Nicholson 1990). A core sample was taken and analysed by Jim Innes for inclusion in the assessment report. These sites are not presented in detail in this chapter, but their most significant conclusions are used where relevant to place activity around the inland mosslands into a wider coastal context.

Geology and topography

The topography of this area today relates quite strongly to the late stages of the last (Devensian) glaciation, which has led to a relatively gentle landscape based on the Shirdley Hill Sands, a coversand deposit which is often less than 1m deep. This overlies the glacial till or boulder clay, which accordingly comes to the surface in small patches across the area, but the sand is generally sufficiently deep for it to determine the characteristic soil and natural vegetation of the area.

The main relief is provided by the two roughly parallel north-west to south-east trending Shirdley Hill Sand ridges, on which stand the villages of Little Crosby and Ince Blundell respectively (Fig 73). The latter is quite wide and of gentle topography, falling slightly to the west into a peat-filled channel to the north of Little Crosby. On the east side the ridge falls more substantially into the Alt valley. The ridge is most pronounced at Ince Blundell park at a height of *c*15m OD. The Little Crosby Shirdley Hill Sand ridge is less substantial. It is *c*300-400m wide, and runs from an area of boulder clay to the south of Little Crosby for a length of *c*1100m, with its high point lying at *c*10m OD. Near the southern edge of the ridge the underlying Triassic sandstone is close to the surface in a small area. The slopes of the ridge are relatively pronounced, falling to the coastal plain at *c*6m OD on the west, and into the peat-filled channel on the east.

The area of Orrell Hill lies at the north-western edge of the Ince Blundell ridge and is associated with a more diffuse, less differentiated area of topography. This falls gently, from a height of *c*6m OD towards the Flea Moss Wood ridge to the south. The 5m OD contour between Flea Moss Wood, Orrell Hill, and a point a little to the south of New Moss Wood, Ince Blundell, marks the approximate present northern edge of the Shirdley Hill Sand, except where it is cut by the Little Crosby/Ince Blundell channel and the river Alt. To the north of this line the dipping contours of the Shirdley Hill Sand have been infilled by Flandrian deposits of peat and clays which have produced a plain stretching from the coast to the Alt valley.

Natural drainage patterns

The Alt flows in one of the four major glacial

Plate 3 Sefton: View of the Coastal plain (Photo R Cowell)

channels which incise this part of the region traditionally known as South West Lancashire (now including Merseyside). Its present mouth lies to the north of Hightown and it is likely that the palaeo-estuary lay in this general vicinity, while the course of the river appears to have been reasonably constant within the Flandrian period (Huddart 1992). To the south of Riddings Moss Wood the western slopes of the Alt valley start to fall down to the river, marking the start of the Ince Blundell ridge. Here the slopes fall gently, but the slope increases south-eastwards, where the fall to the floodplain becomes in the order of *c*4m from the main break of slope. The eastern (Lancashire/Merseyside) side of the valley is wider and of more gentle gradient, although the pronounced Shirdley Hill Sand ridge on which Lydiate is situated forms a third parallel ridge in this area, 2km from the river.

Elsewhere the drainage has been artificially manipulated by a network of ditches, some of which may be quite early in date. The major ditch system, which is *c*2m deep and *c*5m wide, forms a straight boundary between the historic townships of Little Crosby and Ince Blundell (Fig 18), which were vills at Domesday, although it is not known when the boundary became finalised. It possibly takes over the drainage formerly provided by the peat-filled channel that lies mainly to the west of the ditch, although no alluvial clays have been identified in the channel yet. Elsewhere only small remnants of now dried up channels, not traceable for more than a few metres at a time, mark the earlier drainage which has now been subsumed in the system of ditches across the area. A system of roddons, similar to those in the Fens (Hall 1987, 9), whereby water courses cut through the peat leaving raised silt channels when the surrounding peat shrinks after drainage, have been postulated by Huddart (1992), particularly to the east of the Alt but also in the survey area.

Flandrian geology

The Sefton mosses started forming in response to rises in sea-level which raised the water-table, so creating waterlogged areas where reedswamps and fen woodland became established. At certain periods, direct influence from the rising sea-level led to the peat areas becoming inundated with sands and muds as the local environments changed to salt-marsh or dune slack areas. Around 3400-2500 Cal BC the mosses in the present coastal zone were covered by blown dune sand (Tooley 1990; Innes and Tooley 1993), *c*2km landward of the present dune system (Fig 19).

Much of the pre-Flandrian topography of the area north of the coversand ridges is smoothed and levelled by post-glacial deposits that give the area its

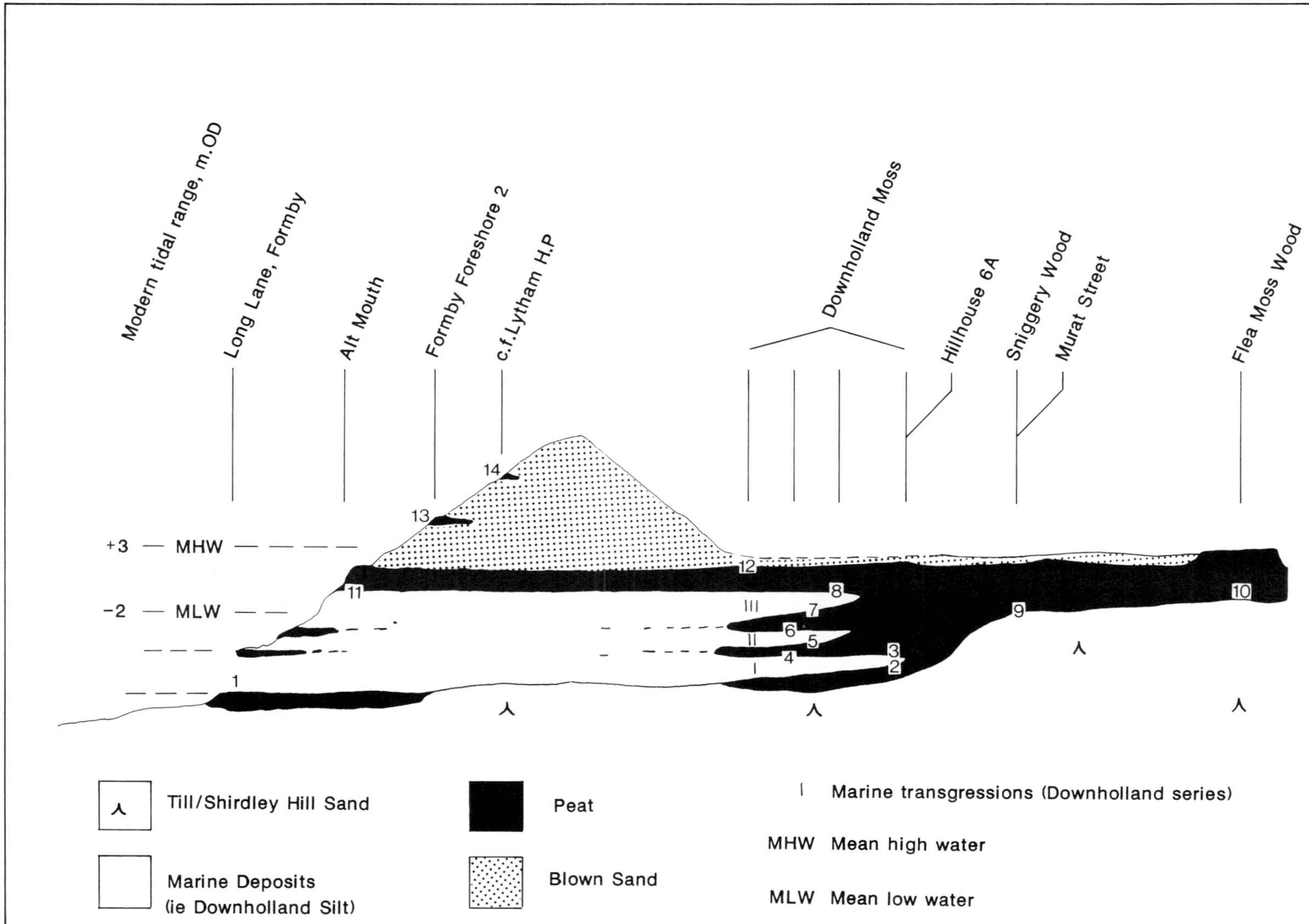

1	c.8000	5	c.6500	9	<5010 (ie less than 5010)	13	2335±120, Hv 4709
2	6980±55, Hv 3936	6	6050±65, Hv 3358	10	<5010 (ie less than 5010)	14	830±50, Hv 3846
3	6760±95, Hv 3935	7	c.5900	11	4545±90, Hv 2679		
4	6750±175, Hv 4706	8	5615±45, Hv 3847	12	4090±170, Hv 4705		

C14 yrs, bp, all dates Tooley 1978a)

Fig 19 Sefton: Altitudinal section through coastal peat

Table 7 Sefton: Marine transgressions summary (after Tooley 1978a)

Transgression	Date (C14 yrs BP)	Date (Cal BC)
DM I	6980 - 6755	5800 - 5650
DM II	?6500 - 6050	5450 - 4900
DM III	?5900 - 5615	4800 - 4500
DM IV	?4800 - 4545	3550 - 3350
DM V	? - 2335	? - 400

character today. Ditch sections and commercial boreholes show that the pre-Flandrian surface falls to *c*5m below present ground level to the north and west of the present coversand edge. This topography has been infilled by a series of deposits of organic (peats) and clastic origin (clays, sands, silts etc), which has formed a plain lying *c*3-5m OD, stretching from the Alt to the coast (Plate 3). The major River Alt channel has also been filled by these Flandrian deposits which partly reflect the influence of sea-level changes in the area.

Tooley (1978a, 105) has recognised five periods of sea-level rise and marine transgression in the north Merseyside area, on Downholland Moss and the Sefton coast, although more research is needed to define them more closely (Plater *et al* 1993). These have been designated Downholland I-V (summarised in Table 7). They have been characterised from relationships between overlapping deposits of clays, silts, sands (deposits of marine origin), and peats (non-marine deposits), with radiocarbon determinations used to date key horizons as sea-level index points (Tooley 1978b). The Downholland terminology is applicable only to the type area and should not be used for correlation with other parts of the region. Not all of the transgressive events are recorded away from Downholland Moss itself. These transgressive overlap sequences supersede the transgression sequence Lytham I-IX used in earlier publications (*eg* Tooley 1974). At the regional scale, twelve periods of positive and negative sea-level tendency have been identified, based upon sea-level index point data, for the eastern Irish Sea coast (Tooley 1982, 1985).

Between *c*12,000-7000 BC sea-level was up to 21m below present OD (Tooley 1982), so that pre-Flandrian sands and clays would have formed a large expanse of land to the west of the present coastline. Sea-level rise was rapid, however, the northern Irish Sea being formed by 7270-7200 BC (Huddart *et al* 1977). Subsequent sea-level rise, including a very rapid phase around *c*4150 Cal BC (Huddart 1992), led to a reduction in the coastal plain, producing marine transgressions around Downholland at least three times before *c*4500 Cal BC. The earliest of these transgressions caused marine deposits (recognised by the deposition of Downholland silt representing former estuarine and saltmarsh environments) to reach *c*2.2km inland of the present coast (Huddart 1992). This was further inland than any of the later transgressions. Each time the sea-level fell biogenic deposits (the coastal peats) were formed. A fourth transgression is also recorded in North Merseyside at the Altmouth, where the regressive overlap is dated to 3592-2926 Cal BC (4545±90 BP; Hv-2679). The second half of the fourth millennium Cal BC was generally a time of marine regression and falling sea-levels in the North West (Tooley 1978a, 35). Extensive areas of former tidal flats were exposed to sub-aerial weathering and colonisation by terrestrial plants. Although oscillations in sea-level continued after this time the transgressions were of limited extent, affecting only small inlets or estuaries, such as the Altmouth or Crossens at the northern end of the Sefton coast (Tooley 1978a, 1982).

In the survey area, the peat in the northern part of the coastal plain began forming by 6215-5960 Cal BC (7230±70 BP; SRR-2697) at Flea Moss Wood (Table 8), and is found directly overlying the coversand. This peat may mark the onset of wetter conditions as marine influences approached the area, although it appears to be earlier than the DM I marine transgression. The onset of reedswamp and fen at this time also affected the Alt Valley, nearly as far south as Sefton village. Midway in the valley at Ince Blundell the peat formation is dated to 6180-5740 Cal BC (7130±110 BP; GU-5210) (Table 10).

In the area north of Flea Moss Wood peat is intercalated with marine silts and clays. The maximum southward extent of these clays lies approximately along an east-west line drawn a little to the north of Flea Moss Wood/Orrell Hill. The marine transgressions fed from the proto-estuary of the Alt into the basin, later to be occupied by Downholland Moss, to the north-east. West of Formby, an area of marine sands had formed a large bar, which existed before the early fifth millennium Cal BC, preventing direct marine influence from the west (Pye and Neal 1993a).

It is difficult, however, to correlate the marine phases across Downholland Moss with events in the area to the south of the Alt estuary, since the clays cannot be dated in the survey area and it is not therefore clear to which transgression(s) they belong. The peat in the mid Alt Valley is sealed by an estuarine clay, with dates for the transgressive overlap ranging from 5204-4791 Cal BC (6040±60 BP; GU-5211) to 5230-4790 Cal BC (6080±90 BP; GU-5212) (Table 10). This would seem to place it right at the very end of the DM II transgressive period. This suggests that estuarine conditions may also have lain close to their maximum to the north of Flea Moss Wood and Orrell Hill. Until the clays in this area are dated more precisely there has, though, to be some doubt in attempting to correlate human activity in the late Mesolithic and early Neolithic to either fresh water, brackish water, or marine environments.

The land to the south of Flea Moss Wood was never affected by marine transgressions. The earliest peat here formed under reedswamps and fen woodlands which accumulated organic material as water-tables rose across the area. If this was a consequence of high sea-level the date for the onset of peat growth of 4780-4510 Cal BC (5770±50 BP; SRR-2699) at Sniggery Wood (Table 9) suggests this was in response to the advance of transgression DM III to the west or north. The clays represent the fact that for parts of the prehistoric period tidal estuarine deposits were being deposited south and east of the present Alt estuary in an intertidal saltmarsh and mudflat zone. The area to the north of Orrell Hill was occupied by saltmarsh during the prehistoric period, with the limit of marine influence lying around the present peat-clay interface *c*100m north of Lady Green hamlet (Fig 73). At several places in the area, such as in a channel fill at Orrell Hill, *c*400m north of Lady Green (Huddart 1992), and most notably in the upper part of the Alt valley, the clays are succeeded by renewed peat growth during the first half of the fifth millennium Cal BC (Table 10). The 4m OD contour marks approximately the upper limit of these deposits in the southern part of the valley. In the northern part of the valley this peat is not apparent, except in thin (*c*0.1m), discontinuous, heavily humified bands towards the top of the marine clay.

Along the Alt Valley the peat is sealed by a surface brown alluvial clay. To the north-west, towards the mouth of the estuary and eastwards to Hightown, this alluvial clay seals the marine clays. The approximate geographical distribution of the deposit is neatly illustrated by a 1778 plan (LRO, DDX 10/3A) showing the lands damaged by the flooding of the Alt (Fig 24). In 1779 an Act of Parliament (LRO, DDBI 24/26) allowed defences to be built to prevent the flooding, represented by the alluvial clay found across the survey area. The limit of the flooded area shown on the map generally extends several hundred metres to the south of the limit of the clay mapped on the ground. This alluvial clay thins out southwards along the Alt Valley until it is only *c*0.1m thick against the 5m contour across the valley, east of Lunt.

The major natural landscape event of the late fourth and early third millennia Cal BC may have been the formation of the sand dune barrier along the seaward fringes of the coast zone (Tooley 1978b; Innes and Tooley 1993), although Huddart (1992) argues for a barrier to the west of Downholland Moss several millennia earlier than this. Dune formation resulted from a lower sea-level which allowed accumulation of material from the freshly-exposed intertidal areas by long-shore drifting and wind-throw. Sand overwhelmed the seaward end of Downholland Moss after 3091-2144 Cal BC (4090±170 BP; Hv-4705) (Tooley 1978a), while the burial of peat deposits at Sniggery Wood probably occurred slightly earlier than this, dated *c*3370-3040 Cal BC (4510±50 BP; SRR-2698) (Table 9). These dunes had the effect of protecting the lowland areas from later marine incursion, but blocked the drainage of coastal mires to the sea. This dune sand is recognised from the palynological sampling site at Sniggery Wood, but the belt of fields to the east have blown sand surfaces with low, narrow, sandy ridges parallel to the modern dune belt, which suggests that dune systems may have existed further to the east than presently. This sand is found overlying the peat to a depth of *c*0.8-0.6m as far as the base of the Woodham Knoll ridge in Little Crosby, although it extends a little further eastwards to the north of the ridge.

The marine transgression DM IV, which extended into the first few centuries of the mid fourth millennium Cal BC (Table 7) caused a general raising of perimarine zone water-levels and the formation of a number of small meres in this coastal lowland zone (Tooley 1978a). These mainly lie to the north of the Alt, in the area yet to be surveyed (Middleton *et al* forthcoming), including Gettern Mere, Black Otter, White Otter, Barton, Formby, and Downholland Meres. Although their existence is uncertain during the low sea-level conditions, it is likely that they became a permanent feature during the late fourth millennium Cal BC, following the accumulation of the sand dunes. A fifth period of high sea-level in this area is inferred during the late first millennium Cal BC, from a fossil dune slack deposit at Formby foreshore dated 800-110 Cal BC (2335±120 BP; Hv-4709) (Tooley 1978a; Innes and Tooley 1993). However, this did not cause a major transgression of the coast because of the protective sand dune barrier. Evidence for it is consequently found only in the survey area along the current dune belt. A second period of dune accumulation is postulated between the ninth and thirteenth centuries AD, with a brief period of dune stability sometime around Cal AD 1042-1270 (830±50 BP; Hv-3846). A number of radiocarbon determinations are available on fossil dune slack deposits in North West Europe (which signify dune stability, high water-tables, and high sea-level) ranging between these dates (*eg* Jelgersma *et al* 1970; Tooley 1978a; Kenna 1986).

Survey techniques

Within an area of *c*20km² there are three main surface exposures of peat in Sefton, which are associated with two types of landscape: the coastal plain, and river channels or valleys. For the purposes of the survey they are treated as a single unit, as they are all but contiguous across the area beneath a variety of later deposits.

Fieldwalking

The Sefton mosslands were the most extensively surveyed part of the county under the auspices of the NWWS. The survey mainly involved fieldwalking, reflecting the relatively large area of dryland/wetland interface available (Fig 74). It was possible to fieldwalk much of the survey area, including both the peat and its pre-Flandrian sand margins, totalling *c*850ha. The fact that the peats are generally seldom thicker than *c*0.5m beneath the topsoil horizon means that the pre-Flandrian subsurface potentially could also have been subjected to damage from ploughing and drainage. The dominant deposit around the mosses, as around the central mosslands, is the Shirdley Hill Sand. In contrast to that area, the relative isolation from St Helens allowed most of the Shirdley Hill Sand from Sefton to survive the programme of extraction for the glass industry in the 1960s.

One problem affecting the survey resulted from the wide expanse of coastal sand which has blown eastwards of the present dune belt. This has buried the peat, consequently protecting it from ploughing and humification. Conversely, it has hidden its surface and adjacent boundary with the pre-Flandrian mineral surface, so reducing the potential for archaeological discoveries. Wherever field ditches cut through the inland blown sand to the east of the dunes it was generally possible to survey limited extents of vertical ditch which showed the blown sand/peat/pre-Flandrian interfaces. Fieldwalking of many fields affected by blown sand, which are largely given over to arable, was also undertaken in an attempt to identify human exploitation of potentially early dune systems.

Pasture fields were also walked and wherever ditch sides were clean a record was made of the stratigraphy. Very few ditches had been cleaned recently and so continuous sections across large areas are very rare, which has made continuous mapping of the peat difficult in some areas (Fig 75). Neither the intertidal nor dune belt zones were surveyed, although peats are known to survive under the latter and outcrop on their seaward side (Pye 1990; Pye and Neal 1993b).

Palaeoecological survey

The work of Tooley (1974, 1978a, 1978b, 1980, 1982, 1985) provides much of the evidence for the effects that changes in sea-level have had on the development of the natural environment. Work in the late 1970s as part of the Liverpool Museum survey programme was specifically targeted at a number of peat areas in Sefton, and pollen analysis of two sites, at Flea Moss Wood and Sniggery Wood, Little Crosby, is presented in the palaeoecological section of this chapter. As the fieldwork for the NWWS developed it became obvious that the lack of detailed palaeoecological work in the eastern part of the mosses would be a hindrance to understanding the archaeological pattern that was emerging. Liverpool Museum therefore sponsored a detailed core from the Alt Valley peats in Ince Blundell as part of the NWWS. The NWWS subsequently provided the funds for a series of four radiocarbon dates on this core, and the details from this site are also presented in the palaeoecological section.

Aerial photography

Aerial photographic survey has been undertaken over a number of years in this area as part of R Philpott's research for Liverpool Museum. In addition C Cox undertook a survey of pre-existing coverage as part of the NWWS. This involved examination of photographs taken mainly in the early 1970s by the County Council. These are now held by the Joint Countryside Advisory Service, who assumed some of the functions associated with environmental matters after abolition of the Metropolitan County.

The mosses

Coastal group: Flea Moss Wood/ Sniggery Wood

The westernmost area of peat in Sefton runs between the present coastline and the low ridge of Woodham Knoll, to the north-west of the village of Little Crosby. The main coastal belt of peat is *c*1km wide and *c*2.5km north to south, with Hightown marking the approximate north-eastern edge, and the outskirts of Great Crosby the southern limits. The total area has been estimated at *c*280ha, although difficulties in the survey (*see below*) make this figure a little artificial. The peat ranges in depth from *c*3m at Flea Moss Wood to *c*0.1-0.3m around the eastern and northern margins. The average depth of the peat in the central part of the coastal plain is *c*0.7m.

The peat could be traced only intermittently across the coastal area, as it is covered by between *c*0.4-0.8m of blown sand from former dunes. This blown sand extends from the railway line, which marks the eastern edge of the present dunes, for a maximum width of *c*1km, its eastern extent lying roughly in a line parallel to the coast. Very few of the ditches in this zone were deep enough, or their sides clean enough, to allow for widespread identification of the stratigraphy beneath the sand. The peat was located along the Far Moss Pool drain, close to the railway line where it was on average a minimum of *c*0.5m deep, beneath blown sand that deepens northwards from *c*0.5-1m. This suggests that the peat probably extends as far as the line of present dunes to the west. An outcrop of peat is recorded from the intertidal side of the dunes, near Hightown (SD 295032) (Tooley 1977, 1978a; Lewis 1982), although this area has not been included in the present survey. This might suggest that the peat could continue under the dunes for another *c*0.5km to the west of the survey zone (this has not been indicated on Fig 75).

The northern limit of the peat is probably an artificial one, as landuse to the north is mainly recreational pasture which did not allow stratigraphic survey. The north-eastern boundary of the peat is marked by the presence of a blue-grey clay. Several kilometres to the north, nearer the present Alt estuary, such clays are of marine origin, representing inundation of the area by estuarine or saltmarsh deposits (Huddart 1992). It seems likely therefore that the clays along the edge of the coastal survey area are also of this origin. These clays reach depths of *c*2m quite close to the northern limits of the peat in the two other survey zones (*see below*), implying that the peat in the coastal zone may be either deeply buried by, or truncated by, the clays to the north. This possibility has led to the projected northern limit for the peat to be drawn running westwards from the most north-easterly identification of the peat. In this section it is *c*0.3m deep and in reasonably good condition beneath its blown sand overburden.

The southern boundary of the coastal peat is marked not by any strong topographical feature, but by a gentle rise in the landscape onto the boulder clay geology skirting the northern edge of Great Crosby. This boundary has also had to be partly estimated as there are few ditch sections available away from the Far Moss Pool drain and Sniggery Wood. In this area the peat is *c*2m deep.

Survey was made difficult in some areas of the coastal zone due to the fact that the effects of the landward migration of sand dunes have led to belts of pasture or recreational landuse dominating on the thin sandy soils. West of the railway line, between Hightown and Great Crosby, the sandy area consists of dunes and golf courses which were not included in the survey area. Extending to the east of the railway, south of Hightown, there is additionally much recreational grassland. In the areas of peat, the surface mostly consisted of blown sand, which postdates the late Neolithic. Fieldwalking was also extended onto the coversand Woodham Knoll ridge (Fig 74), as it is topographically the most distinct feature in the area. Approximately 270ha of arable land was walked in this coastal zone.

To the east of the Woodham Knoll ridge a narrow channel extends south-eastwards for a distance of *c*900m. Those parts of the channel close to Flea Moss Wood contain peat which is *c*0.3m deep, but its greater part is now filled with a peaty sand, as drainage and farming has led to the humification of the former peat for most of the length of the channel (Fig 75).

To the north of the channel, a narrow band of Shirdley Hill Sand separates it from a slightly more extensive area of peat that forms a north-eastward extension of the coastal peat to the north of the coversand ridge. In the more westerly part of this area the surface is littered with small Shirdley Hill Sand islands and ridges lying in the remnant peat which fills the hollows between. In this latter area it is difficult to identify how deep the peat is, as there are no suitable ditch sections in the area, but it appears to be fairly shallow.

The channel is located in mainly arable land and

fieldwalking was extended to both sides of its line. Most of the area was fieldwalked, encompassing the moderately steep eastern slopes of the Woodham Knoll ridge and the more gentle gradient leading to the brow of the Ince Blundell ridge *c*1km to the east (Fig 74). The peat to the north of the channel occurs in an area where a number of fields are given over to pasture. This made the precise delimitation of the peat here more difficult, although survey in the surrounding fields suggests that the actual area is unlikely to be much more extensive than that shown on Figure 75.

Orrell Hill

The second area of peat in Sefton lies *c*800m to the north-east of the coastal peat. It is separated from it by the low, diffuse topography of the Shirdley Hill Sand running up to Orrell Hill. This slightly elevated area lies at a height of *c*6m OD and marks the southern edge of an area of peat of *c*40ha. The area surveyed around the peat amounted to *c*100ha, the survey area being approximately defined by a rough oval area, with the small hamlet of Lady Green midway along its southern circumference and North End Farm near to its northern limit (Fig 74). The east-west limits of this peat deposit were more difficult to trace (*see below*).

About 25% of the surveyed area consisted of pasture and woodland, all of it concentrated in the western and north-western part of the survey area. The peat to the west of the bypass lies within this pasture, but as the ditches were almost all too vegetated to survey, tracing the edge of the peat was extremely difficult here. It could only be recognised in a small area near the south-eastern edge of the dryland, where it is a minimum thickness of *c*0.3m beneath the topsoil. The western border of the peat is therefore only approximate in this area on Figure 75. In the extreme north-west of the surveyed area the peat was only *c*0.1m deep and was mixed with small patches of grey clay.

The limited field data for this area of mossland can be augmented by a survey along the main ditch that runs north-west to south-east across the survey area, undertaken when a section was exposed during ditch cleaning operations by the National Rivers Authority in 1982-3 (Huddart 1992, 229-231 and fig 8). This showed that the surface peat extends *c*400m to the north of Lady Green Farm and is underlain by gradually thickening grey silty clay, which is found in two bands and represents former marsh environments. The clay overlies a soil formed over the Shirdley Hill Sand coversand. In some places a thin peat, *c*0.1m thick, had formed over this soil before the clay deposition. The work also showed that small areas of peat close to Lady Green reached depths of 1.5m, while the peat to the north-west, although generally shallow, was variable in depth and reached almost 1m deep in a channel formed in the underlying clay near its northern limit.

The peat is identifiable more easily to the north-east of the bypass, where a large area of arable land is found and much of the peat lies at the surface. Much of the peat in this fieldwalked area is quite shallow, generally being less than *c*0.3m thick, although there are some small channels close to the bypass which are up to *c*0.5m deep. The general shallowness of the peat has caused a number of small islands and ridges of Shirdley Hill Sand to be interspersed with the southern edge of the peat here. To the west of the Ince Blundell bypass a number of similar ridges and islands are also apparent in the pasture fields at the southern end of the peat area.

Around the northern and north-eastern edge of this deposit the peat thins to *c*0.1m of very humified peat. Within *c*400m north of Lady Green hamlet the peat disappears and clays deepen to over *c*2m around North End Farm. North of here they deepen towards the Altmouth where they reach *c*5m in depth (Wray and Cope 1948). The average depth of the surface clays identified in ditch sections over the northern part of the surveyed zone is *c*1.5m. The field survey showed that the upper part of this clay differed from the lower half, the latter being a light bluey-grey colour, and probably represents marine or estuarine deposits, as identified by Huddart (1992), to the east of the bypass and further to the north nearer the Alt estuary. The overlying clay is browner and may relate to the regular flooding of the area recorded in the post-medieval period (Fig 24). Huddart (1992, 231-232) shows that *c*1km to the north, at Alt Bridge, the upper clays are of estuarine origin.

The area between North End Farm and the estuary of the Alt to the north-west, which is mostly arable land, was not extensively fieldwalked. It soon became clear after a small number of fields had been sampled here that they were clean of all but a few very recent artefacts. This reflects their former character of regularly flooded lands which were not reclaimed until quite late in the post-medieval period. The ditches in this area were also not deep enough to provide more stratigraphic evidence than could be got from the clay surface.

The eastern boundary of the Orrell Hill deposit is very difficult to identify because of the presence of these clays and a general lack of deep, clean ditch faces in the area. There is a gap *c*200m wide east-west shown on Figure 75 between this deposit and the peat found in the Alt valley proper. This may be artificial, but more evidence is needed before the peat distribution can be shown to be continuous across the area and into the valley.

Alt Valley

The two bands of silty clay noted at Orrell Hill extend eastwards, recorded to an average thickness of *c*1m, into the Alt valley. As in the previous peat deposit, the lack of clean ditch sections in this zone has meant that the peat has only been noted intermittently under this clay. It has been traced as a thin, *c*0.1m, deposit in a narrow belt running eastwards to the floodplain of the Alt, south of New Moss Wood. Charcoal was noted in one ditch section close to the wood. The northern limit to the peat, which thickens to the south-east, is shown as an artificial line on Figure 75, as the estuarine clays deepen to the north-west and make identification of peat impossible from ditch sections.

To the south-east of the wood, where the valley slope becomes more pronounced, the peat becomes thicker, *c*0.5m on average. This band of peat is between 300-400m wide, and runs along the western side of the Alt for a distance of *c*4km. For much of its southern distribution a basal peat, *c*0.5m deep, is overlain by a band of light blue-grey estuarine clay between *c*0.1-0.3m thick. The accumulation of this clay has been dated through the NWWS to shortly after *c*5200-4800 Cal BC. Sealing this layer is a second band of peat, of about the same thickness as the clay, beneath the ploughzone soil. Along a strip *c*200m wide adjacent to the course of the Alt, and wider over the clay further north, the peat is overlain by a brown clay. Its extent roughly corresponds to the distribution of post-medieval alluvial clay shown on a map of 1778 (LRO, DDX 10/3A) (Fig 24). The peat in the valley accounts for *c*150ha, although the total area may be larger than this as it is hidden by the clays further to the north.

The western boundary of this wetland is marked, from a point east of Orrell Hill, by the gentle sandy slopes of the Ince Blundell ridge. The brow of the ridge has attracted modern settlement, consisting mainly of scattered farms and the small hamlets of Homer Green and Lunt at its southern end. South-east of Lunt the limit of the survey zone is marked by a large domestic refuse tip which covers the width of the floodplain as far as Sefton village (Fig 74).

The River Alt marks the eastern limit of this survey, although the peat does extend to the east of the river, in a belt stretching from *c*1km wide at its southern end to about 3km wide towards the north near Lydiate. From here it widens out into the large expanse of Downholland Moss in West Lancashire.

Virtually the whole of the valley and adjoining slope is given over to arable farming and it was possible to walk *c*225ha of both wetland and adjacent dryland. The depth of presumed alluvial clay capping the upper peat makes some areas of the floodplain flat, and not very receptive to recovery of artefacts. From a point approximately opposite Carr Houses, however, the clay thins a little and the undulating Shirdley Hill Sand topography beneath it intermittently comes to the surface in a series of small islands and ridges.

Archaeology and palaeoecology

Early Mesolithic period *c*8000 BC – 6800 BC

The late glacial coversands that form the dryland over most of the survey area emanate from aeolian and possibly fluvio-glacial sources (Wilson *et al* 1981; Wilson 1985; Huddart 1992, 231). In the first millennium after their formation they may have remained relatively unstable as soil formation was retarded (Innes 1986; Innes and Tooley 1993). Possible archaeological evidence for human activity contemporary with this environment remains limited and tentative. One struck flint piece from the survey, found on the northern edge of the ridge at Woodham Knoll, Little Crosby, adjacent to the Flea Moss Wood peat (Fig 76; Me 039), has technological characteristics unlike any other flintwork known from the county (Fig 20.19). These might tend to place it in a late Palaeolithic context, a view since tentatively strengthened (J Wymer pers comm), although one qualification to this interpretation is based on the fact that the raw material, which is a very dense, creamy, speckled chert or even sedimentary type rock, is similar, though not identical, to the raw material used for a struck piece of late Neolithic or early Bronze Age date, found *c*900m to the south-west (Me 007).

If the former artefact is of such an early date then it seems unlikely to have been associated with activity based on exploitation of the wetland, as Flea Moss did not start forming until a little before 6215-5960 Cal BC (7230±70 BP; SRR-2697). There are some indications, from gravelly sand deposits, that a channel predating the peat may have flowed to the west of the ridge, which could have provided a focus for human activity, although such deposits were only recognised in a limited area in a few ditch sections.

By the beginning of the Mesolithic period rising sea-level had led to the formation of the Irish Sea, with the connection between the Isle of Man and mainland Britain being severed at least by 7270 BC (9270±2000 BP; Birm-141) (Tooley 1980, 83). The extensive presence of wetland in the area during this period, however, seems unlikely as sea-level would still have been relatively low. Identification of human activity from the artefactual record relating to the period is problematic because early Mesolithic material is difficult to distinguish from that of the later period, unless the distinctive microliths appear in the surface assemblages. There is some evidence that at this time in southern England blades, which are the second main characteristic of Mesolithic assemblages, are of larger proportions than in the later Mesolithic (Pitts and Jacobi 1979), but whether this is also a feature of the local Mesolithic is not clear. When work on the quantification of excavated assemblages from early Mesolithic sites on the Wirral (Cowell 1992c) is complete, and can be compared with assemblages from probable late Mesolithic sites at Ditton and Croxteth Park (Cowell in prep b), such a local framework may exist.

One indication of early Mesolithic sites on the Wirral is the use of a range of grey-brown or black cherts. Chert is very rarely found to the east of the Mersey, and the few examples are of a much poorer quality, possibly representing *ad hoc* use of the occasional pebble found in the boulder clay. At one site in Lunt in the Alt Valley (Me 094), however, chert artefacts of a similar kind to those found on the Wirral occur in association with a blade assemblage, which is tentatively attributed to a site of this early period. The seven relevant pieces of struck flint from Me 094 exhibit some technological features associated with on-site curation of flint found in the North Wirral assemblages, including core rejuvenation (Fig 20.4) and dressing of the striking platform. These features are rare in the late Mesolithic assemblages so far identified in the county. At the Greasby site on North Wirral, the surface flint assemblage from fieldwalking is notable for the small proportion of the grey-brown and black cherts present in relation to light coloured cherts, the latter group being shown by excavation to be stratigraphically later. This may be due to some extent to the fact that lighter material is more visible in the ploughsoil, but the opposite pattern prevailed in the pits and general layers beneath the ploughsoil. This suggests that the small amount of possible early Mesolithic material on the surface at Lunt could under-represent the actual importance of this group of material on the site as a whole.

Other examples of material referable to this early date are even more difficult to identify. One possible focus lies on the northern edge of the sand ridge at Flea Moss Wood, Little Crosby (Me 039), where there is an almost continuous scatter of flint over *c*120m by 80m, mostly of probable later Mesolithic date. A few pieces, however, including a large core with long blade removals, and a large point (both a blue-white patinated flint) may belong to this early period (Me 038).

Both locations are close to river channels. Me 094 is on the floodplain of the Alt Valley where a low sand ridge protrudes from the peat. The site lies *c*150m from the present course of the Alt, but it is not clear how far this may have meandered across the valley floor since the early Mesolithic. Me 039 lies almost on the northern edge of the sand ridge overlooking the Woodham Knoll peat channel. If this material is of early Mesolithic date, then it predates the rise in groundwater levels across the area around the late seventh/early sixth Cal BC millennium, which led to the formation of reedswamp in the vicinity of both these locations. These sites are therefore unlikely to have been bases for exploiting widespread fen landscapes. The main factor determining location would therefore probably have been related to riparian exploitation.

Little can be said about the nature of the sites represented by this putative early Mesolithic evidence, as it amounts to no more than *c*10 pieces in total. If the chert discovered is from the same Welsh upland sources as that on the Wirral, it might imply that this part of the county lay within the annual cycle of the hunters who were ranging across this area; part of the county would have been accessible from the Wirral across dryland that is now under Liverpool Bay, a distance of *c*15-20km. The Little Crosby area would have provided the nearest belt of light sandy soils for the people using the low sandstone hills around Greasby and North Wirral, and the Alt Valley might also be expected to have provided attractive riparian resources, as well as a route into the sandy areas to the east. The typology and raw material of a point from the Little Crosby site (Me 038) is very closely matched by a piece from Rainford (Fig 32.7; Me 106). This was found on the sandy ridge overlooking Holiday Moss, *c*18km to the east (*see Chapter 5*). The site in the Alt Valley may be partly covered by later peat, and further evidence with the potential to aid understanding of the landuse during this period may still be hidden.

Later Mesolithic/earliest Neolithic *c*6800–4000 Cal BC

This part of Sefton has the richest prehistoric evidence in the county, most of which, on the basis of

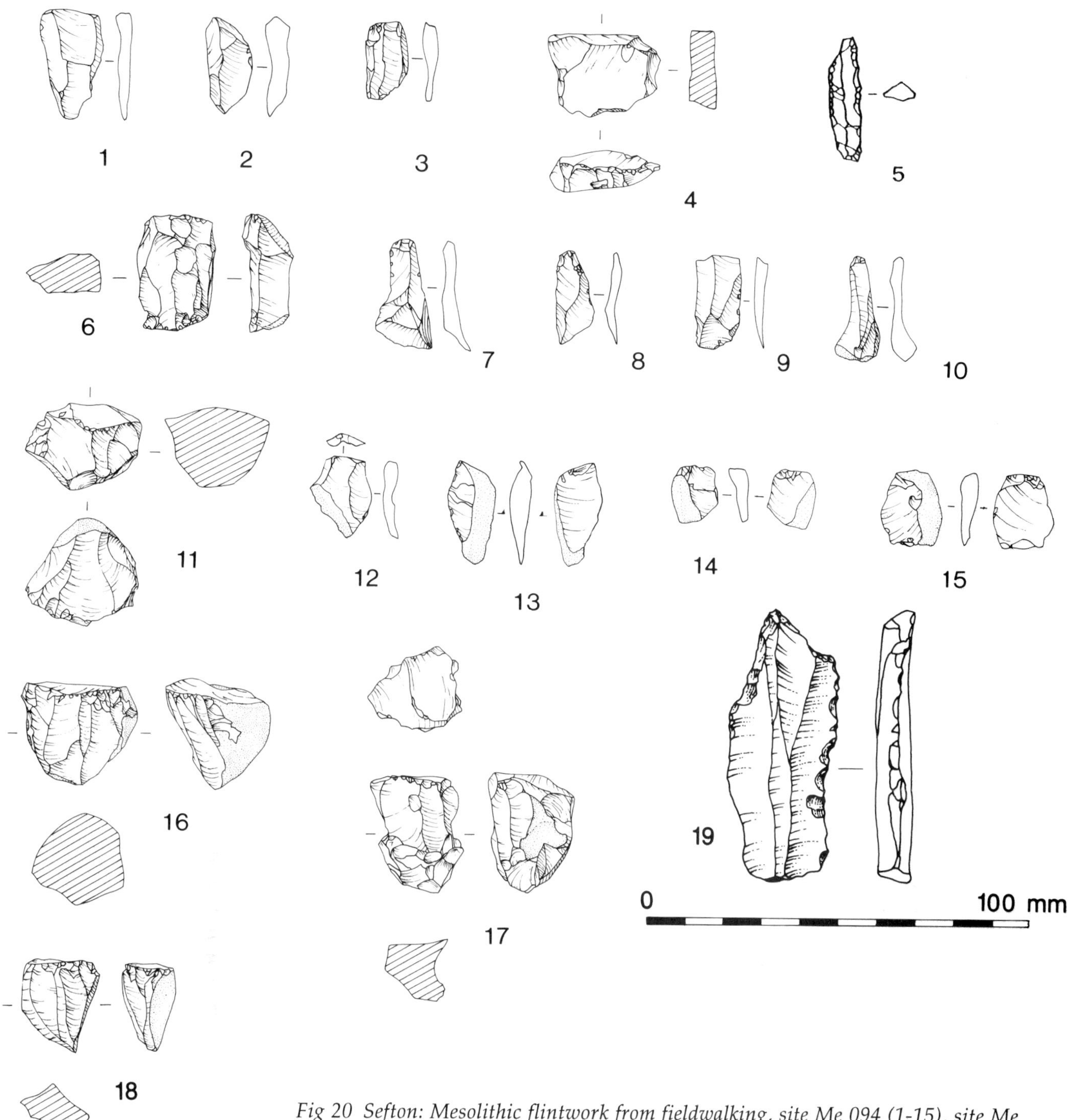

Fig 20 Sefton: Mesolithic flintwork from fieldwalking, site Me 094 (1-15), site Me 080 (16-18), Ince Blundell; site Me 039 (19), Little Crosby

the flint technology, is attributed to the Mesolithic (*see Chapter 1*). Although there is no excavated evidence to provide a chronological framework for this material it would seem likely, based on dates from other coastal sites in the lowlands (Bonsall *et al* 1986), that most of it may postdate the inception and spread of wetland in the area, *ie c*6215-5960 Cal BC (7230±70 BP; SRR-2697). The lack of an excavated framework also makes it difficult to identify the introduction of early Neolithic culture into the area through flint assemblages alone, and the end of the Mesolithic period in this area may therefore be quite vague and could be marked by assemblages that differ little in many technological details.

Sea-level rise between *c*7000 BC and *c*5000 Cal BC was rapid but thereafter slower, and was marked by a series of oscillations (Tooley 1985, 106), so that the late Mesolithic saw the first real effects of these changes on the present coastal areas. These effects were felt prior to the spread of marine environments across the present coastal zone, as areas of wetland started to form behind the advancing coastline. Reedswamp was established around Flea Moss Wood at *c*6215-5960 Cal BC (7230±70 BP; SRR-2697) (Table 8), and shortly afterwards fen carr woodland had become widespread in the area. To the east, waterlogging is evident in the Alt floodplain at about the same time. By the middle sixth millennium Cal BC widespread reedswamp and fen carr lay across the valley (Table 10).

Although freshwater wetlands were widespread around the Shirdley Hill Sand ridges in the south of the survey zone, it lay close to different conditions a little to the north. Here the marine transgression of DM I at the beginning of the sixth millennium Cal BC, and that of DM II in the later sixth millennium Cal BC (Table 7), resulted in estuarine conditions in the area east of Hightown. With its flanking saltmarsh this landscape extended into Downholland Moss to the north-east of the estuary (Tooley 1978a; Huddart 1992), but it is not known how far south of the Alt estuary these estuarine deposits extended. What does seem clear, though, is that during the sixth millennium Cal BC many parts of the northern coastal zone maintained a generally semi-aquatic landscape. Coastal conditions are likely never to have been far away, probably mostly within several kilometres of the fens which were interspersed amongst the areas of sandy dryland around Ince Blundell and Little Crosby.

At some stage (the dating evidence is not yet precise enough) saltmarsh reached to within 100m or so of the dryland areas at Orrell Hill and Woodham Knoll. This had probably been established before the end of the sixth millennium Cal BC when, in DM II, tidal estuarine conditions replaced fenland in the Alt Valley as far south as Lunt.

The estuarine/marine clays never reached as far south as Sniggery Wood, so that during the sixth millennium Cal BC the area was a sandy plain as far as Woodham Knoll, with fen, and later, towards the middle of the millennium, cotton-grass bog interspersed with the alder fen lying along the plain to the north. The onset of wet conditions in the extreme south of the plain did not occur until the early fifth millennium Cal BC, in DM III. It is dated after *c* 4789-4510 Cal BC (5770±50 BP; SRR-2699) at Sniggery Wood (Table 9). This area subsequently became dominated by reedswamp, freshwater pools, and fen carr woodland.

The archaeological evidence of the late Mesolithic and earliest Neolithic period therefore has to be seen against this background. During the sixth millennium Cal BC coastal conditions lay close by for most of the time as fen and swamp lay around the main sandy ridges and in the Alt Valley. In the first half of the fifth millennium Cal BC, apart from a period in the first few centuries, coastal conditions lay a little further away and fen became more prevalent, spreading south along the coastal plain, probably colonising the former marshland around Orrell Hill and becoming widespread again in the Alt Valley.

The presence of humans in the area at this time may be inferred from the pollen record. Some slight disturbance to the vegetation is noted during the earlier sixth millennium Cal BC when coastal conditions were prevalent south of the Alt estuary. At Flea Moss Wood there is a short phase when, though woodland cover remains fairly constant, broken ground weeds are evident. This is followed by a lengthy period of woodland regeneration (Table 8, phase 2).

It is, however, with the spread of fen during the DM III marine phase at the beginning of the fifth millennium Cal BC (Table 7) that the main effects potentially attributable to human impact are noted. On the dryland areas oak woodland was dominant, with other taxa such as elm, pine, lime, and ash less well represented, and with a variety of shrubs also present. At this time a short period of open ground is recorded at Flea Moss Wood, associated with a reduction in tree cover and the presence of *Triticum* (cereal-type) pollen, dated to 4939-4720 Cal BC (5920±50 BP; SRR-2696) (Table 8, phase 4). Subsequent human activity in the vicinity of the fen carr at Flea Moss Wood is not recognised in the vegetational record which ends after *c*3622-3340 Cal BC (4670±50 BP; SRR-2695).

At Sniggery Wood, however, *c*1.5km to the south-west, several episodes of woodland reduction and subsequent regeneration do occur during the rest of the fifth millennium Cal BC. The final clearance episode is associated with the elm decline (Table 9, phases 2, 4, 6). There is some evidence that the woodland further away from the wetland may have been more open by this time, although this could be the effects of the coastal influence rather than human interference.

The current problems in dating flint material found around these mosslands accurately within the Mesolithic, make for difficulties when associating it with the stratigraphic and pollen evidence from the adjacent peat sites. Coastal activity in Cumbria

Table 8 Sefton: Flea Moss Wood, palaeoecological summary

cms	Date (Cal BC)	Local Pollen Zones		Ecology of Moss	Regeneration/Disturbance Phases	
0		FMW-e	Characterised by sycamore. Pollen of other trees higher.	Humified Mor humus. Possible woodland floor.		
5		FMW-d	Birch values high. Wider variety of other trees and shrubs in low %. Wide range of herbs.	Increased acidification and drying out of bog surface with birch carr.	FMW-6	Major woodland disturbance with evidence of local cereal cultivation.
30	3622-3340		No pollen	More acidic wet bog conditions with some alder carr.		
			Elm falls at top of zone from low but consistent values .	Cotton grass pollen replaces aquatic pollen.	FMW-5	Local tree regeneration. Canopy undisturbed with sporadic records of weed pollen.
120	4939-4720	FMW-c	Characterised by oak and alder pollen. Pine and hazel reduced.		FMW-4	Small-scale clearance with some cereal-type pollen recorded.
130			Other trees and shrubs recorded. Some herbaceous pollen (dryland or wetland) present.	Alder carr and freshwater communities with rich wetland shrub and herb flora and aquatics.	FMW-3	Woodland with little or no disturbance.
185			Cereal-type pollen recorded at 120cm.			
210		FMW-b	Oak and alder characterise AP. Aquatic and wetland herb pollen common. Occasional ruderal types.	Reedswamp	FMW-2	Tree pollen still dominate but pollen of ruderal weeds may indicate small clearings in woodland.
240			Grasses and sedge pollen remain significant		FMW-1	Closed canopy mixed hazel and pine woodland replaced by oak and alder carr. Wetland herb types present.
295	6125-5960	FMW-b	Oak, pine, and hazel pollen. Alder rise at top of zone			

is dated to the sixth and fifth millennia Cal BC (Bonsall 1981), and together with the pollen evidence for clearance at this time from the local mosses it seems reasonable to suppose that some, if not most, of the local flintwork should date from this period also.

The sites found in Sefton are restricted to small clusters of material which are found mainly in two areas (Fig 76). They are distributed firstly around the northern edge of the Woodham Knoll ridge at Little Crosby, where they are found on the brow of the ridge and on the eastern slopes, overlooking the peat channel to the east. The flint material on the west, overlooking the coastal plain, is more dispersed.

Concentrations of struck flint are also quite strongly

represented in the Alt Valley, both on the floodplain and on the lower part of the western terrace slopes. The local vegetation record provides no indications of disturbances that could be the result of human activity in the valley.

The intervening mossland between these two areas of Mesolithic activity, centred around the Orrell Hill wetland, has produced less archaeological material. The stratigraphic evidence is also less clear, although it seems likely that reedswamp did not become established until the late sixth millennium Cal BC, and that prior to this the area was never far from the estuarine conditions that were widespread in this northern area during the late Mesolithic. The lack of Mesolithic material possibly reflects the less advantageous resources of the saltmarsh for settlement, with its exploitation perhaps being undertaken from sites further to the south. Only one possible small site (Me 058) has been located, set *c*150m back from the present edge of the moss, and a single blade has come from the edge of a small, shallow, peat area a little to the south (Me 048). A small blade from within the peat, found on the edge of a narrow peat-filled channel (Me 060) might suggest that at least one site is covered by peat, and that more may exist around the moss, although careful walking of a number of just emerging sand islands around Me 060 did not provide any further evidence. The discovery of Me 094, however, shows that ground conditions can be absolutely crucial in locating sites that are only slightly ploughed out, and therefore this possibility should not be dismissed for the area of Orrell Hill.

The sites with concentrated material are generally represented by 5-35 pieces of struck flint. Of a number of single find sites, recurring along the 4m OD contour of the Alt terrace, two consist of single blade cores (Fig 20.16, 18; Me 073, Me 088). One other site on the same contour (Me 080) provides an association between similar cores and knapping debris, which suggests that differences in surface visibility may have led to an under-representation of the quantity of concentrated sites in this topographic zone.

The raw material in most of the sites ascribed to the late Mesolithic is generally a light grey or blue-grey patinated flint that is likely to have come from the coastal margins. It could have been acquired quite easily at most times during the sixth and fifth millennia Cal BC.

One site on the Little Crosby ridge (Me 039), which has produced four cores, covers a larger area than many of the others, and may represent either several reoccupations of the same area or possibly a more intensively settled locus. Me 093, in Lunt, has many similarities to Me 039, except that cores are not a feature of the assemblage at the former site. Without excavation it is not possible to determine if this is just an accident of recovery (although the site was walked at 2m intervals) or whether it represents a functional difference, *ie* the 'industrial' element (*see below*) is lacking and knapping is geared more towards maintenance of tools and domestic activities. The flint assemblage on the Alt floodplain in Lunt (Me 094) is of similar size (Fig 21.B) and has a similar morphology to that at the Woodham Knoll site (Me 039). It includes cores, flakes, and blades, including primary knapping debris and flint pebbles, many of which are less than 30mm in diameter. This suggests that raw material was being worked from nodules on the site.

The implication behind the occurrence of cores at these sites is that they represent locations where raw material has been collected and then worked to provide the basis for the toolkits to be used for hunting and domestic activities. The sites referred to above may be solely 'industrial' areas where the raw material is collected, roughly prepared, and the waste left, with the more usable pieces taken to other types of settlement representing hunting or base camps. The latter type of site, which is less specialised, has not, however, been found elsewhere in the region. It is the coastal or estuarine locations that would have provided a range of natural environments whose resources could sustain larger groups for longer periods of time, as is suggested for Mesolithic sites in the Cumbrian coastal area (Bonsall 1981). In the absence of larger surface assemblages, from which it would be easier to make interpretations, it seems possible that the 'industrial' element at these Sefton sites may have been included within a range of domestic activities. These areas could therefore form the home range while, as is argued in Chapter 7, the interior of the county was the location mainly for specialised hunting and foraging activities, which were based in river valleys and around the western edge of the central mosslands.

The other types of Mesolithic site around the Sefton mosslands are more restricted in their area and density of finds. Interpretation of these sites is difficult as it is rarely clear to what extent post-depositional factors or survey conditions hide their true character. They are found in the same areas as the larger, more complex flint assemblages, on the slopes of the Woodham Knoll channel, and the terrace and the flood plain of the Alt. Nine sites, other than the large site Me 094, are located on small sand islands in the peat in the floodplain (*eg* Me 076-081, 082-085). All of these have small flint assemblages,

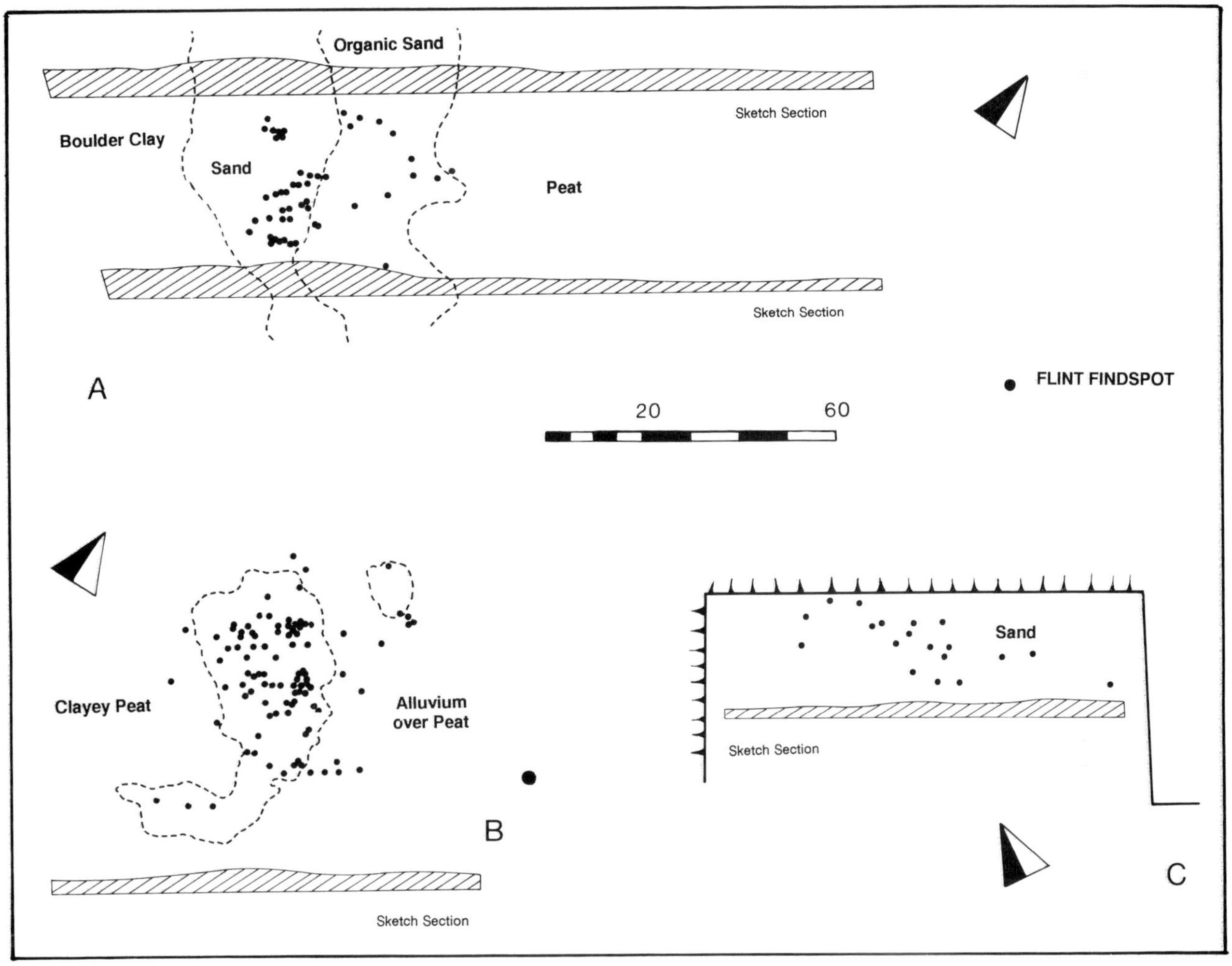

Fig 21 Sefton: Nature of a selection of surface flint sites in Ince Blundell, site Me 093 (A), site Me 094 (B), site Me 080 (C)

which includes a blade element, and use the same raw material as is found at the latter site. The small size of these sites may result from the fact that much of their extent is protected from the plough by the post-medieval alluvial clay. At the larger Lunt site (Me 094) this has thinned out considerably, so allowing more plough damage of the pre-Flandrian levels to take place. Until more evidence is available these sites could therefore be interpreted as small specialist camps or as similar in function to the large site at Lunt (Me 094).

A number of single findspots (*eg* Me 071) are more commonly found on the top of the Alt Valley ridge, to the north of Carr Houses, Ince Blundell, in an area where no convincing scatters are located. The group of sites along the 4m OD contour of the Alt terrace (Me 073, Me 088) also falls into this category, although survey conditions may be a factor here (*see above*). More of these sites are probably also located lower down the valley to the north but will be totally covered by the alluvial clay, and the peat below it, as the alluvial material deepens and smooths out the hummocky coversand topography. The western side of the Woodham Knoll ridge also produced mainly single pieces, although the later blown sand cover potentially hides more Mesolithic material. This type of site need not be contemporary with the larger concentrations of material, as the pattern of exploitation of the landscape need not have been a static one over a thousand years. At some times the area could have been used as a home range area, at others for specialist camps and activities.

Although the palaeoecological evidence is not totally complementary, the archaeological distribution of sites in Sefton suggests that in the coastal zone during the Mesolithic the evidence indicates that the margins of the river valleys and channels

Table 9 Sefton: Sniggery Wood, palaeoecological summary

cms	Date (Cal BC)	Local Pollen Zones	Ecology of Moss	Regeneration/Disturbance Phases
10	3370-3040	**SW-d** Characterised by Oak and alder. Sharp rise in heather. Birch values lower. Low values of wide variety of herb types. Pine and bracken rise at top of zone. Fewer fern and *Sphagnum* spores.	Rising water table increases nutrient level. Fen carr with open water pools replace acidic mire conditions.	**SW-9** Regeneration of more open woodland. Difficult to separate unstable coastal conditions from clearance indicators.
32			*Sphagnum* major peat component.	
		SW-c Birch pollen dominant type with oak, alder & hazel. Low elm, increase in ash pollen, small peaks of ruderal weeds. Less fern spores but high Sphagnum peak & rise of heather at top.	Drying out and acidification of moss with birch carr. Cotton grass major peat component.	**SW-8** Extensive clearance. AP falls sharply twice. The elm decline. High values of ruderal weed pollen, that of cow-wheat, common on sandy acid soils, may suggest use of fire for clearance.
56				
64	4034-3760*			**SW-7** AP recovers. Mixed oak woodland.
		SW-b Birch dominant with less oak & alder. Isolated increases in heather, willow & ash pollen. Sedge main herb pollen. Fern spores important. Elm decline at top of zone.		**SW-6** Significant clearance (mainly oak). Ruderal weeds, hazel/*Myrica*,willow and heather well represented.
72				
100				**SW-5** Open oak/ alder woodland recovered.
				SW-4 Small clearings in mixed oak woodland (mainly oak). Bracken and ruderal herbs increase.
112				
		SW-a AP 50% TP. Alder and oak major tree types with hazel and birch. Elm low but consistent. Grass major herb type with wide range of wetland herbs. Freshwater pollen throughout zone with coastal ones early.	Eutrophic fen carr	**SW-3** Local woodland regeneration. Open habitat indicators recorded may suggest a slightly broken canopy.
148		Upper boundary marked by rise in birch.	Reedswamp with freshwater pools	**SW-2** Limited forest clearance (oak and alder). Clearance indicators and ruderal herbs recorded. Grass pollen rises. Peaks of willow and hazel pollen. Heather recorded. Cherry and ash suggest open woodland .
160			Salt marsh	
				SW-1 Dense mixed woodland with only hazel as understory. Herbs indicative of open conditions probably on unstable sandy soils.
174	4780-4510			
			Shirdley Hill Sand	

* Date relative to FLII/FLII boundary at Park Road Meols (see Ch 3 *this volume*)

were exploited for settlement. It is likely that the coastal fens were also an important resource, although later sand deposition may be obscuring this relationship a little. The peat in the Little Crosby channel lies directly on the coversand, wherever it has been seen in section, which implies that the channel was not primarily alluvial, but subject to gradual waterlogging which led reedswamp to colonise the depression gradually. Here, and in the Alt Valley floodplain, the location of swamp and fen would have meant that the collection of plants, fishing, and the hunting of birds and woodland animals, would be relatively easy. The adjacent, higher, drier, light sandy soils of the Little Crosby ridge and the Alt Valley terrace may therefore have provided conditions where larger groups, in residential or base camps, could have been sustained for relatively long periods.

The Mesolithic sites seem placed to exploit the wetland whilst remaining at a short distance from it. Those sites already mentioned on the slopes of the Alt Valley also share this characteristic, including the potential base camp at Lunt (Me 093), and the smaller sites or findspots (Me 073, 080, 088), which all lie at least 100m from the contemporary fen edge. This distinctive location pattern might reflect the desire to allow animals to water unhindered at the edge of the swamp or fen and so increase the opportunities to hunt them.

If some of the larger sites represent residential or base camps, a number of the sites represented by single findspots of cores or implements may indicate specialist task sites, which were less intensively occupied, within the fen areas. This does not allow for the possibility of biased recovery due to post-depositional factors, which it is not possible to isolate at this stage.

The presence of marsh and marine environments within a day's easy travel by foot or boat from these suggested residential locations would have provided further opportunities for specialised activities to be undertaken by smaller, more highly mobile groups, associated with wildfowling, fishing, and raw material collection. Sites within these environments are likely to be very small, and as the coastal and alluvial deposits are widespread and deepen to the north of the main sand ridges in Little Crosby and Ince Blundell, they are very difficult to identify through surface survey.

The woodland disturbance phases at Sniggery Wood during the late fifth millennium Cal BC coincide with the transition from Mesolithic to Neolithic culture across the country as a whole. The nature of the landscape changes, suggesting that short-lived, small-scale woodland clearance did not alter during this period. This is a feature which is common to all the mossland areas in the county to the east of the River Mersey. The economic change from the Mesolithic to the Neolithic is consequently almost impossible to distinguish in this region from the pollen record alone, in contrast to many areas of the country where dated sites are found with domesticates and Neolithic cultural material before the end of the fifth millennium Cal BC (Zvelebil and Rowley-Conwy 1986).

Earlier Neolithic *c*4000-3200 Cal BC

It has already been seen that the fifth millennium Cal BC saw a series of small woodland reduction episodes around the western, coastal mossland. The earliest of these, at Flea Moss Wood (Table 8, phase 4), suggests that cereals may have been involved. The extent to which this activity represents native hunters rather than farmers, whether they be incomers or acculturated natives, is beyond the scope of the evidence at the moment, but the pattern described for the earliest Neolithic on the Wirral (*see Chapter 3*) may also be relevant for the Sefton coastal areas.

After the elm decline the pattern of woodland disturbance is difficult to identify in the pollen record at each of the main sites. At Flea Moss Wood the period is one of general woodland regeneration. At Sniggery Wood (Table 9, phase 8) dryland tree cover is reduced on two occasions before the mid fourth millennium Cal BC, although there is no evidence for cereals being associated with this phase. After this the marine influence of DM IV in the second half of the fourth millennium Cal BC (Table 7) affected the area. The effects of this make it difficult to distinguish human manipulation of the landscape up to *c*3370-3040 Cal BC (4510±50 BP; SRR-2698), after which the peat profile is terminated by blown sand.

The archaeological material from the early Neolithic is very difficult to identify within the county. Neolithic technology, in the form of leaf arrowheads and stone axes, had presumably, on the national chronology, been introduced into the area by this time. There is no contemporary local pottery from this part of the region and none was discovered during the survey. The identification of sites of this period therefore has to be achieved mainly from flint scatters. This is made difficult, both by the small size of sites located, and by the fact that there is no framework from locally dated, excavated sites against which surface material can be compared.

Two stone axes have previously been recorded from

the intertidal zone slightly to the south-west of Hightown (Fig 76; A 18, A 19) and it is possible that they may be connected with clearance of woodland during a period of lower sea-level. Six flint sites have been tentatively interpreted as being of early Neolithic date, allowing for the difficulties of dating such material. Raw materials for this flintwork are more varied than in the Mesolithic, with a patinated beach pebble variety, not apparent on Mesolithic sites, quite common, as well as the use of boulder clay flint. Even so, the material included within this classification still only totals *c*50 pieces. Until more evidence is available it remains a possibility that other pieces belonging to this period may be hidden within assemblages classified as Mesolithic.

Coastal wetland areas appear to have remained as important foci for human exploitation during this period, whilst areas of the county to the east see intensification of neither the archaeological nor palaeoecological record during this period. The Little Crosby ridge includes little material that can be placed with any confidence in this category, although a previously recorded axe roughout (A 18) on the brow of the ridge may signify settlement here. It might also suggest that some of the flint material located here may be of unrecognised Neolithic date. One site on the ridge (Me 039) includes a series of split pebbles with primary flakes and blades common, and the occasional bi-polar core, the pebbles almost certainly having been brought from the beach. The activity is difficult to date, but is technologically distinct from the material so far associated with the Mesolithic. Neither does it appear to be part of the late Neolithic/early Bronze Age material in the area, which is also very distinctive (*see below*). A tentative Neolithic date might therefore be assigned to the group. Flint technology at the Anglesey chambered tomb of Trefignath was almost all coastal flint, knapped in a bi-polar manner (Healey 1987).

Another site in this general location (Me 032) comes from the surface margins of the peat in the Woodham Knoll channel, *c*350m from the brow of the ridge; no Mesolithic flint has been found as low down the ridge as this. It has probably been disturbed by a recent drainage episode, suggesting that the flint material probably predates the peat formation. Unfortunately the assemblage is made up of only seven pieces, which makes dating difficult. The use of chert and the presence of a core rejuvenation piece could suggest an early Mesolithic date in the light of other evidence from the Wirral (*see above*). Such a date would have preceded the onset of wet conditions at Flea Moss Wood, explaining why this site lies beneath the peat when all the other Mesolithic sites are higher up the ridge. The retouched piece in the assemblage, however, might suggest an early Neolithic date. If this is the case then it suggests that the site could have been deliberately placed within the wet area, or the fen conditions did not cover the site until after the early Neolithic.

The possibility also exists that unrecognised elements of Neolithic date are included in the two large sites in the Alt Valley at Lunt (Me 093, Me 094) which have been assigned primarily to the Mesolithic, but which include raw materials identical to Me 064.

Although this suggests that the areas settled in the Mesolithic continued to be used, there is a little evidence which may suggest signs of a slight shift or extension of the pattern. As the marine conditions of DM III retreated from the vicinity after the first half of the fifth millennium Cal BC (Table 7), Orrell Hill may have become more attractive for exploitation, as fen woodland spread across the area. Some limited late Mesolithic material is found here (*ie* Me 058, Me 061), but a number of sites which are found to the south-east of the mossland are more noteworthy, located on almost imperceptible localised sand ridges in a generally featureless landscape. This includes Me 065, which contains a leaf arrowhead, the adjacent Me 066, which has pieces with similar technological traits in the same raw material, and Me 064, which has almost identical characteristics to Me 050 to the north of Flea Moss Wood. The single implement find Me 057 is also probably Neolithic. With the exception of the latter these are all set *c*400m off the current edge of the peat.

The uncertainty of attribution of this material to the early Neolithic period, and the small numbers of flints involved, makes it difficult to identify the nature of these sites. If the material is not Mesolithic, then its similarities to material locally attributed to that period suggests little change in exploitation in the area. The pollen evidence also suggests that there was little change in the nature of vegetation disturbances during the late fifth and early fourth millennia Cal BC. The implication therefore is that widespread clearance and arable farming was not a feature of the early Neolithic in this part of the region, in contrast to North Wirral (*see Chapter 3*). Although cereals may have been introduced into this region at the beginning of the fifth millennium Cal BC, as early as anywhere in the country, hunting and gathering appears to have remained a more significant part of the economic base with woodland clearance being fairly small-scale and limited in time.

Table 10 Sefton: Alt Valley, Ince Blundell, palaeoecological summary

cms	Date (Cal BC)	Local Pollen Zones		Ecology of Moss	Dryland Vegetation
			Oak pollen characteristic with alder and hazel important.	Freshwater and fen carr communities develop as marine influence declines.	Mixed oak woodland on drier ground.
			Elm remains high.	↑	Less alder carr as reedswamp develops in valley.
		IB-c	Ash and lime recorded.	Reedswamp	
			Grass and sedge pollen high with a range of wetland herbs.	↑	
18				Saltmarsh	Saltmarsh as sea retreats.
20	5230-4790				
				Marine incursion	
30	5204-4791				
32			Characterised by alder pollen with oak and hazel important.	Saltmarsh and estuarine conditions increase	Mixed oak woodland with alder carr predominant within valley. Oak more abundant away from the watercourse
			Elm remains high.		
		IB-b	Pine falls early in zone.		
			Rise in oak with fall in alder marks top of zone.		
74			NAP rises due to grasses and wetland herbs.	Extensive fen carr and reedswamp	
			Top of zone marked by rise in alder.	Rising water tables.	
		IB-a	Pine with major values of alder, oak, and hazel.		Mixed deciduous woodland with pine.
			Elm pollen high.		
			Herbaceous pollen very low.	Probably waterlogged areas.	
80	6180-5740				
82					

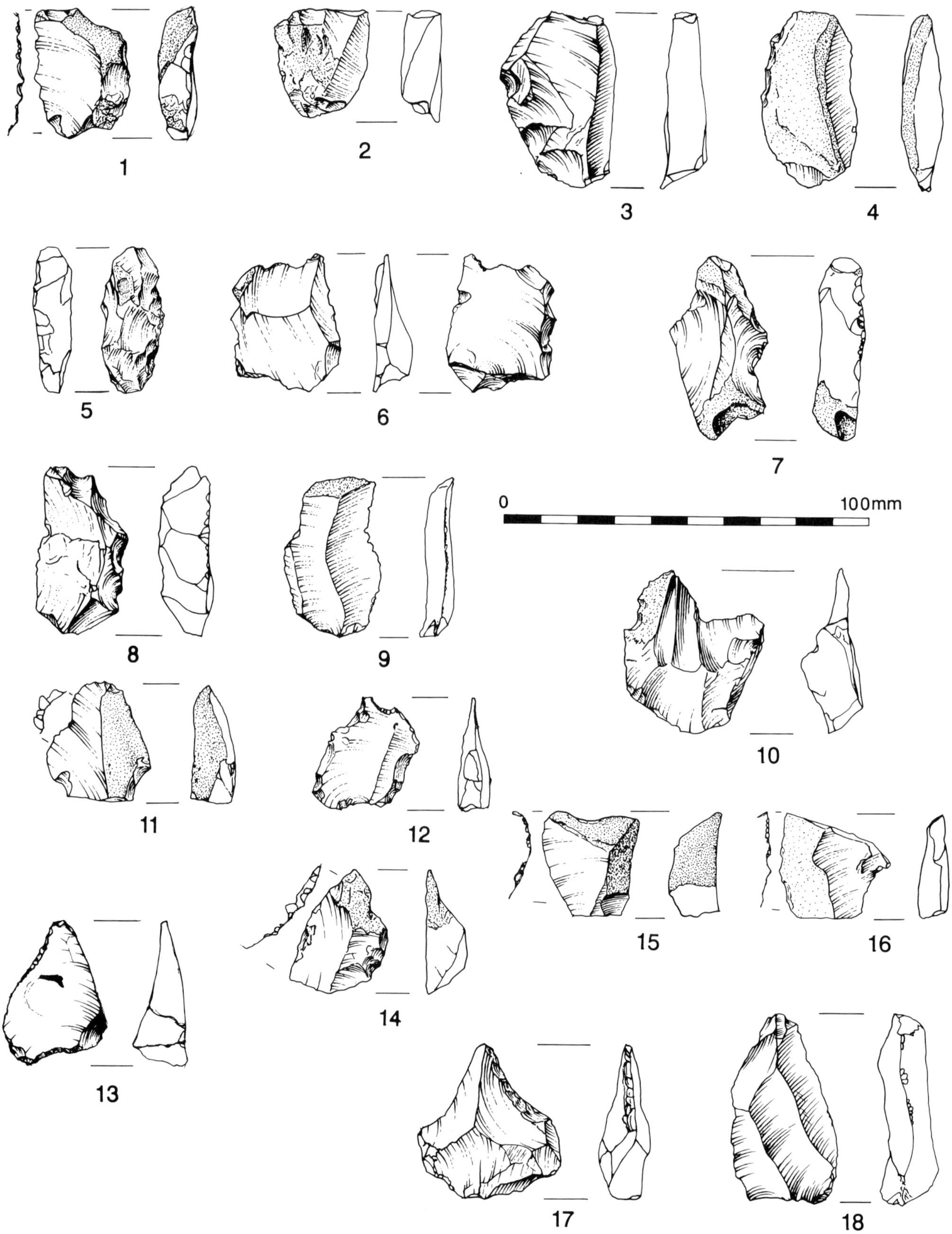

Fig 22 Sefton: Late Neolithic/early Bronze Age flintwork from Woodham Knoll, Little Crosby

Late Neolithic/early Bronze Age *c*3500–1650 Cal BC

The middle Neolithic coincided with a marine transgressive episode (DM IV) finishing in the late fourth millennium Cal BC in South West Lancashire. The environs of the moss at Sniggery Wood had been dominated by coastal influences during this time, the culmination of which was the widespread dispersal of blown sand, covering the Sniggery Wood site after *c*3370-3040 Cal BC (4510±50 BP; SRR-2698). This suggests that sand dunes slightly to seaward of the present fore-dunes, which had become established in DM IV, shifted. Blown sand reaches as far inland as the foot of Woodham Knoll, with low ridges across the plain possibly representing ploughed out former dunes.

As in North Wirral, the later Neolithic period marks the beginning of a dry phase, which occurs generally across the North West (Tooley 1978a). A peat bed outcropping on the beach south of the Altmouth at Hightown, sealing estuarine clays, has provided a date of *c*3592-2926 Cal BC (4545±90 BP; Hv-2679) for the extension of fen woodland beyond the present coastline (Tooley 1977). This suggests that marine conditions had also retreated over parts of the Sefton coast during this time. Although none of the peat in the survey zone is as late as this period, it seems likely that to the east and north of the inland limit of the blown sand the estuarine environments had retreated northwards. It would also appear that the fen at Flea Moss Wood became drier, to be followed by the spread of *Sphagnum* moss into the area after *c*3622-3340 Cal BC (4670±50 BP; SRR-2695).

Although it is easier to attribute certain classes of flintwork to this period, there are still difficulties in its finer chronological resolution, as a regional context for technological change from excavated sites is lacking. A relatively large group of material has, however, been located during the NWWS fieldwork which can be placed in this period (Fig 22), the typological characteristics of which have already been documented (Cowell 1991b). There is some doubt as to its more exact attribution within the period, since it shares characteristics with published sites of both late Neolithic and early Bronze Age date from the Fens near Peterborough (Pryor 1976). This writer leans towards a late Neolithic date on technological grounds, before or at the mid third millennium Cal BC, and a tentative agreement has also been provided (J Gardiner pers comm). Other views, based on familiarity with the Fenland material, suggests an early or even middle Bronze Age date (R Middleton pers comm). At present there is not really enough evidence to be sure.

The flintwork is found in a more restricted area around the mosslands than previously (Fig 76). A little comes from around the Orrell Hill peat area, mainly as single finds (Me 063, Me 068), and none was found in the Alt Valley, although one site is known from previous survey *c*3km away at Lydiate, on the west-facing ridge overlooking the Alt Valley (Lewis 1982). The bulk of the material, though, comes from the Woodham Knoll ridge, where five small to medium concentrations of flintwork have been located (Me 018, Me 019, Me 007, Me 036, Me 037). These consist almost totally of implements and large flakes, with little indication of on-site knapping (Cowell 1991b).

This suggests specialist activities in this location, with the flint prospecting and reduction sequences having taken place elsewhere. The raw material, which may be coastal in origin, and of reasonable quality, is from pebble types that are not used in any of the other technologies so far described, implying either a source far distant from the previous ones, or a local source that may not have been previously available, possibly due to changes in sea-level. The coastal wetland area was covered by dunes, approximately in their present position, although it is not clear if the blown sand found inland of the present dunes had reached as far as the Woodham Knoll ridge by the late Neolithic. No flintwork of this type has yet been found overlying the blown sand on the coastal plain to the west of the ridge, but other types, identified as later in date, do occur in the sand (*see below*). This may suggest that the blown sand had not spread inland across the whole coastal plain by the late Neolithic or early Bronze Age, and that there may still have been fen or heathland around the ridge where the sites are located.

In the south-eastern end of the channel at Woodham Knoll burnt flint findspots and sites are found, which are difficult to date. Nowhere is the late Neolithic/early Bronze Age type of flintwork found with the burnt material, the former being restricted to the coversand soils, mainly well above the peat in the channel.

The lack of sites around Orrell Hill implies that the Little Crosby ridge was the more favourable situation for human activity at this time. Here the small, reasonably tight concentrations of material along the ridge, might suggest repeated visits to the area, rather than one large contemporary settlement, particularly as there is no evidence that flint knapping was taking place at any of these sites. Although this question can only be properly clarified through excavation, some possible supporting evidence may come from a late Neolithic/early

Bronze Age fen edge settlement at Fengate near Peterborough. Here, areas with struck flint, pottery, and burnt bone which were interpreted as domestic, had more than 67% of the total flintwork as by-products from knapping (Pryor 1978, 155). If the flintwork can be interpreted in this way then it suggests that the Little Crosby occupation may have been on a seasonal basis. One objection to this is that very little of this type of material has been found in fieldwalking away from these coastal mosslands (Cowell 1991a).

Pastoralism is difficult to define archaeologically, although there is a general trend to identify it as a common component of Bronze Age landuse, from ethnographic parallels and studies of sites such as burials (Fleming 1971; Green 1974). In the vicinity of Flea Moss Wood there is evidence for a large-scale clearance associated with cereal farming in the late Neolithic, after 3622-3340 Cal BC (4670±50 BP; SRR-2695) (Table 8, phase 6). The mire at this stage had become acidic bog which was drying out, with birch carr prevalent. The palaeoecological evidence does not extend into the Bronze Age at this site. As the flintwork is difficult to date with any accuracy, it is not clear if it should be associated with the clearance phase or with a later period for which there is no palaeoecological evidence. It would seem likely though that the adjoining mire had ceased to be largely aquatic fenland by the time the flintwork was in use. In the Somerset Levels, the mineral nutrients and moss peats at the edge of marshes are said to have provided ideal conditions for fodder or for seasonal intensive use by animals (Coles 1978). It would seem likely that in winter the bog would get very wet underfoot and so might be avoided, and that therefore good grazing land might have been available around the foot of the Woodham Knoll ridge during the summer months.

This area of Sefton may thus represent a focus for mobile pastoralists who grazed the not too distant dunes and nearer carrs surrounding the dryland ridge. It might represent free ranging summer grazing grounds, with a more restricted pattern in the winter. The evidence that might represent such winter areas, where presumably women and children would live for most of the year, if the menfolk were mobile during the summer, is rare anywhere else in the county. It may also be that it was these areas where cereal farming was more likely, away from the fodder areas (Coles 1978). Base camps or flint knapping sites are lacking for this period, with inland sites being represented only by the occasional scraper or flint knife implement. Cereal-type pollen, and indicators of arable farming such as ribwort plantain, are represented in a clearance phase, dated 2560-2295 Cal BC (3910±50 BP; SRR-2931), from a coastal site *c*8km to the south at Mount Pleasant, Waterloo (Innes and Tooley 1993, 39). This site was overwhelmed by sand after that date. The evidence from the Rimrose valley to the north of this site also implies clearance activity taking place close to the valley at about this time (Cowell and Nicholson 1990).

Later prehistoric period *c*1650 Cal BC – AD 43

In the Sefton area the palaeoenvironmental background for the late prehistoric period is largely absent. At Sniggery Wood peat growth had ceased during or after the late Neolithic through sand accumulation, while at the other two sites later disturbance makes the post-Neolithic evidence difficult to interpret. The area was largely free from direct marine influence during this period, with the coastline likely to have been in roughly its present position. A well marked phase of dune stability is recorded from the coast a little to the north, at Formby Point, dated c300 Cal BC (2335±120 BP; Hv-4709) (Innes and Tooley 1993, 39), with the formation of a dune slack soil in the present dunes.

A recurring type of site assigned to this period is fairly widely represented across the survey area, although as with other post-Mesolithic types, it has no excavated parallels and so therefore dating must only be tentative. The flint assemblages involved are very small, between two to five pieces, and are mainly identified through their lack of formal knapping characteristics. They consist of small irregular natural pebbles and pebble fragments, and in some cases they are probably struck waste chunks, which have been retouched along a serviceable edge to form tools that were presumably used as scrapers or cutting implements. They include a whole range of local flint pebble types.

The main locations (Me 005, Me 010, Me 011, Me 027) are on the Woodham Knoll ridge and adjacent former dune areas which overlie the peat to the west in the coastal plain. They also appear to have an association with areas of peat, forming the main type of material (Me 024, Me 030) from the west-facing slopes of the peat channel adjacent to Woodham Knoll, where they are located close to the present peat interface. Fewer examples are found around the Orrell Hill peat, and very few from the slopes of the Alt terrace and floodplain overlying the upper peat. They sometimes occur (Me 072, Me 079) in the general area of the Alt floodplain as thin scatters of burnt flint nodules which are difficult to date, but are often assigned a Bronze Age or late prehistoric date (Silvester 1991).

This represents a fairly *ad hoc* type of flint industry, with the activity represented being wide ranging and less structured than that recognised in other areas of the survey zone. This kind of extensive distribution places the emphasis around the coastal blown sand areas, and in river valleys where seasonal flooding might be expected. The question remains as to whether it is chronologically distinct from the material discussed earlier, or whether it represents functionally different sites within one of the periods already covered. It would seem unlikely to be Mesolithic or early Neolithic, as material derived from controlled knapping techniques is found in almost the same areas. Its distribution across the former dune areas suggests it may have been discarded after the main dune formation after the late fourth millennium Cal BC. On stratigraphic and technological grounds, therefore, it might tentatively be placed in the Bronze Age, reflecting a deterioration in flint knapping techniques which may be associated with more marginal landuse. A middle Bronze Age site on the edge of a peat bog, at Bonds Farm, Pilling, in the Lancashire Fylde, dated to the middle to later Bronze Age, however, includes an assemblage of fine flint scrapers as well as other prestige items (Edwards 1991). This could be interpreted as a higher order site in the landscape, while the flint material from Sefton seems to represent a different kind of landuse, possibly not even associated with settlements as such.

Romano-British period AD 43–410

The Romano-British palaeoenvironmental background is largely lacking, as no contemporary peat growth has yet been identified. A general sea-level rise is inferred for North West England around 410-180 Cal BC (2270±65 BP; Hv-2916) and between 390 Cal BC – Cal AD 680 (1975±200 BP; Hv-5215) and Cal AD 540-853 (1370±85 BP; Hv-4708) (Tooley 1978a, 1980). Although the present coastal area was not directly affected, local water-tables may have been raised.

Archaeological evidence for this period, as for the preceding Iron Age, was not recovered from around the Sefton mosslands. This continues the general pattern for the county, with little firm evidence for settlement in the sandy western or northern areas. More evidence, in the form of cropmark enclosures and occasional excavated sites, comes from the firmer soils in the more geologically mixed areas of the interior (Philpott 1991).

The lack of evidence in the west may result from the light soils not providing the right conditions for aerial photography, but chance finds are similarly lacking from the area. The only positive evidence comes from the Altmouth, where artefacts, now lost, recorded in Liverpool Museum included two pieces of *mortaria* and two possible Roman bronze items. The important Roman trading settlement of Meols lay *c*10km across Liverpool Bay on the North Wirral coast. It is possible, therefore, that some coastal settlement lay on the eastern side of the bay also. Whether or not the areas a little inland, where blown sand, mossland, and light sandy soils were extensive, were as attractive for settlement must remain in doubt.

Early medieval period AD 410–1066

Those parts of the county producing evidence for Romano-British settlement also seem to have become the focus for the pre-Conquest and later medieval clearance and settlements, *ie* the southern and south-eastern parts (Philpott 1991). As there is little evidence for the native settlement of this area prior to the influx of Anglian settlers in the sixth or seventh centuries AD, it is not possible to discern the relationship between the two periods. In Sefton the English settlement which, as in other areas in the county, has to be identified from place names, lay in an arc to the south-east of the coastal and mossland townships. One of the earliest settlements identified, as in other parts of the county, may be around the seat of the parish church. This lies in Sefton village and is one of the few churches in the area to have surviving medieval fabric. The later settlement, with the moated site of Sefton Hall opposite the church, grew up around the church site on heavy clay land. The village lies only several hundred metres from the Alt, where field names suggest carr vegetation may have been found on the mossland, which could have been used for grazing. Part of the river valley lay within the demesne land in the medieval period. It seems likely that the pre-Conquest settlement lay in the same location as the late medieval one, but other early sites in the mossland areas are difficult to identify.

It was not until the tenth century AD that the pattern of settlement can be seen a little more clearly. Norse settlers from Ireland began to colonise the area and their presence is attested by a concentration of place names along the coastal fringe, the major names suggesting that this settlement took place mainly in two zones: along the coastal strip, from the modern Southport in the north to Bootle, on the outskirts of Liverpool, in the south, and along the River Alt (Lewis 1982). The extension of Scandinavian names along the Alt Valley reaches as far as the western fringes of the central mosslands (*see Chapter 5*). Two alternative interpretations may be given for this pattern: either settlement was taking place in areas that were underpopulated during the

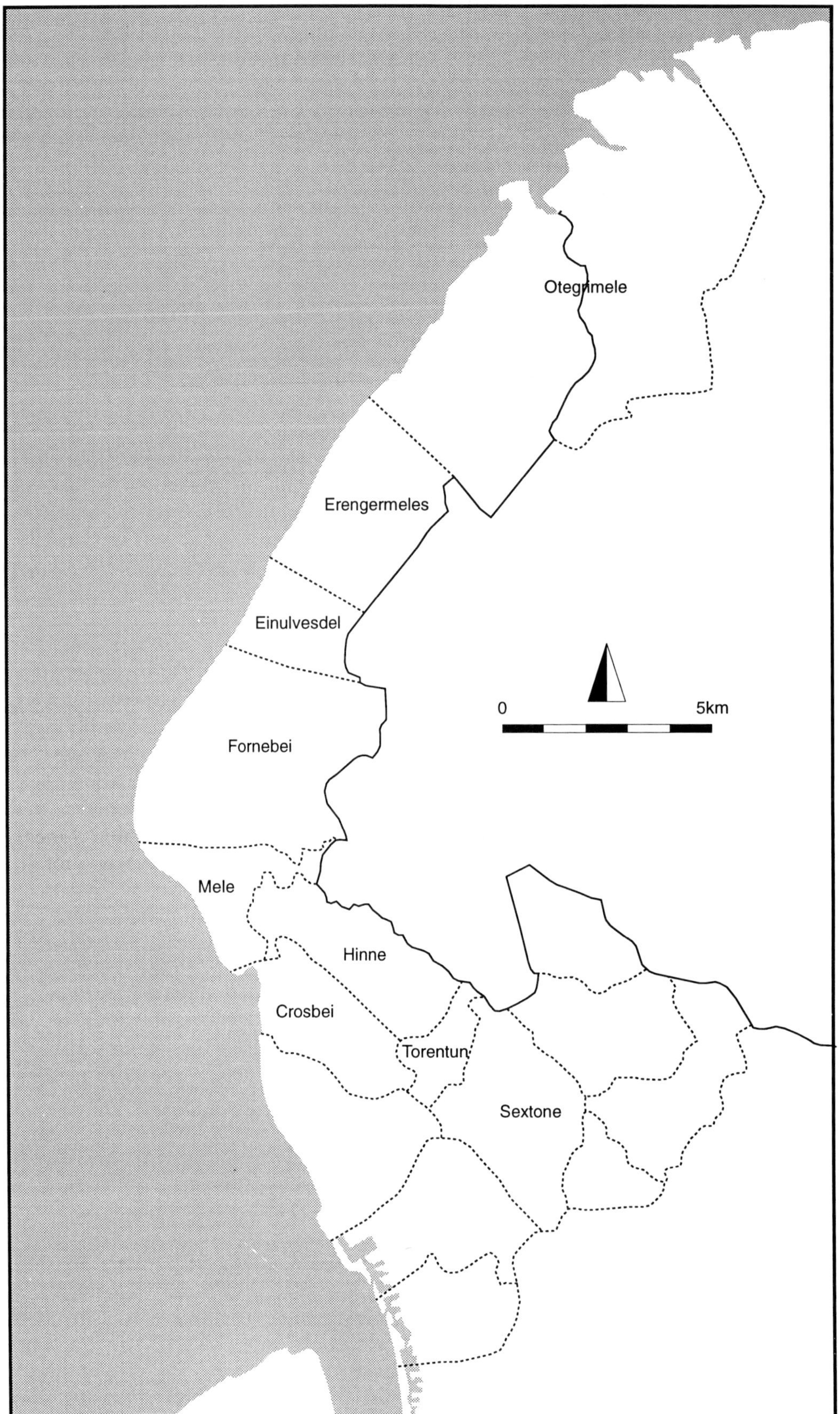

Fig 23 Sefton: Domesday manors (after Lewis 1982)

early medieval period, roughly coinciding with the coastal wet areas, therefore allowing Scandinavian influence to spread easily, or alternatively settlement was strong in these areas through choice, perhaps linked to landuse in the dune and wet areas.

Whatever the reason, the arrival of the new population then quickly led to existing settlements changing the character of their language and place names. The fact that most of this land contained the wetlands suggests that, unless the incomers were forced to remain on the marginal land, they were prepared to subsist with a high reliance on grazing, using the dunes, slacks, and marshes prevalent in the area. Within the survey areas of the coastal zone, Ince Blundell and Little Crosby both had well developed villages with associated open arable fields in the medieval period, which might imply early foundations, and at Little Crosby, a Viking coin hoard and possible chapel, the Harkirk, lay to the south of the present village (Lewis 1982), well away from the mossland to the north of it.

The place-name *Hinne* (Ince) may be Celtic, and its meaning of 'island' must represent its topographic position whereby the sandy ridge was surrounded on three sides by marshland and fen or seasonally flooded clayland. The open field system and the village lie on the upper part of the sandy ridge, similarly well away from the Alt Valley marshes. Lunt derives from the Scandinavian *lundr*, meaning grove (Ekwall 1922). The probable medieval and post-medieval precursors of the present settlement lie on the flat part of the ridge away from the Alt Valley.

At Domesday the Sefton area, along with the townships associated with the western half of the central mosslands, lay within the Hundred of West Derby (Fig 23). This was one of the six large units of pre-Norman administration *inter ripam et Mersham* (between Ribble and Mersey) which were treated as large royal manors for the assessment in 1086 (Farrer 1899).The Hundred as a whole contained a large amount of woodland, most of which probably lay in the vills around the central mosslands. It also extended into the vills of Lydiate and Maghull on the higher ground, based on sandstone and boulder clay, to the south-east of the Sefton mosslands. Elsewhere in Sefton, woodland at Domesday seems not to have been as prevalent as to the east, which in the twelfth century formed the basis of the hunting areas in the Royal Forest of Lancaster (Cunliffe Shaw 1956) (*see Chapter 5*).

The areas to the south of the main Sefton woodland vills, which are all characterised by being located in areas of heavier clay or sandstone soils, such as Sefton and Thornton, appear to have had the largest areas of cleared land given over to arable in the district at Domesday (Farrer 1899); as such they probably represent the earlier settlement foci. In contrast, vills such as North Meols (now modern Southport), on the coast, had relatively little arable land (Lewis 1982), which must be a reflection of the fact that the sandy wetland areas were more suited to grazing.

Later medieval period AD 1066-1500

By the late medieval period most of the coastal townships, as well as the ones south of the Alt, carried nucleated villages with some arable land almost everywhere. There is little medieval documentary evidence for the mossland areas within the system of landuse, although subsequently much of the coastal belt north of the Alt was used extensively as grazing land, with Cistercian monasteries setting up granges in a number of townships (Lewis 1982). In the thirteenth century, a degree of reclamation of the coastal marshlands probably took place as part of the grant of land, including turbary rights, in the north part of Ince Blundell to the monks of Stanlaw abbey at Whalley, Lancashire, who were settled at the site known today as Grange Farm (Farrer and Brownbill 1907, 79).

Nucleated settlements did exist in the coastal zone, for in the fourteenth century there is evidence that a number were overwhelmed by the sea, or at least abandoned because of unstable dunes. *Argarmeols* (Birkdale, south of Southport) was totally lost, while both *Ravenmeols* (Altcar township, adjacent to Hightown) and Little Crosby lost part of their arable lands, probably reflecting these environmental changes (Lewis 1982). The townships in the coastal area south of the Altmouth appear to have contained a mix of arable and grazing land during the late medieval period. This was associated with scattered, small, nucleated villages and hamlets. Nineteenth century tithe maps make it clear that the mossland areas, often called moor, were placed on the edges of the townships and were used as common grazing land.

Post-medieval period AD 1500–1850

More gradual reclamation of the marginal land seems to have taken place from the sixteenth century onwards, as arable areas expanded into the margins of the townships; post-medieval field names refer, for example, to fields such as Riddings (clearing) in the early eighteenth century in Ince Blundell. Mossland areas were enclosed for grazing and were also used for turf cutting. Turf was primarily used for fuel, a detailed account being

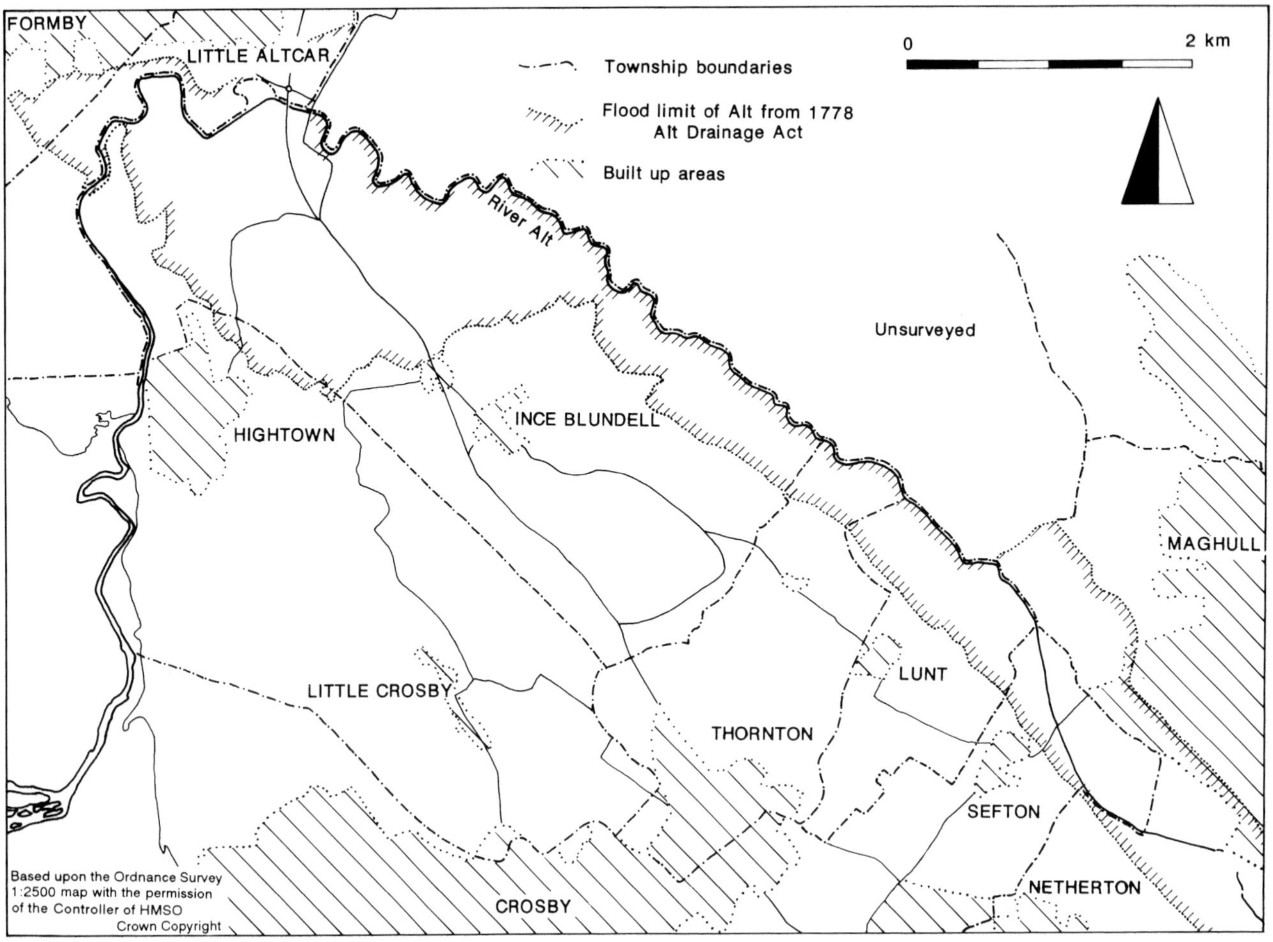

Fig 24 Sefton: Plan of damaged lands adjoining the Alt, 1778

given in the 'Great Diurnal of Nicholas Blundell' (Tyrer 1968) for the period 1702-1728, in Little Crosby township. Earlier court rolls dating back to the sixteenth century also laid down stringent conditions for turf extraction and the use of the mosses, with, for example, the digging of new moss pits being forbidden in 1604 (LRO, DDX 109/36, 27).

In 1778 the Alt Drainage act was passed (LRO, DDBL 24/26) which straightened the course of the Alt and allowed for the cutting of new ditches to prevent the flooding that was a common feature of the area. The clay that was deposited probably allowed arable farming to spread onto these areas formerly covered by mossland. Until drainage took place, the results of the flooding were often dramatic, whereby some areas by the Alt were 'almost drowned in winter... the families were frequently in such distress as to flee from their home and seek refuge, and yet in summer seasons, this country is distressed for want of water, and that to a degree, as to require driving the cattle the space of a mile to drink, the springs being exhausted' (Holt 1795, 141). The flooded lands reached from the Alt estuary to Orrell Hill and down into the Alt Valley as far as Sefton. The silty clay deposited by this flooding can be traced today on the ground, although it is less extensive than the plan of the affected lands shows (Fig 24). The clays are still *c*0.6m deep in places, and along the valley it provides a useful sealing layer to protect the peat from plough damage. This also probably protects parts of prehistoric settlements which lie around the lower slopes of the small sand islands that are protruding slightly from the peat and alluvial clay.

Palaeoecological survey

Flea Moss Wood (FMW) (SD 311027)

Stratigraphic reconnaissance was undertaken in the region of Flea Moss Wood, Little Crosby, to establish peat depths and choose the best sampling site for pollen analysis. Flea Moss Wood was selected as it forms a wooded covert in an area of peatland reclaimed for cultivation, and it was thought likely that deposits of more recent age may have survived there, protected from the effects of recent arable farming techniques. The surface of the wooded area stands more than 1m above the surface of the surrounding cultivated fields. The site was also interesting because the edge of the marine Downholland Silt lies within the wood, and so the site lies very close to a stratigraphic and ecological boundary of major importance. A peat core 3m deep was recovered from a point in the central area of the wood, landward of the marine clay boundary. The altitude of the sampling site was 4.09m OD.

Pollen analysis

Samples were taken for pollen analyses at 100mm intervals throughout the core and the results are shown in Figures 25 and 26. Pollen preservation was good. Five pollen assemblage zones have been recognised, and are described below.

FMW-a 3-2.95m The basal zone is characterised by *Pinus, Quercus,* and Coryloid, while *Alnus* and *Betula* are present in low frequencies. Gramineae and Cyperaceae are the main non-arboreal pollen types. Few other herbaceous pollen occur. This zone (2.9-2.98m) has been dated to 6215-5960 Cal BC (7230±70BP; SRR-2697).

FMW-b 2.95-1.85m This zone is characterised by *Quercus* and *Alnus,* with lesser amounts of *Pinus,* Coryloid, and *Betula. Ulmus* maintains moderate frequencies, and herbs of wetland environments such as *Nymphaea, Filipendula,* and *Ranunculus* are common. Some weed pollen types like *Artemisia, Rumex,* and *Melampyrum* occur sporadically, and Gramineae and Cyperaceae values remain significant.

FMW-c 1.85-0.40m *Quercus* and *Alnus* dominate this zone, while Coryloid and *Pinus* frequencies are much reduced. Other trees and shrubs including *Tilia, Fraxinus, Ilex, Prunus,* and *Viburnum* are now represented, and *Ulmus* is still present in moderately high values. Cyperaceae and *Pteridium* are low but consistent, while Gramineae is very low indeed. Wetland herbs, as well as others like *Rumex* and Chenopodiaceae which may be dryland weeds, are present. Cereal-type grains were noted at 1.2m and this level (1.16-1.24m) was dated to 4939-4720 Cal BC (5920±50 BP; SRR-2696).

FMW-d 0.40-0.05m *Betula* is the main tree pollen type, but with a very wide range of other types in lower frequencies. *Quercus, Alnus,* and Coryloid are present in moderate values throughout, but *Pinus, Ulmus, Fagus,* and *Acer* increase towards the end of the zone. Filicales are high, while *Calluna* and Gramineae increase. A wide range of ruderal weeds or dryland herbs occurs, including *Plantago lanceolata, Rumex,* Cruciferae, *Taraxacum*-type, and *Artemisia.* Wetland herbs are still recorded and *Sphagnum* is important for the first time. Peat from near the base of this zone (0.28-0.36m) has been dated to 3622-3340 Cal BC (4670±50 BP; SRR-2695).

FMW-e 0.05-0.00m The uppermost zone is dominated by *Acer,* with higher values for *Pinus, Fagus,* and *Ulmus. Betula,* Coryloid, *Alnus,* and Filicales all fall sharply, and the assemblage of dryland herbs is reduced in both variety and frequency. After the decline of the previous zone, total tree pollen values are much increased. Wetland types are low, and Cyperaceae are no longer being recorded.

Ecological history

The three radiocarbon dates from Flea Moss Wood provide an age context for the ecological history of the site which agrees well with the ages inferred from the pollen data. The date of 6215-5960 Cal BC (7230±70 BP; SRR-2697) for the base of the peat profile is closely comparable with the age of the transition from Flandrian I to II (the Boreal to Atlantic transition) which has been dated to 6180-5720 Cal BC (71070 BP; Q-916) at Red Moss (Hibbert *et al* 1971). The replacement of *Pinus* by *Alnus* at the end of pollen zone FMW-a therefore does seem to represent the regionally occurring rise of alder pollen frequencies at the start of chronozone Flandrian II. The date of 3622-3340 Cal BC (4670±50 BP) for the lowest counted levels at the base of pollen zone FMW-d also agrees with the pollen evidence, as *Ulmus* frequencies are low but consistent below the FMW-c/d boundary, but temporarily fall above that horizon. It is unfortunate that no pollen was countable at the 0.4m level due to poor preservation, but it seems almost certain that the *Ulmus* decline of about 4000-3650 Cal BC may be safely regarded as lying at a depth of 0.30-0.50m. The average sedimentation rate between the upper two radiocarbon dates in the profile is about 10mm every 13 radiocarbon years, which would fit such an interpretation well. It appears unlikely that any hiatus in sediment accumulation or loss of sediment by erosion has occurred here; a temporary drying of the mire surface around the time of the elm decline seems to have led to oxidation and loss of the pollen record. The date of 4939-4720 Cal BC (5920±50 BP) for mid-Flandrian II sediments at 1.16-1.24m depth accords well with both the pollen data and the other two dated levels.

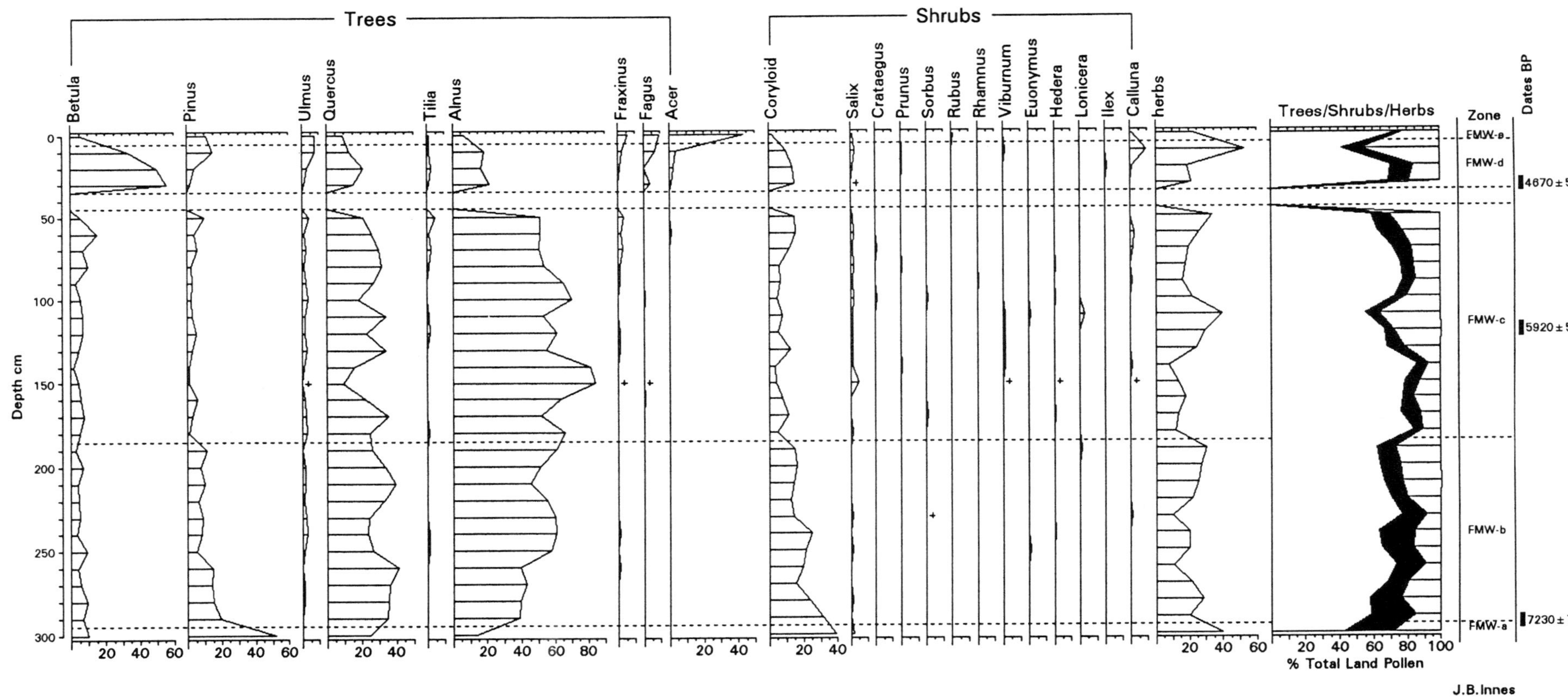

Fig 25 Sefton mosslands: Flea Moss Wood, percentage arboreal pollen diagram (trees plus group)

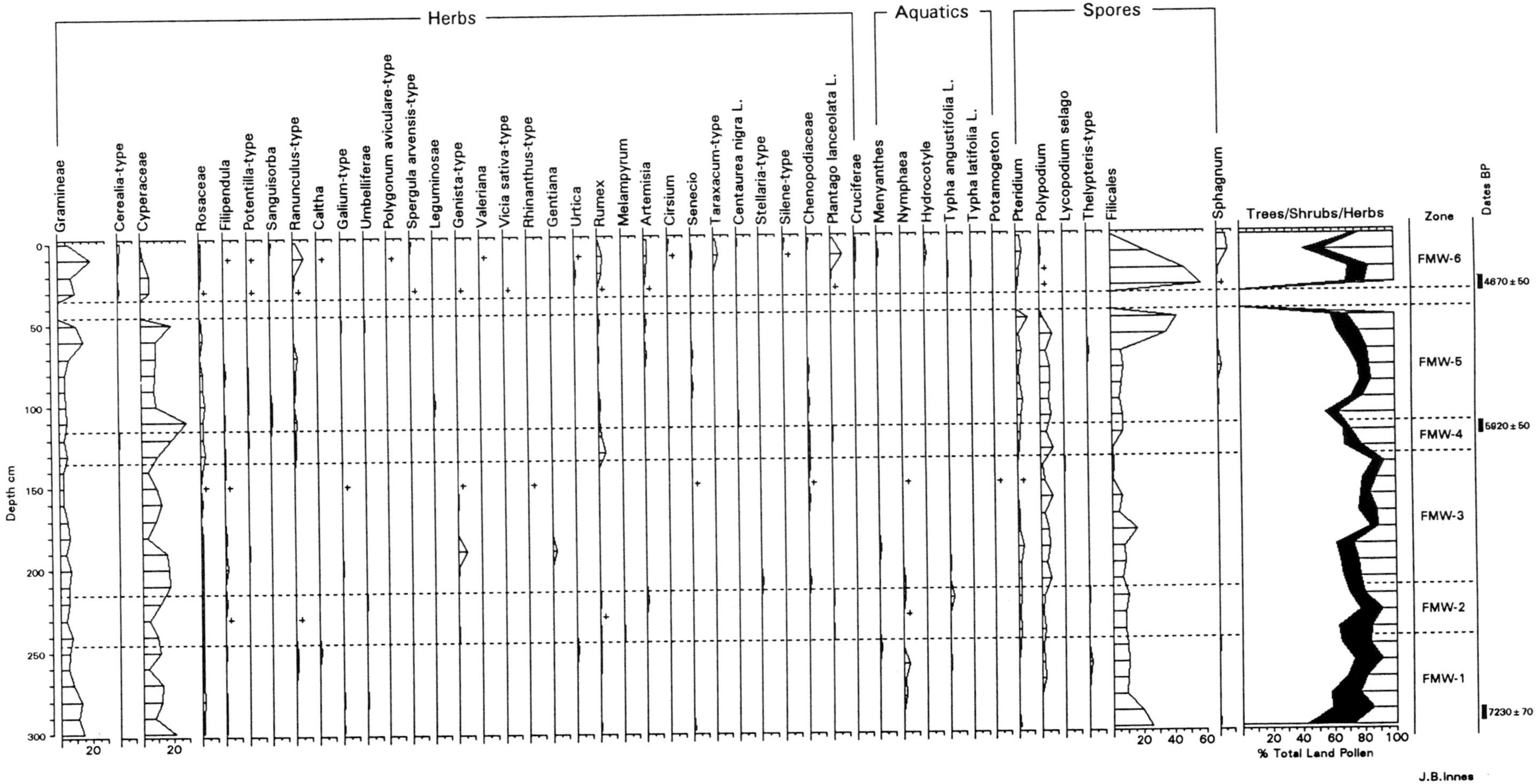

Fig 26 Sefton mosslands: Flea Moss Wood percentage herb pollen diagram (trees plus group)

Peat formation began at Flea Moss Wood above Shirdley Hill Sand at the very end of chronozone Flandrian I, probably in response to surface water-logging caused by raised water-tables as sea-level rise affected the Sefton coast. Peat formation began at Flea Moss Wood 6215-5960 Cal BC (7230±70 BP) almost 1500 radiocarbon years earlier than at Sniggery Wood, at 4780-4510 Cal BC (5770±50 BP), but this is due to the much lower altitude of the sub-peat sand surface at Flea Moss Wood, almost 2m lower (+1.09m OD) than at Sniggery Wood (+3.00m OD). As peat formation began the local woodland was dominated by *Pinus*, *Quercus*, and Coryloid, but this mixed oak-pine woodland was soon locally replaced by *Alnus* carr as conditions became wetter. Few wetland or aquatic pollen taxa occur in the initial peat formation period, suggesting that organic inception may have been quite a slow, gradual process without aquatic habitats available. Once the site had moved through the early marsh and reedswamp phases into the alder carr or swamp carr of early zone FMW-b, however, there were many aquatic habitats which supported a rich wetland shrub and herb flora. These alder-dominated wetland conditions may have been widespread in the area as dryland pollen types are generally low. Much of the *Quercus* pollen may have been coming from oaks from drier areas of the fenwood communities, or more regional pollen. The aquatic nature of the local environment is supported by the detrital nature of the peat itself, incorporating alder wood, the identification of pollen of floating aquatics like *Nymphaea*, and the presence of freshwater diatoms. These include the genera *Pinnularia*, *Cymbella*, and *Navicula*. The replacement of fen-carr conditions by more acidic bog environments occurs in zone FMW-c, when the pollen of aquatics such as *Nymphaea* declines and the fibres of *Eriophorum* occur in the stratigraphy, replacing the detrital peats of the lower profile. The reduction in *Pinus* at the end of zone FMW-b may be linked to the spread of bog conditions and the covering of higher areas of sandy soils which may have maintained pine populations. Pine is, however, quite capable of finding suitable habitats in peat areas, and this fall in pine may be part of long term edaphic changes starting at the end of Flandrian I and its continued replacement by alder during the wetter Flandrian II.

The ecological changes which occurred during the formation of the sediments of pollen zone FMW-d include the replacement of *Alnus*-dominated communities by *Betula*. This is likely to reflect the drier, and perhaps more acidic, nature of the bog surface which, in its most extreme form, may have caused the lack of pollen preservation at the 0.4m level. The well humified nature of the peat itself supports this. The presence of many dryland weed pollen types confirms the change to a more open and drier local habitat. The final stratigraphic change is that of the upper counted pollen level very close to the surface, where the peat becomes almost a well humified or humus material similar to that of a woodland floor. Natural colonisation of the drier ground surface is likely to have taken place throughout these upper two pollen zones, with a wide range of shrub types present. The domination of the upper level by *Acer* pollen could reflect a natural succession in early Flandrian III. It is more likely, however, that the humus sediment at and near the surface may be part of, or at least contaminated by, the modern woodland floor deposits at Flea Moss Wood, where planted trees including sycamore, beech, and ash predominate. The sudden removal in this upper counted level of several pollen types such as Coryloid, *Alnus*, *Betula*, and Filicales which were important in zone FMW-d, suggests that some hiatus may exist here.

Landscape history and forest disturbances

There are indications of possible human impact on the vegetation at a number of pollen levels at Flea Moss Wood. The following phases in forest history are recognised at this site:

FMW-1 3.00-2.45m There are no real indications of woodland disturbance during this phase, as the pollen record is dominated by trees and shrubs which suggest a closed canopy environment, firstly comprising mixed oak, hazel, and pine forest, then oak/alder carr woodland. Herb pollen is restricted to wetland types associated with the local fen carr vegetation.

FMW-2 2.45-2.15m During this phase the pollen of ruderal weed species indicative of disturbed or open ground is recorded, although not in high frequencies. These include *Plantago lanceolata*, *Rumex*, *Artemisia*, Chenopodiaceae, and *Melampyrum* as well as spores of *Pteridium*. Such records could be consistent with the creation of small clearings within the local woodland in the drier parts of the area. The overall dominance of tree types is little reduced, however, and any disturbed areas must have been of limited extent and duration.

FMW-3 2.15-1.35m The assemblage is once again without indications of woodland disturbance, with only a low *Pteridium* curve present which could represent open conditions. Bracken could as easily be associated with shady woodland environments, however, and without the ruderal weed assemblage can not be interpreted as evidence of open ground.

FMW-4 1.35-1. 15m A second phase of disturbance occurs which may more securely be attributed to human influence as likely open ground types *Rumex*, *Plantago lanceolata*, Chenopodiaceae, and *Pteridium* are joined by cereal-type (*Triticum*) pollen grains. Disturbance seems to have been very limited in extent, for these records are in low frequency. Their occurrence together, however, suggests the creation of small woodland clearings in the vicinity. This phase is approximately dated to 4939-4720 Cal BC (5920±50 BP; SRR-2696).

FMW-5 1. 15-0.35m While sporadic records of weed pollen occur in this phase they are not sufficient to form evidence of renewed woodland opening. It is very probable that the local tree canopy remained undisturbed during this time.

FMW-6 0.35-0.0m This phase contains considerable evidence of woodland disturbance, particularly in its upper part where total tree pollen values are temporarily heavily reduced. Cereal pollen records suggest cultivation near to the site, and this is supported by a very wide range of dryland herb pollen which includes some arable-type weeds like *Centaurea nigra*, *Taraxacum*-type, and *Polygonum aviculare*. *Plantago lanceolata*, *Rumex*, Cruciferae, *Artemisia*, and a number of other weeds also occur. This is a major phase of woodland disturbance, apparently for agriculture. The date of 3622-3340 Cal BC for the 0.28-0.36m level allows it to be attributed to the activities of Neolithic farmers. Some recolonisation of the land by trees and shrubs occurs at the end of the phase, but many open ground indicators are still recorded.

Sniggery Wood (SW) (SD307015)

The Sniggery Wood site lies immediately to the west of the small wooded area from which it takes its name, on the seaward fringe of Little Crosby township. The site lies in arable fields and the stratigraphy has been somewhat truncated by the effects of drainage. This has not affected the peat sediments at the site, however, for they are covered by an overburden of blown sand which forms a sheet across this western part of Sefton district. This sand sheet varies in thickness, but in general attenuates to landward and thickens to seaward, where it merges with the sand dune system which fringes the Sefton coast (Pye 1990; Innes and Tooley 1993). A site was chosen for analysis where the peat was undisturbed and the sand cover was relatively thin so that the sediment succession could be sampled easily. The location was also chosen because the sand dune fringes may well have provided favourable conditions for early human settlement and agriculture, combining organic peat soils with well drained sands. The site also lies close to the boundary of the local Downholland Silt (marine clay), and was thought likely to yield information relating to coastal change in the area near the landward limit of marine conditions during the period of its deposition.

Reade (1908) studied the stratigraphy in this area during excavations for the construction of an outfall sewer and showed that the peats at Sniggery Wood are part of the same general sequence which outcrops on the beach at the Altmouth near Hightown (Reade 1872b; de Rance 1877), where pollen analyses have been under taken previously (Travis 1926; Tooley 1977).

Pollen analysis

Pollen analysis was undertaken on the peat bed at intervals of 40mm, and the resulting pollen diagrams are shown as Figures 27 and 28. Pollen preservation was good. Four local pollen assemblage zones were recognised, as described below.

SW-a 1.74-1.08m The lowest zone is characterised by *Alnus* and *Quercus*, with lower frequencies of Coryloid and *Betula*, and low but consistent values for *Ulmus*. Tree pollen accounts for about 50% of total pollen. Gramineae is the main herb pollen type present but a wide range of other wetland herbs occurs. Some coastal herb types, including Chenopodiaceae, *Artemisia*, *Plantago maritima*, and Compositae Tubuliflorae of *Aster* type, occur in low percentages at the base of the zone. A variety of shrub types occur, but all in low values. Freshwater aquatic taxa are present throughout the zone, and Filicales spores increase markedly near its end. The base of the zone (1.70-1.74m) is radiocarbon dated to 4780-4510 Cal BC (5770±50 BP; SRR-2699). The upper boundary of the zone is defined by a rise in *Betula* frequencies.

SW-b 1.08-0.56m This zone is dominated by *Betula*, with lesser amounts of *Quercus* and *Alnus*, while Coryloid and *Ulmus* values are unchanged from the previous zone. Cyperaceae replaces Gramineae as the main herb type. *Calluna* shows isolated increases in frequency, as do *Salix* and *Fraxinus*. Aquatic and wetland herb types occur, but in small amounts. The occasional ruderal herb pollen occurs, such as *Plantago lanceolata*, and Filicales remains important. *Fagus* occurs, but other shrub types are reduced in number. The upper boundary of the zone is defined by a fall in *Ulmus* pollen frequencies.

SW-c 0.56-0.32m The zone is characterised by *Betula* with considerable frequencies of

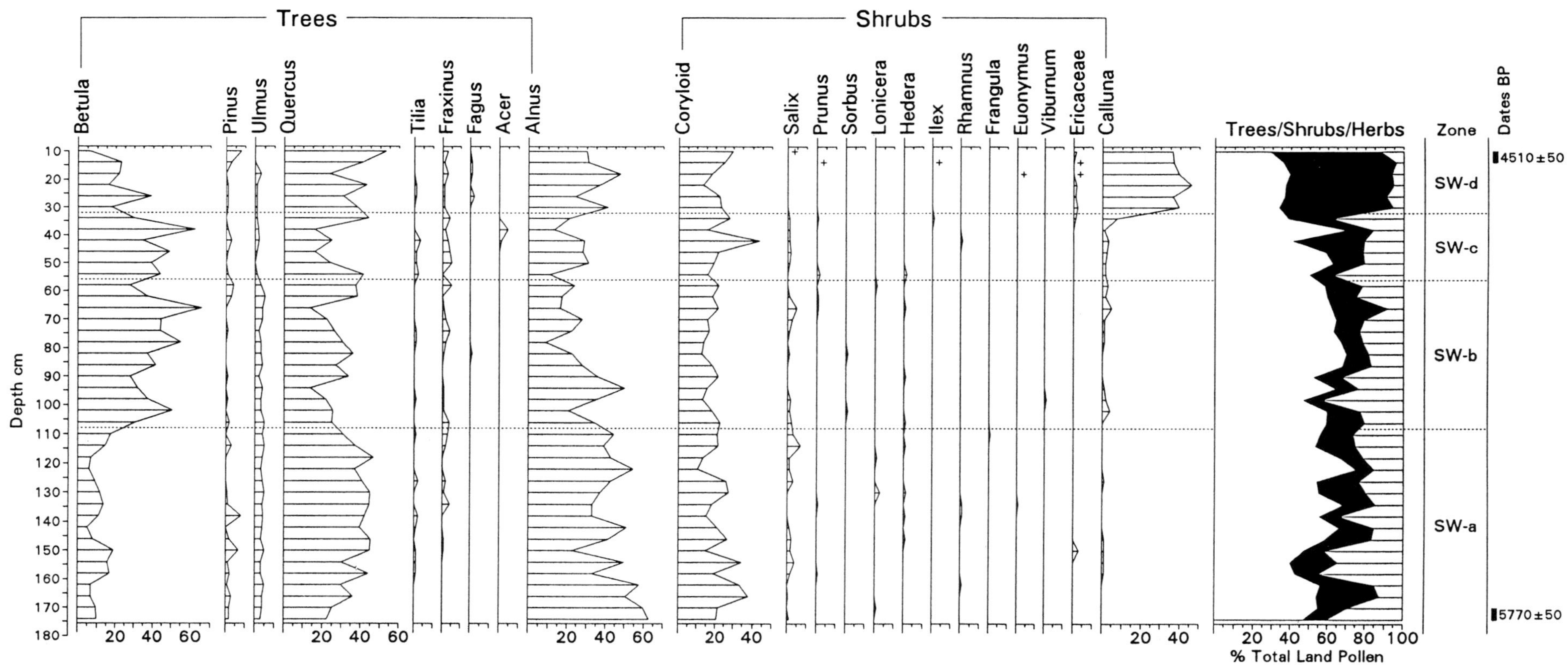

Fig 27 Sefton mosslands: Sniggery Wood percentage arboreal pollen diagram (trees plus group)

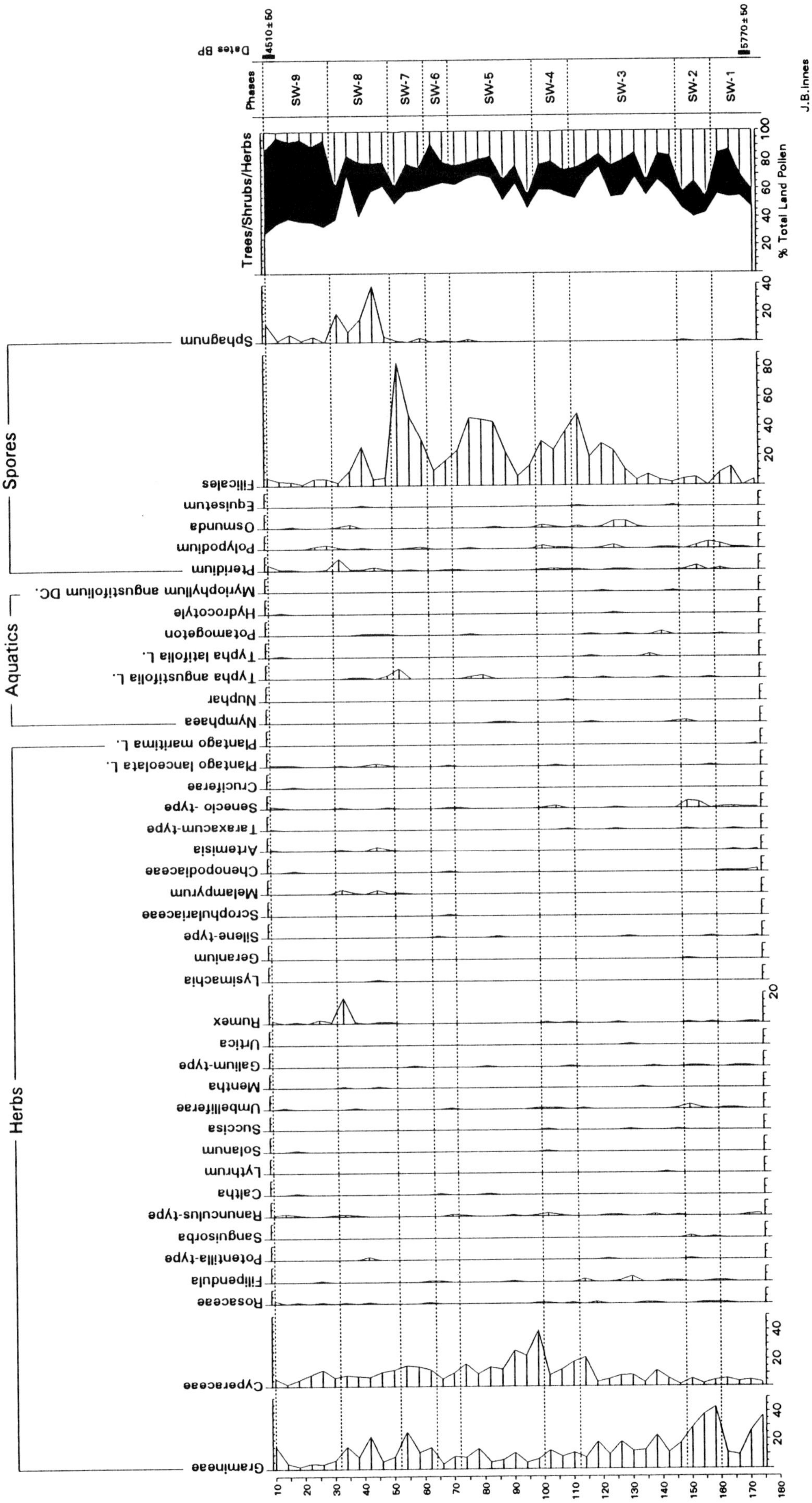

Fig 28 Sefton mosslands: Sniggery Wood, percentage herb pollen diagram (trees plus group)

Quercus, Alnus, and Coryloid. *Ulmus* values fall from moderate but consistent levels to become very low and sporadic. *Fraxinus* increases in importance, and a peak of *Acer* occurs. Herb values remain generally low although small peaks of ruderal weed pollen occur, including *Plantago lanceolata, Artemisia,* and *Melampyrum. Pteridium* percentages rise through the zone, but Filicales spores are much less abundant. A high *Sphagnum* peak occurs late in the zone. The upper boundary of the zone is defined by a sharp rise in *Calluna* frequencies.

SW-d 0.32-0.10m The highest zone is characterised by *Quercus* and *Alnus,* and by a very sharp increase in *Calluna* which remains abundant through out the zone. *Betula* declines from the high values of the previous zone and is present, with Coryloid, in moderate percentages. *Pinus* rises to a slight peak at the end of the zone, and *Fraxinus* is present consistently. Herb pollen frequencies are low but a wide variety of types is present, especially late in the zone when ruderals are common. Filicales and *Sphagnum* are greatly reduced. The upper boundary of the zone (0.10-0.14m) is dated to 3370-3040 Cal BC (4510±50 BP; SRR-2698).

Ecological history

The two radiocarbon dates provide a chronological framework for the ecological changes at the site, and were chosen to date the beginning and end of peat formation. Pollen zone boundaries themselves have not been dated, but the duration of peat formation between *c*4780-4510 Cal BC and *c*3370-3040 Cal BC suggests that the regional elm decline of the Flandrian II/III transition should be present within the pollen record, unless there is an hiatus in the peat profile. There is no reason to believe that any hiatus exists and so the decline in *Ulmus* frequencies at the zone SW-b/c boundary may be regarded with confidence as the regional elm decline and so dated to about 3900 Cal BC by analogy with other dated profiles in Merseyside (Innes and Tomlinson 1991). Approximately consistent sediment accumulation rates through this peat bed (about 7.7 radiocarbon years per 0.1m) would place peat of elm decline age at very close to the SW-b/c level in the diagram, providing support for its identification as this important pollen stratigraphic feature. *Ulmus* percentages below this level are lower than at other Merseyside sites but are consistent, and it is probable that elm frequencies were naturally lower in this area in Flandrian II, due perhaps to edaphic factors associated with the local Shirdley Hill Sand soils.

Landscape history and forest disturbance

Indications of disturbance of the woodland which may have been caused by human activity occur at intervals in the pollen diagram from Sniggery Wood. The history of forest clearance at the site is described by the following nine phases.

SW-1 1.74-1.60m Very few indications of possible human activity occur in this phase, and the open habitat herb taxa are probably due to the proximity of coastal environments, or to naturally unstable sandy soils prior to their stabilisation by rising water-tables and peat formation. Drier areas inland were covered with oak-dominated broad-leaf forest fringed with alder fenwood. *Ulmus* was a quite minor constituent of the forest, with few understorey shrubs recorded except for Coryloid.

SW-2 1.60-1.48m Limited clearance of the oak woods occurred with values for *Quercus* and *Alnus* reduced. Ruderal herbs which are recorded include *Plantago lanceolata, Rumex,* and *Taraxacum*-type, in addition to a high peak of *Senecio*-type. *Salix* and Coryloid rise sharply and *Calluna* appears for the first time. *Fraxinus* and *Prunus* pollen show the more open nature of the woodland. A small peak of *Pteridium* also occurs in what seems to record a period of small openings in the woodland canopy.

SW-3 1.48-1.12m This is a phase of forest regeneration with most of the previously recorded open habitat indicators removed from the diagram, although complete canopy closure may not have occurred. Some slight indications of continued open conditions may be seen in the occasional weed or heliophyte shrub taxon, but not sufficiently to suggest significant breaks in the forest cover.

SW-4 1.12-1.00m A second phase of forest disturbance occurs, with dryland tree pollen, particularly *Quercus,* again in decline and *Plantago lanceolata* returning to the assemblage. Increases in other disturbed habitat weeds and *Pteridium* occur, and renewed creation of small clearings within the forest seems the most likely cause.

SW-5 1.00-0.72m The vegetation history of this phase is again of open *Quercus* and *Alnus* woodland on the drier areas around the site. There is little evidence of significant interference with the vegetation cover.

SW-6 0.72-0.64m Several open habitat indicators are recorded in this phase. *Quercus* falls sharply while Coryloid, *Salix,* and *Calluna* all increase and *Plantago lanceolata,* Chenopodiaceae, Caryophyllaceae, and *Senecio*-type all occur. *Pteridium* spores also increase. A significant recession of woodland took place at this time.

SW-7 0.64-0.56m Pollen indicating forest clearance is no longer recorded, and tree pollen values recover their former magnitude. Clearings created during the previous phase appear to have reverted to mixed woodland and remained undisturbed.

SW-8 0.56-0.32m Extensive clearance of woodland is recorded during this phase, which begins with the *Ulmus* decline and so can be dated to approximately 3900 Cal BC. Tree pollen frequencies are twice reduced sharply while ruderal taxa are present in high values. *Pteridium* percentages increase, and a continuous *Plantago lanceolata* curve points to the creation of areas of open grassy ground. *Rumex* is prominent and rises to a major peak, while *Senecio*-type and *Artemisia* are present in quantity. Chenopodiaceae and Cruciferae pollen occurs, and particularly significant are high and sustained values of *Melampyrum*, which is common on sandy acid soils and may be diagnostic of fire disturbance of woodland.

SW-9 0.32-0.10m Dryland tree pollen values are high in this phase, and regeneration of woodland seems to have occurred after the disturbance of the previous phase. The continued presence of dryland herb pollen, with significant frequencies of *Fraxinus, Fagus, Acer,* and Coryloid, may be interpreted as reflecting a more open kind of woodland. This is as likely to be due to the effects of long term ecological changes as to renewed clearance. In this phase, however, the possible effects of human activity are complicated by vegetation instability induced by the movement of blown sand across the area around 3200 Cal BC. Higher herb and bracken values, as well as high *Calluna, Pinus,* and Coryloid values, are probably attributable to natural blown sand fringe environments.

Alt Valley, Ince Blundell (IB) (SD 342031)

This site lies south-east of Carr House Farm in the upper Alt Valley, near to the altitudinal limit of the marine Downholland Silt deposit which marks the maximum penetration of estuarine conditions during the period of mid post-glacial high sea-level (de Rance 1872a; Tooley 1978a; Huddart 1992). The location was chosen to determine the timing of this major palaeoenvironmental event in the area, to investigate the changes in environmental conditions which accompanied it, and to discover whether evidence existed of human impact on vegetation communities during this critical period of ecological change. Peat deposits lie both above and below the estuarine silt layer in the upper Alt Valley, and this sequence rests upon Shirdley Hill Sand. The archaeological evidence from the survey for the presence of late Mesolithic people in this area also determined the need for the sample.

Pollen analysis

Samples were taken at 0.4m intervals throughout the core. Pollen preservation was good and the resulting pollen diagrams are shown as Figures 29 and 30. Three pollen assemblage zones are recognised and are described below.

IB-a 0.80-0.74m The lowest zone is characterised by *Pinus*, with major frequencies of *Alnus, Quercus,* and Coryloid, while *Ulmus* is consistently high. Herb pollen values are very low indeed and non-tree types are represented mainly by spores. The base of the zone (0.80-0.82m) is radiocarbon dated to 6180-5740 Cal BC (7130±110 BP; GU-5210). The end of the zone is defined by a sharp rise in *Alnus* pollen and a fall in *Pinus*.

IB-b 0.74-0.30m This zone is characterised by *Alnus*, with *Quercus* and Coryloid also important. *Pinus* declines early in the zone, while *Ulmus* remains high. Non-tree pollen increases in frequency, mainly contributed by a wide range of herb types of wetland habitat. Gramineae rises very sharply at the end of the zone (0.30-0.32m), which is radiocarbon dated to 5204-4791 Cal BC (6040±60 BP; GU-5211). The end of the zone is defined by a fall in *Alnus* and a rise in *Quercus*.

IB-c 0.20-0.00m *Quercus* dominates the zone, with *Alnus* and Coryloid also important. *Ulmus* remains high, and *Fraxinus* and *Tilia* are recorded. Gramineae and Cyperaceae are present in very high frequencies, and other wetland herb types are common. The start of the zone (0.18-0.20m) is radiocarbon dated to 5204-4791 Cal BC (6080±90 BP; GU-5212)

Ecological history

The pollen evidence supports the radiocarbon dating in showing this profile to have formed during early to mid Flandrian II times. *Ulmus* frequencies are high throughout the diagram. The relative proportions of *Alnus* and *Pinus* pollen in zone IB-a suggest a date very soon after the Flandrian I/II boundary and the radiocarbon date of 6180-5740 Cal BC (7130±110 BP; GU-5210) agrees very well with this, the transition being dated at nearby Flea Moss Wood to 6215-5960 Cal BC (7230±70 BP; SRR-2697).

The early part of the diagram, zone IB-a and the start of zone IB-b, shows the local landscape to be closed woodland with very little evidence of any

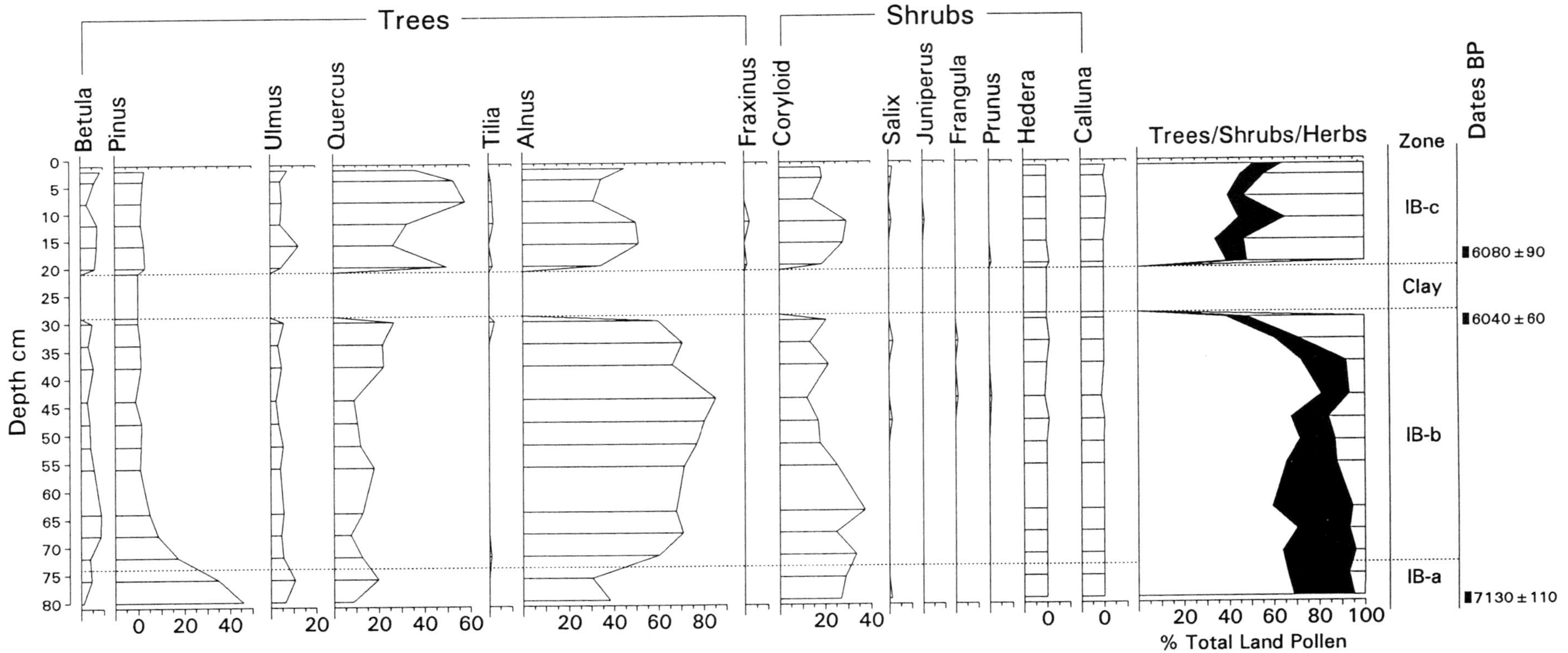

Fig 29 Sefton mosslands: Alt Valley, Ince Blundell, percentage arboreal pollen diagram (trees plus group)

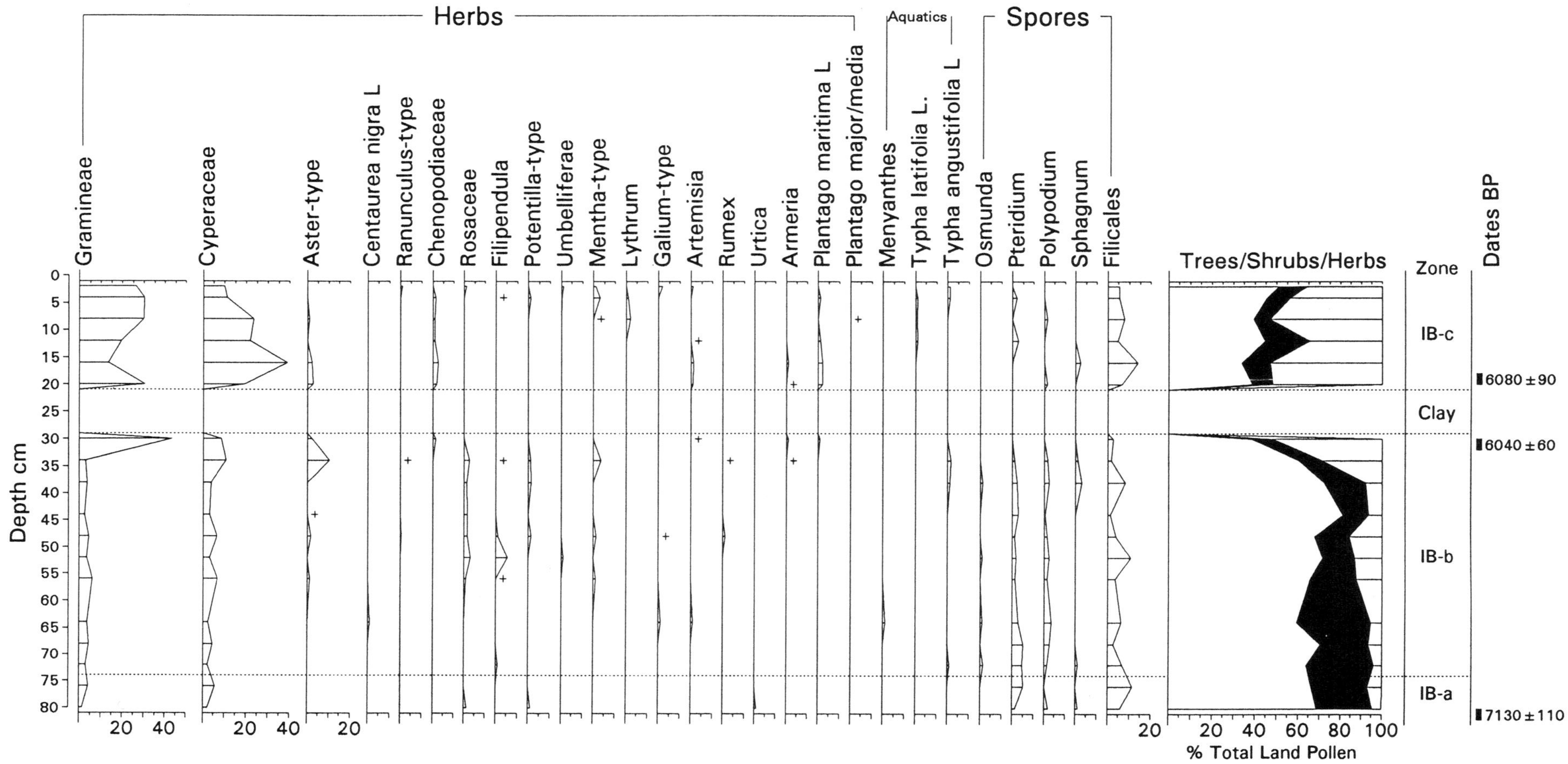

Fig 30 Sefton mosslands: Alt Valley, Ince Blundell, percentage herb pollen diagram (trees plus group)

open ground. Mixed deciduous woodland with a lot of pine was replaced by one dominated by alder. This alder community was probably of local importance within the Alt Valley and away from the watercourse it is likely that oak was more abundant. Waterlogging must have occurred for peat to form at the site, but not until halfway through zone IB-b do wetland and aquatic herb frequencies rise sufficiently to suggest that water-tables were very high over a considerable area, so that fen carr and reedswamp environments developed extensively. Taxa such as *Mentha*-type and *Filipendula* become common, and several other wetland herbs occur.

These freshwater aquatic habitats persisted until the end of zone IB-b when many herb taxa indicative of saltmarsh and general coastal environments are recorded, including *Plantago maritima*, Chenopodiaceae, *Armeria*, *Aster*-type, and *Artemisia*. These, and the sharp rise in Gramineae, indicate approaching estuarine conditions which culminate in the inundation of this site and the deposition of the silty clay layer in the profile between 0.30-0.20m. This clay layer contains many brackish and marine diatom taxa which prove the estuarine origin of the clay and support the pollen evidence. The radiocarbon dates which date the duration of this marine transgression event are almost identical at a little before *c*4900 Cal BC. They are in close agreement with the date of 5210-4792 Cal BC (6050±65 BP; HV-8649) obtained for the lower boundary of this same clay layer further down the Alt Valley (Huddart 1992).

A relative fall in sea-level after this time caused the resumption of peat accumulation, but the continued presence of saltmarsh taxa in the pollen record in the early part of zone IB-c suggests that estuarine conditions remained near. A change in peat type occurs, with high Cyperaceae and Gramineae pollen values suggesting a *Phragmites* and sedge-dominated local flora. These pollen changes reduce the dominance of alder in the assemblage, and the more regional oak pollen increases in importance. A continued decline in sea-level influence causes the gradual removal of saltmarsh types from the assemblage, and their replacement by freshwater wetland herbs of fen and carr conditions. *Typha latifolia*, *Mentha*-type, and *Lythrum* are typical of these environments.

Landscape history and forest disturbance

There are no indications in the pollen record of woodland disturbance which may have been caused by human activity. Only *Pteridium* occurs, which may be considered an indicator of open conditions, but in the absence of other disturbance indicators it is likely that *Pteridium* was growing locally within shady woodland or carr environments. The mainly local nature of the pollen data from this site has not recorded any vegetation change caused by the late Mesolithic inhabitants of this area.

Conclusions

The main points of archaeological and palaeoecological significance from the survey in Sefton that have particular relevance to research questions in the region or beyond are illustrated here, while the wider and more detailed discussion of these issues is left to Chapter 7, where they are seen in the context of patterns within the county. The main locations that may have the potential to develop this research are described here, as well as the threats that might destroy the information. This evidence can be used to set the criteria for preservation of wetland areas through management strategies, or for developing its research potential, through rescue archaeology if necessary.

Prior to the NWWS there was little archaeological evidence for the prehistoric period in Sefton. The NWWS fieldwork in the Sefton mosslands has now produced potentially some of the most important prehistoric sites in the lowlands of North West England. Sefton contains the greatest density of flint sites and the largest surface assemblages in the county, only the previously surveyed areas centred on the sandstone ridges of the North Wirral having a similar density of finds. In contrast, the Sefton survey has also produced approximately 20 sites that may partially lie in associated organic deposits. The associated palaeoecological work from three sites in the district has also provided much significant evidence for the earlier prehistoric period to set beside the archaeology.

In keeping with the results of previous survey work undertaken in the county, most of the new archaeological evidence in Sefton probably relates to the Mesolithic period. Two sites, which may have been located adjacent to river channels, have produced a small amount of flint material that may represent the first early Mesolithic material from the county east of the Mersey.

By the time fens had formed around the coastal plain and in the Alt Valley in the late Mesolithic period (after approximately the beginning of the sixth millennium Cal BC), flint sites become reasonably common. They are located on the coversand ridge overlooking the coastal plain and associated channel at Little Crosby, and in the Alt Valley, where they are found both on the floodplain and on the terrace slightly above the maximum extent of former wetland. The intervening peat area, at Orrell Hill, has less evidence for settlement, but this may be to some extent due to the fact that it lay closer to the estuarine and saltmarsh environments that were prevalent in the northern part of the survey zone during significant parts of the late Mesolithic.

The survey has also produced a small number of sites assigned tentatively to the Neolithic. These are found in the same areas as, and mirror the form of, the Mesolithic flint sites. Palynological evidence suggests that small woodland clearances seen in the sixth millennium Cal BC continued after the elm decline (approximately marking the horizon of Neolithic cultural material nationally) in the coastal area, with little sign of intensification. It is assumed that the coversand soils were providing well drained woodland for sites on the edge of the fens which were being exploited for a wide range of resources, with coastal resources also in close proximity for parts of the period, and that this altered little into the fourth millennium Cal BC of the Neolithic. It has been suggested that this coastal area was a 'home range' where mobility may have been more limited than in other parts of the county to the east, because of the range of concentrated resources available. Occasionally, as at Flea Moss Wood, cereal-type pollen is present. The first instance of this, as at the Bidston Moss site on North Wirral, is at the beginning of the fifth millennium Cal BC, in the chronological Mesolithic.

The NWWS fieldwork has also produced the largest assemblage of late Neolithic/early Bronze Age flintwork from the county on the coversand ridge at Little Crosby. Palaeoecological evidence is limited for this period, but there may have been a general trend in the coastal area to acid bog, which was subsequently (after the end of the fourth millennium Cal BC) overwhelmed by blown dune sand. A closer dating of these natural events and the flintwork is necessary before a clear pattern of landuse and settlement can be drawn up, but the general landscape trends and the nature of the flint assemblage might suggest that the area was important for pastoral farming. Cereal farming is, however, recorded in the coastal dune area a little further to the south, in the mid third millennium Cal BC.

Subsequent evidence, both archaeological and palaeoecological, is very limited for the periods after this. Some flintwork from the surface of the coastal plain blown sand, and along the main channels, may date to the later Bronze Age, but by the late second and first millennia Cal BC of the late Bronze/ Iron Age evidence is absent. Although settled farmsteads are identified for the Roman period elsewhere in the county, there are none in the coastal areas of Sefton, with the evidence for this period restricted to the occasional chance find from the intertidal area close to the mouth of the Alt. In the medieval period the wetland areas all form marginal land lying within the system of townships that had grown up, forming grazing land that was held in common. The focus of medieval and later settlement lay at some distance from the wetland areas in each of the main townships.

Significance of the Sefton wetland evidence

If some of the flintwork from Me 094 and Me 038 is of early Mesolithic date, this will be the first evidence for the period located to the east of the Mersey, complementing the richer evidence from the North Wirral sandstone areas (Cowell 1992c). This provides the basis for an understanding of the relationships between sites of this period in the North West lowlands and the larger body of evidence from the uplands, including the Pennines (Jacobi 1978) and, to a much lesser extent, North Wales (Jacobi 1980).

Concentrations of late Mesolithic flintwork discovered during the course of the NWWS Sefton fieldwork have produced the first evidence east of the River Mersey for a type of site so far very rare in the lowlands. Several locations around the coastal wetlands and the Alt Valley suggest that during the late Mesolithic this area may have formed a 'home range', where possible base camps, which included 'industrial' activities, were located. This suggests a number of family groups were living in the vicinity for part of the year within a generally mobile system. This contrasts with those areas away from the coast, in the eastern part of the county, where the only type of Mesolithic site so far discovered appears to be the small specialist camp (*see Chapter 7*).

The late Mesolithic sites and the potential Neolithic sites identified in Sefton may also be important in the context of understanding the local technological and chronological framework for the Neolithic which is lacking for the region. These sites may also contain evidence which will aid an understanding of the economic transition from hunting and gathering to agriculture, as may be suggested by the possible presence of cereals at the beginning of the fifth millennium Cal BC near Flea Moss Wood. This is enhanced by the fact that the same phenomenon is seen at Bidston Moss on the North Wirral (*see Chapters 3 and 7*).

The pollen evidence from the survey suggests that there was little change in the nature of the clearance activity between the Mesolithic and the late Neolithic (after which most pollen evidence is absent). This suggests that changes in the nature of settlement were also slight, with hunting and gathering remaining important and cultivation playing a limited role. On the Wirral, however, there is a stronger contrast seen towards the end of the fifth millennium Cal BC in the pollen evidence, suggesting that economic change was more dramatic and sustained here during the early Neolithic. This provides for the possibility of elucidating the nature and chronology of the adoption of agriculture by research in two areas of potentially contrasting economic activity during a cultural transition period.

The evidence from the flint scatters of later Neolithic/early Bronze Age date, and that found in the overlying blown sand in the coastal plain, which may be a little later, provide the potential for identifying the nature, chronology, and development of settlement and landuse during this period. This is important, as elsewhere in the county this period is marked only by a little burial evidence and a series of scattered chance finds.

Archaeological and palaeoecological potential

Alt Valley

All three of the main Sefton mossland areas have produced sites that are located within peat deposits. The most important area is the Alt Valley, where six sites have been located on sandy islands or ridges that are emerging as the peat shrinks. A further five locations are recorded where single findspots are found on the surface of the peat, without any obvious link with eroding sand islands or ridges. This flintwork is mainly of Mesolithic date, although Me 093 could be Neolithic, while the single pieces are very difficult to date. The archaeological potential of these sites, particularly the former group, is on the whole good, as most may still have some of their area protected from ploughing, as they have largely been preserved by peat deposits, generally capped by alluvial clay. This may therefore allow the survival of *in situ* organic material.

A number of the Sefton sites which are characterised by burnt flint nodules (*eg* Me 024, Me 026, Me 079), and which are difficult to date, occur in or on the edge of wet deposits. They are mainly found in the Little Crosby channel but also occasionally in the Alt Valley. The northern half of the valley floor has a 1m thick blanket of alluvial and marine clay overlying the peat, which serves to protect it but also makes the discovery of sites difficult. The clay narrows across the valley floor the further south it is found, leaving a narrow band of peat on the surface along the lower part of the valley slopes, which is often highly humified.

Survey of the area as far south as Searchlight Wood has located few sites, probably because of the depth of the alluvial and marine clays found there. Most of the single finds come from this area, mainly located towards the edges of the peat near its junction with the sandy river terrace, where alluvial clay is absent. It is possible that the single finds may represent the initial erosion of larger sites, instead of being isolated pieces associated with 'off-site' activity. The relatively high density of artefactual evidence from the surrounding dryland and the southern half of the valley (*see below*) suggests it is likely that clay and peat hide further prehistoric archaeological evidence in this zone.

In the southern half of the valley, where the alluvial clay thins in extent and depth, the main concentrations of flint lie on the valley floor or on the valley slopes, and it is in these areas that the largest sites are located. This is probably due to the fact that overlying sediments are in a more advanced state of erosion. The main part of Me 093, on the terrace overlooking the river, is found on Shirdley Hill Sand (Fig 21.A), and although once covered by peat, little organic material is likely to have survived on the ridge, even though archaeological features in the sand may have done so. A part of this site, found at the foot of the slight ridge which marks the present edge of the organic material, may have a little more potential. The peat is badly humified here and is unlikely to be much deeper than *c*0.3m. The occasional flint piece is also found a little further out into the peat, where physical and organic preservation may be somewhat better. At the other large site in Lunt, Me 094, it seems very likely that although *c*1ha of the site is partially subject to plough erosion, other elements are still largely protected by peat and clay (Fig 21.B).

The terrace slopes around Lunt appear to have some potential for the location of more sites, as the 4m contour seems to have been crucial for sites in the Mesolithic period, possibly representing a well drained dryland soil set just off the fen in the valley floor. These slopes were under-represented in the fieldwalking, due to large amounts of land here being left uncultivated close to the landfill tip, while unsuitable field conditions prevailed in a number of other fields. The potential for the discovery of more sites in this zone is therefore high. The crucial contour is, however, generally a minimum of *c*20m from the edge of the peat, which severely reduces their potential for the survival of associated organic material.

Nine smaller sites lie to the north of Me 094 in a similar topographic situation (Me 076-079, 081-085), as well as a number of dispersed single findspots around them. These may either be functionally different from the two larger sites, where less debris has been discarded, or they may be similar but better protected from ploughing. In most cases the smaller groups of material are located right at the edge of the peat/clay and sand interface, rather than on the top of the sand islands. This implies that they are still mostly unaffected by ploughing and are covered by layers of alluvial clay and peat. In this southern part of the valley floor it is difficult to assess the depth of the peat associated with these sites, as they generally lie 50m or more from the ditches where the sections are available for recording the peat. In the ditches the peat is found in two bands of *c*0.4-0.6m deep with a similar thickness of marine clay, whose inception began towards the end of the sixth millennium Cal BC, separating them. The peat is in reasonable condition in the central part of the valley where the best sections were available, and pollen preservation was good in the sample undertaken close to Me 089. The closeness to the River Alt also makes it seem likely that the lower peat, which mainly lies beneath the field drains, is still waterlogged. The potential for organic preservation associated with these valley floor sites (which are likely to be mainly Mesolithic in date) is consequently quite good.

Orrell Hill

The Orrell Hill area is more difficult to assess, as no palaeoecological work has been carried out here and archaeological finds are limited. Only one site (Me 060) appears to have some potential. It is represented only by a broken blade, but it lies within a small peat-filled gully in an area where the peat is thinning out and small sand islands are becoming visible to the north-east. It is possible that this find may represent the beginnings of a larger site preserved to some extent by the depth of the organic deposit, which is likely to be more than 0.4m deep. In general, however, wherever the peat is visible in the ditch sections it is quite humified and never reaches a greater thickness.

Much of the rest of the Orrell Hill peat is either covered by post-medieval alluvial clay, to the north, or, to the south-west, is mainly pasture land. This landuse suggests that the potential for the discovery of more sites is reasonable. This may be particularly so in the slightly higher area that is given over to pasture to the west of the Ince Blundell bypass, where a number of grassed ridges and islands can be seen on the surface. Sites of all periods surround the peat here. Two small Mesolithic flint concentrations lie to either side of this pasture area (Me 058, Me 061), set slightly back from the peat, and Neolithic and late Neolithic/early Bronze Age findspots also come from the peat margins (Me 059, Me 062). The latter are likely to be quite shallow and probably fairly well humified.

Coastal plain

In the coastal plain, west of the Woodham Knoll ridge, a fairly well developed peat lies beneath *c*0.6-0.8m of blown sand. This area of peat has produced evidence of archaeological material from the blown sand surface, which probably reflects occupation and landuse in former dune areas. It is not clear how far plough damage has affected the distribution of flintwork on the surface, but the depth of the sand may indicate that sites near the base of the sand could have survived in reasonable condition. The peat beneath the sand is fairly well preserved as it lies close to the water-table, and so could provide a palaeoecological context for the origins of settlement in the blown sand areas, as well as for the occupation that was concentrated on the adjoining Woodham Knoll ridge flanking the peat to the east.

The flint material from this Shirdley Hill Sand ridge has produced one of the densest concentrations of surface flint from the NWWS in the county (Fig 76). These sites cover the period from the Mesolithic to the late Neolithic and early Bronze Age. The latter group of material is particularly important owing to its rarity in the county, although it may postdate the accumulation of sand across the peat. The Mesolithic, and possibly Neolithic, occupation on the ridge, and along the slopes of the small channel to the east of it was, however, contemporary with the peat. Although the ridge is under arable farming, the potential for survival of occupation evidence, from results of excavations on other Mesolithic sites in the county (Cowell in prep b), is reasonable. The proximity of the adjacent peat, which *c*100m to the north is associated with marine clay deposits, enhances the importance of this dryland material.

The occurrence of this material on the ridge, and along the organic soil of the lower slopes of the small channel to the east, implies that the potential for archaeological sites on the coversand surface beneath the main body of peat close to the ridge is reasonably good.

Threats

Development

A number of threats found in many other wetland areas in the county are not as marked here. As the area is a designated Green Belt, housing and industrial development are restricted and there is no commercial peat cutting. The main development threat comes from the continuing road programme and, in particular, the plans currently being discussed for a new road through Ince Blundell. The two alternative routes both cut through the peat areas, one affecting the Alt Valley peat north of Carr House Farm, where a number of small Mesolithic sites have been identified on sandy islands in the peat; the other, more favoured route, will cut through the area of sandy islands and peat deposits on the west side of the Orrell Hill area. It was not possible to survey this peat thoroughly as most of it is given over to pasture.

Agriculture

A large expanse of land is under arable cultivation and includes much reclaimed mossland. This is consequently threatened by extensive drainage and ploughing, which constitutes a longer term and more subtle threat to the remaining peat. This is particularly so as much of the peat is less than 0.5m thick. On average the survey found many areas where the top *c*0.3m of peat was humified, often leaving little peat unaffected. One sample taken from a drainage trench in the Little Crosby channel, close to a number of sites, including Me 032, produced *c*0.2m of almost completely humified peat beneath the topsoil, which proved barren of identifiable organic remains.

The main areas where this has happened, and will have clearly affected archaeological survival, are in the Little Crosby channel and the archaeologically important area immediately to the north of it and Flea Moss Wood. Other areas are the north-eastern side of the Orrell Hill moss, where a little Mesolithic material has been found, and the south-western fringes of the Alt Valley peat, where it meets the sandy terrace and is unprotected by alluvial clay.

Not all areas, however, have suffered to the same extent. The peat, with a later deposit capping it, has usually been protected from the effects of ploughing, and in these areas also, the water-table is naturally fairly high due to the peat deposits lying at about sea-level.

Natural erosion processes

The dune belt along the Sefton coast is partly developed for recreational purposes, with golf courses being largely restricted to the south of the Altmouth, while to the north, the dunes are managed for wildlife and conservation. Coastal erosion is, however, a continuous threat in Sefton and the dunes are being eroded at a rate of on average *c*3ha per year around Formby Point (Cowell *et al* 1993). This poses a threat to the Neolithic peat deposits in the dunes at Hightown, which may be an extension of the peat deposits in the coastal plain to the east.

Future research

The Sefton area has the greatest potential for future work in the county. The two major threats are the projected Ince Blundell bypass and the continuing agricultural erosion. A decision has not yet been made on the continuation of the former project, but if it does go ahead then rescue work should precede development. This would serve to determine whether sites exist on the small islands in the pasture fields at Lady Green, and sampling of the peat and marine clay should take place to identify the date and nature of the environmental developments here. This would not only provide local evidence for any archaeological sites found, but would have a wider application for understanding the site evidence from around the peat deposit as a whole.

The most urgent area to safeguard against agricultural erosion is perhaps in the southern half of the Alt Valley. Here the largest sites, which are likely to be Mesolithic, are found (Me 093, Me 094), suggesting plough damage may be greater than lower down the valley, but the main sites are still likely to have potential for organic survival because of the surrounding peat and alluvial clay. Such sites should be tested to identify survival of features and associated organic material, to provide better dating for the flintwork and so relate it to the depositional history of the bands of peat and marine clay in the Alt Valley, and to identify if suitable peat deposits survive from which local pollen may provide an idea of the nature of activities being carried out on these sites.

The smaller, probable Mesolithic, sites a little lower down the valley (Me 076-079, 081-085) should also be tested for the same questions, and to determine whether they differ in function from the two larger sites, or are similar sites that are better preserved. These also are very likely to have organic survival.

The valley terrace slopes in the area to the south-east of the large sites (Me 093, Me 094) have also not been thoroughly surveyed due to field conditions, suggesting that fieldwalking when possible, and test pitting of the 4m OD contour, should take place to see how common these types of sites are here, and whether peat has survived in the unploughed areas.

A similar programme of research would probably be productive in the area of flint sites, sand islands, and peat channels to the north of Flea Moss Wood (Me 050, Me 052, Me 054), although here the peat seems less well preserved than in areas of the Alt valley. Some of the flint material from this area may be Neolithic, which would increase its importance in the context of the poor regional evidence for such sites.

The sites from the Woodham Knoll channel are, in the main, probably not associated with significant organic deposits, as much of the former peat in the channel has now turned to an organic sand. One Mesolithic site, on the dryland slope just above the channel (Me 022), and the large scatter of Mesolithic material on the upper part of the slope of the channel overlooking the Flea Moss Wood peat (Me 038, Me 039), are good candidates to identify and date the nature of Mesolithic settlement in the area, and relate it to the pollen evidence from the latter site. One site (Me 032) may lie beneath or within the channel deposits where the peat is deepest (*c*0.4m). Although there seems little potential for associated pollen from the peat in the vicinity of the possible site, the peat is the wettest in the area in winter, which might allow for limited organic survival.

The dryland late Neolithic/early Bronze Age sites on the Woodham Knoll ridge (Me 018, Me 036, Me 037) are important because of their rarity, and it would be very useful to test a sample of these to identify the nature of the settlement they represent, and to provide radiocarbon dating evidence that might allow the flintwork to be attributed more accurately. There is little hope of organic survival from these sites, but samples from the peat adjacent to the ridge on the west may provide evidence to date the encroachment of sand over it, and allow a better idea of the palaeoenvironment contemporary with the flintwork. This may accordingly allow a better interpretation of the nature of activities carried out at these sites on the ridge than is presently possible.

5

THE CENTRAL AND EASTERN MOSSLANDS

The central mosslands lie in an intermediate topographical position between the coastal mosslands at *c*5m OD and the Carboniferous ridge, rising to *c*120m OD, that bounds the Lancashire plain to the east (Fig 4). Topographically they can be divided into two areas: a belt of eastern mosslands lying at the foot of this ridge, and an extensive, almost contiguous area of mossland occupying the plain to the west. These western mosslands occupy an area of *c*1100ha, and are found in the historic townships of Kirkby, Simonswood, Knowsley, Windle, and Rainford (Fig 31). They stretch in a belt *c*7km long from north to south, widening from *c*3km in the south between Knowsley Park and the western edge of St Helens, to *c*5km between the north-eastern edge of Kirkby and the north-western edge of the town of Rainford. A small area of peat in Bickerstaffe, at the head of the Rainford Brook valley, *c*1km to the north of the main complex, has also been included within this western group of mosses.

The eastern group consists of a belt of mossland lying in the eastern half of the historic township of Rainford, on the north-eastern edge of the county and spreading northwards into Bickerstaffe in West Lancashire. In the south, the peatland complex of Holiday and King's Mosses occupies *c*160ha within a basin between the town of Rainford to the west and the slopes of the Carboniferous Billinge ridge to the east. This basin is separated by a ridge from a larger peatland complex to the north which encompasses Holland Moss, and in Lancashire, White Moss and the eastern of the two Bickerstaffe Mosses. The area of this group of mosses totals *c*265ha. The physical bounds of the Holland/White Moss complex are marked on the east and north by the modern spread of the industrial estate of Skelmersdale and the M58, while the western edge is marked by the break of slope down to the valley of the Rainford Brook. To the west, Bickerstaffe Moss, also in Lancashire, stands isolated from the peatlands already mentioned and is not included in the current survey. Instead it will be surveyed as part of the West Lancashire programme of NWWS fieldwork (Middleton *et al*, forthcoming).

Landuse

Since the time of Domesday, when the area was heavily wooded and sparsely settled, the central mosslands have remained little exploited for settlement and industry, and so largely retain their distinctive pattern of scattered farms, cottages, and the occasional hamlet. The belt of urban and industrial development, where many of the key features of the Industrial Revolution occurred, has been concentrated on the surrounding heavier clay and sandstone derived soils. To the south is St Helens, developing from its early glass, transport, and chemical industries, and Prescot/Whiston, where pottery and watchmaking industries were of prime importance in the region. To the east is a belt of industrial towns, such as Leigh and Wigan, now on the fringes of Greater Manchester, which were based largely on coalmining, while to the west lies the expansion of modern Liverpool which has subsumed a series of small, post-medieval villages stretching northwards from West Derby to Kirkby. The only nucleated settlement to develop within the mosslands-dominated landscape was the medieval village of Rainford, which is now a small town.

These mosslands are therefore characterised by an agricultural landscape with arable farming predominant, although some areas of poor pasture also occur on ground created by the commercial extraction of the Shirdley Hill Sand for the use of Pilkington's Glass in St Helens during the 1960s and 1970s. An important feature of the rural landscape of the area has been the growth of Knowsley Park on the western edge of the mossland belt, around the ancestral seat of the Earls of Derby. Much of the surrounding land in the mosslands still belongs to the Derby Estate, and this has traditionally been associated with hunting. Many of the woodland plantations that are dotted across the landscape were created for this pursuit during the last few hundred years.

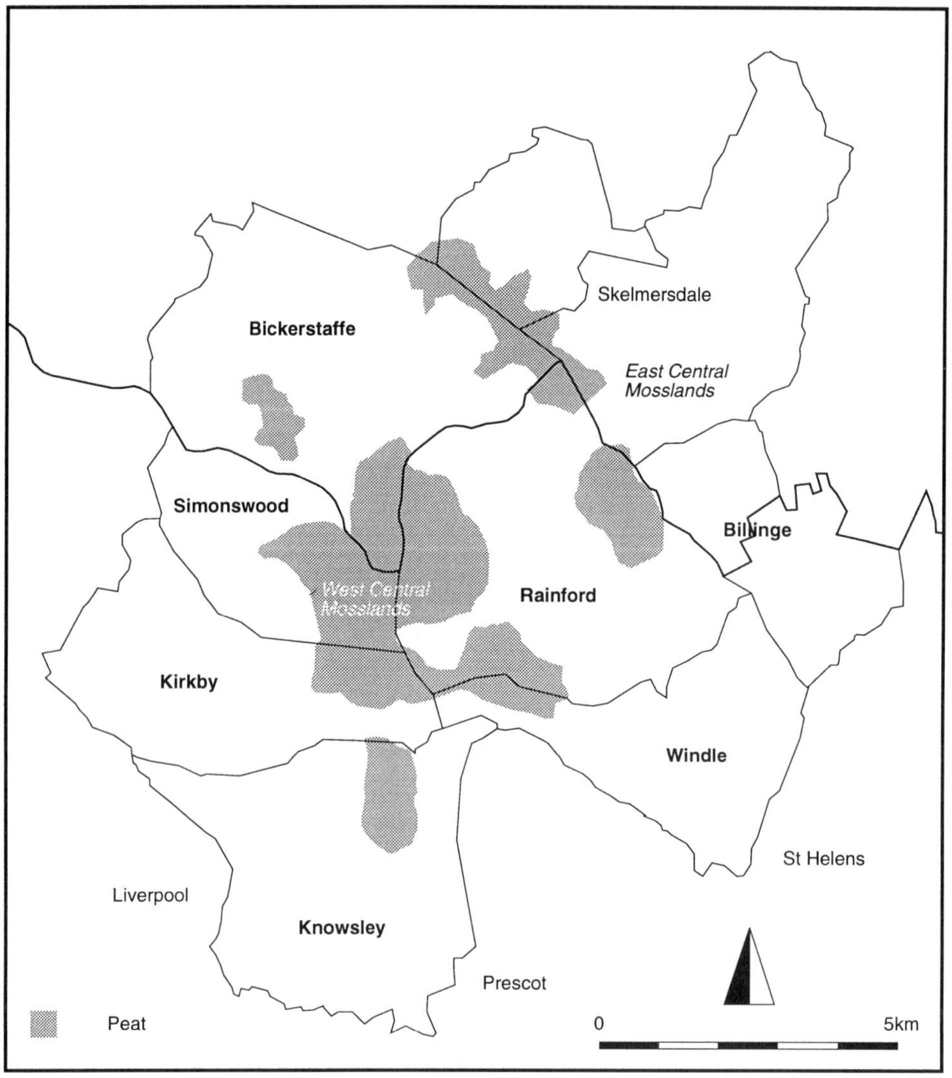

Fig 31 Central mosslands: Location of survey area

The nineteenth century growth of St Helens avoided the mosslands' wet terrain and surrounding light sandy soils. Instead these areas were drained and enclosed to provide the urban population with much of its food grown on rich reclaimed mossland soils. However, modern urban spread has begun to have an impact on the mossland areas. Despite this (in contrast to the situation on the Wirral), urban sprawl has accounted for little loss of the peat, apart from an industrial estate on the southern edge of Skelmersdale which has started to affect the edge of White Moss.

Commercial extraction of the peat has had a far more direct effect on the central mosslands. The chief area of exploitation lies at Simonswood Moss on the west side of the system (Plate 4), where over 120ha is being commercially milled. White Moss is notable for playing host to what is probably the last commercial peat works in England which still operates entirely by hand cutting. The White Moss peat works, which covers an area of *c*10ha, has all the appearance of an open air museum with the tiny workforce still using the traditional range of peat spades and barrows which have long disappeared from the rest of the country. Another small commercial operation of 8ha lies 1km to the north, based on a localised deep basin. Further south at Holiday Moss, over 30ha has already been commercially mined, and the site is now used for landfill.

The loss of much of the sand from around the mosses through commercial extraction for the glass industry poses further problems for archaeological field survey in the central mosslands. Although this has not affected the peat areas directly, it has probably removed archaeological evidence from the dryland/wetland interface; a zone which survey, particularly in Sefton, has shown to be important during the prehistoric period. A few pieces of

Plate 4 Central mosslands: Simonswood Moss, peat cutting, with sand island appearing in foreground (Photo R Cowell)

isolated lithic material were found on the mined areas as part of the NWWS, but it is doubtful how far this represents an accurate picture of the original pattern of artefact distribution.

To set against this loss there are some areas in the central mosslands that, echoing the general late development of the area, are still undrained and unfarmed. The best is in the Lancashire part of Holland Moss, where over 30ha is largely left in a semi-natural state, standing proud of the surrounding moss farmland. There are other smaller areas around Holiday Moss and Bickerstaffe Moss where semi-natural mossland is found in small belts and most of the small woodland copses have localised areas of uncut peat around them.

Previous research

Prior to the commencement of the NWWS this area had seen little scientific research, nor had it been noted for the discovery of chance finds. Some limited fieldwalking was undertaken by Liverpool Museum around the west side of the central mosslands as part of the creation of the Merseyside SMR in the early 1980s, which located some prehistoric material, mainly representing Mesolithic sites and landuse. The historical development of the landscape in a number of townships containing peat deposits within their bounds was also studied (Cowell 1981). A number of peat cores was also taken for palynological study as part of this research programme, including examples from Simonswood Moss and Knowsley Park Moss. Since that time radiocarbon dates have been acquired through a variety of sources for the pollen diagrams.

The eastern mosses had received little academic attention prior to the present survey, no systematic archaeological fieldwork ever having been undertaken, although occasional stray finds, including stone axes (one from the survey area: A 29), and occasional struck flints from the area further to the south along the slopes of the Billinge Ridge, had been reported (Cowell in press). Some limited palaeoecological work was undertaken in the Liverpool Museum research programme. A peat core was taken from Holiday Moss for pollen analysis, although it remains stored due to lack of resources to study it. University College of Wales, Aberystwyth, also undertook some work in collaboration with the Liverpool Museum survey. This included coring at Simonswood Moss, and of the eastern group of mosses, Holland and Holiday Mosses, which provide the main specific vegetational information for this area (Baxter and Taylor 1981). In comparison to the other mosslands treated in this volume, however, the eastern complex remained the least understood in both archaeological and palaeoecological terms.

Geology and topography

The west-central group: Kirkby, Simonswood, Knowsley Park Mosses (Figs 77, 81)

These mosslands lie in a large depression which deepens to the east. Field observations during the survey suggest that, from variations in the depth of the peat, there may be a series of smaller basins which have probably acted as separate loci for the growth of the whole wetland complex. The large depression is bounded on its southern side by a Carboniferous ridge in Knowsley Park, where the undulating landscape of the Coal Measures sandstone starts to rise relatively suddenly from the mossland surface at *c*60-95m OD.

The central group: Windle, Reeds, Rainford, Catchdale Mosses (Figs 77, 81)

From the Carboniferous ridge in Knowsley Park the ground falls markedly to the east to *c*45m OD, where a shallow basin formerly contained Catchdale Moss, and then more gradually to *c*20m OD across the boulder clay plain upon which St Helens has grown up. To the north-east lie the deeper basins in which Reeds and Rainford Mosses lie. The eastern limit to the mossland complex is marked by the boulder clay ridge marking the western side of the valley of the Rainford Brook. This lies at *c*45m OD in the undulating terrain around Eccleston at the southern end of the valley, and terminates at the northern end near Bickerstaffe, at a height of *c*60m OD. The north and western edges of the depression, which are in Bickerstaffe, Simonswood, and Kirkby, present no topographical distinction, with the peat terminating on sandy surfaces that are mostly lower than the central areas of the moss.

The pre-Flandrian surface of the central group of mosslands is dominated by a late glacial coversand, the Shirdley Hill Sand. This is generally found at its thickest to the west and north-west, where it lies in a band a little under 1m thick. This sand extends northwards into West Lancashire, where it is quite extensive around the major wetland complex in this part of the region. Westwards, it becomes much more localised within the larger areas of glacial till and low sandstone outcrops on the eastern fringes of suburban Liverpool.

To the south and east the sand thins and boulder clays become more prominent. The latter forms low, localised ridges and hills surrounded by peat on the southern side of the western moss complex. This is most noticeable in the south-east between Mossborough Hall and Knowsley Park, where boulder clay separates Reeds and Windle Mosses from Kirkby Moss.

The east-central group: Kings, Holiday, Holland, White, Bickerstaffe (east) Mosses (Figs 77, 81)

The eastern mosses lie in two main basins, with the surface of the peat lying at a height of between *c*60m OD on Holiday Moss to *c*70m OD on White Moss. Topographically, the main defining feature on the eastern side of both basins is the steep Carboniferous ridge running from Billinge in the south to Upholland/Skelmersdale in the north, and this forms the highest relief in this part of South Lancashire/Merseyside, rising to *c*120m OD on average. The Holiday/Kings Moss complex lies between this high ridge and the leeward side of the ridge on the eastern side of the Rainford Brook valley. The western group of mosses lies in the large basin terminating on the western ridge of the valley, *c*2km away. Holiday Moss lies at the head of the Black Brook valley running south-eastwards towards Billinge Chapel End, and is separated from Holland Moss to the north by the slopes of the Rainford Junction ridge, which is less than 1km wide. Fieldwork showed that the soils on the lower slopes of the Billinge ridge are largely formed from heavy glacial clays around the northern end of the moss, although around the other slope margins the clays have been capped by Shirdley Hill Sand.

The Holland moss complex is more extensive, stretching from the brow of the Rainford ridge on the west, where it overlooks the edge of the central mosslands, to the Upholland Carboniferous ridge. The mossland largely overlies a landscape of late glacial coversands, although the margins along the slopes of the Rainford Brook ridge and the Carboniferous ridge to the east are formed by clays.

Natural drainage pattern

The main drainage of the central mosslands has probably remained fairly constant through the Holocene period. In the east this is formed by the Black Brook/ Sankey Brook, which flows along the foot of the Carboniferous ridge to the east of Rainford, before cutting through the boulder clay plain to the south to join the Mersey north of Warrington. To the west of Rainford the northern part of the Rainford Brook flows through a well defined valley before it joins the Black Brook/Sankey Brook east of St Helens. The other main drainage system lies several kilometres to the west of the central mossland complex, and is formed by the River Alt. This flows into Liverpool Bay north of Little Crosby from the central sandstone ridge around Huyton (*see Chapter 4*). The same glacial channel links to join the Mersey in the Ditton Brook to the west of Widnes. Near the western edge of the central mosslands the Alt has a major tributary, the Croxteth Brook.

Flandrian geology

The surface of the peat in the western complex rises imperceptibly from *c*40m OD at the northern end to *c*60m OD in the south. Much of the current peatland area probably once formed a discrete mire macrotope which is now truncated and fragmented. The present sites, however, retain extensive sections of deep (*ie* >5m), well preserved peat, and at Knowsley Park considerable depths of peat (>7m) still remain.

As the subsurface survey across these mosses was not continuous, the subsurface topography of the main basin is difficult to identify in detail. However, at a more general level it seems that Simonswood and Kirkby Mosses appear to lie within one extensive, undulating basin complex in the west, which is separated from Rainford Moss, to the east, by a low surface sand ridge extending roughly northwards from the Mossborough Hall clay ridge. Stratigraphic survey (Figs 43, 45, 47) indicates that Simonswood Moss probably lies within two shallow basins separated by a Shirdley Hill Sand east-west subsurface ridge. Knowsley Park Moss to the south lies within two interconnected, elongated basins, the southern one being the deeper (>6m). The evidence of ditch sections and farmers' comments also suggests that Reeds Moss lies in a shallow basin in the south-east. The subsurface topography of the east-central mires suggests that each of the main mires occupies a shallow basin, although at some stage in the past they probably formed a discrete mire macrotope.

The only estimates for the date of accumulation of organic material in the basins comes from the pollen cores from samples taken in the early 1980s from the Simonswood and Knowsley Park Mosses. The former has produced one date that places inception of the mire in the later fifth millennium Cal BC (Table 12), while Baxter (1983) has reported a slightly earlier date for a basal peat from the moss. Knowsley Park Moss has a late glacial origin (*ie* *c*7500 BC), derived from palynological evidence (Fig 39). No radiocarbon dates are available for the origins of the eastern mosses, although Baxter and Taylor (1981) have suggested approximate periods of inception from the pollen assemblages in the diagrams they studied at Holiday Moss. Here, peat growth may have started early in the Flandrian period (*c*7000 BC), while a basal pollen sample from Holland Moss provided a mid-Atlantic estimate (late sixth millennium Cal BC) for its inception.

The development of these mosses followed a similar course, with peat accumulation beginning in the discrete hollows within the underlying sand and clay. Extensive reedbeds and carrs developed during the Mesolithic period, but were eventually replaced by ombrotrophic communities which continued to dominate the systems until their reclamation for agriculture in historically recent times.

Survey techniques

Fieldwalking

As much of this area has been avoided by modern development it was possible to survey over 85% of the mossland, and this has been spread fairly evenly across the moss complex (Figs 78, 82). A total of *c*1650ha of arable was surveyed and a further *c*1000ha of pasture, woodland, etc over the area of the central mosslands. One problem associated with the fieldwalking programme in this area is the widespread mining of the Shirdley Hill Sand around the mosses for the glass industry of St Helens during the 1960s and 1970s. This involved severe disturbance to the upper part of the subsoil where any archaeological features which had survived ploughing would have been located. The topsoil, which would have contained ploughed out artefacts in some areas, has been removed, typically leaving an undulating surface with hedges set on baulks around the edge of the field. A proportion of this type of land was walked to test if topsoil material was retrievable (Figs 79, 83), with very limited results, which would at least point to former locations from where the mosslands might have been exploited. The areas shown as affected by this quarrying on Figures 79 and 83 are only approximate guides to its full extent. Only fields walked, or adjoining fields that could be seen to have been obviously affected, are recorded. Many more, where landuse, lack of access, or other difficulties made identification impossible or uncertain, have been omitted from the maps.

Palaeoecological survey

The western group of mosses received some attention from the Liverpool Museum palaeoecological research programme, although the eastern group was relatively under-represented in the sampling strategy. Although in other areas of the NWWS region a comprehensive stratigraphical survey of surviving peats was attempted (*eg* the Over Wyre peatlands further to the north (Middleton *et al* forthcoming), this was not possible in Merseyside, and a selective approach was adopted in response to constraints imposed by limited time and access difficulties.

Four key areas of mosslands were selected for stratigraphic investigation, *viz* White Moss, Holland Moss, Simonswood Moss, and Knowsley Park. These areas contain extensive areas of well preserved deep peats, and all have received previous palynological investigation by other workers (Cundill 1981; Innes and Tomlinson 1983a, 1983b). However, little stratigraphical work had been undertaken to provide a context in which to place the pollen diagrams. Because of this, priority was given to recording stratigraphy from the palynological study sites and to areas which still retained extensive profiles of peat, rather than the indiscriminate approach adopted hitherto. It is important to emphasise that extensive areas of truncated and residual peats and organic soils surround the areas studied, and that the degree of preservation of stratigraphy represented in the profiles cannot be regarded as typical of the region as a whole.

The NWWS also arranged for further radiocarbon dating of pre-existing cores from Knowsley Park Moss.

Aerial photography

The existing aerial photographic coverage for the area, including the collection held at Liverpool Museum recently assembled by R Philpott, was also assessed and replotted by C Cox as part of the NWWS (Cox 1992). Much of this was done during the course of the field survey and so little checking of features on the ground has been possible.

The mosses

Although mossland is not continuous across the northern central part of Merseyside, for the purposes of discussion it has been found easier to group sites as units. A number of natural divisions are evident within the complexes which are used to distinguish differences in the human exploitation of the area.

West-central group: Kirkby Moss, Simonswood Moss, Knowsley Park Moss

The two former adjoining mosses are not distinguishable topographically and occupy the western half of the whole western complex, roughly being defined on their eastern side by Simonswood Brook and the Mossborough Hall ridge. The mossland in general survives to a depth of *c*2-4m across the area. It is divided between mainly arable farmland and commercial extraction/landfill. The mossland occupies an area of *c*520ha, while a total of *c*680ha of mossland and dryland fringe was fieldwalked, including the peat mining areas. The Shirdley Hill Sand has been removed from many areas around the western and southern dryland fringe. A sample of the land returned to arable was fieldwalked to test for surviving unmined areas and the possibility of artefact recovery. In the north, a little over 200ha is being commercially extracted from Simonswood Moss. It was possible to monitor all of this commercially worked area by inspection of surface peat and ditch sections. The latter were particularly amenable to survey around the perimeter, where they are kept clean by the Simonswood Moss Co to a depth beyond the peat-mineral soil junction. The company also maintains a conservation policy for a small part of the moss, leaving a shelter belt around the cut area, varying in width from 10-20m, which has been abandoned to nature. This peat baulk is therefore *c*0.5-1m higher than the internal cut area, and carries a vegetation of birch scrub and *Molinia* grassland.

The dryland/wetland interface to the north-west side of the moss is occupied by a further *c*75ha of industrial development adjacent to the railway line. A lack of time, and a complicated land ownership pattern for the belt of fields immediately to the north of the industrial complex, prevented fieldwalking here. A guide to the archaeological potential of this northern mossland edge comes from the results, incorporated in the present work, of a previous survey (Cowell 1981) of 50ha of fields near the moss. To the south-east, on Kirkby Moss, a further *c*25ha of dryland/wetland interface was inaccessible as waste tipping had taken place directly onto the peat. As there was no extensive series of ditches around this area it was difficult to define the location of the peat interface, which has therefore here been estimated by interpolating data from surrounding areas.

In the south-west of the survey area, Knowsley Park covers a total area of a little over $2km^2$. The moss, occupying *c*85ha, lies at the northern end of the park, in the part given over to arable land and plantation, although the latter is gradually being reduced for arable land. The moss is known to be over 7m deep from stratigraphic survey. The southern end of the peat is marked by the course of Mill Brook and is covered by grassland. The geology of this section of the park is characterised by Carboniferous sandstone and boulder clay. Some areas of this landscape were fieldwalked in the early 1980s and the results from the nearest fields have been included here. The moss is separated from Kirkby Moss by a belt of sand 150m wide, which was only partially walked as it has suffered quite heavily from sand extraction along the route of the East Lancashire road.

Central group: Catchdale Moss

This moss survives only as a place name, 800m south of Windle Moss in a gentle, north-south trending dry valley formed in boulder clay. It is mentioned in documents of the seventeenth century (Bailey 1937), but is not shown on nineteenth century maps nor on modern Soil Survey maps (Hall and Folland 1970). The area was surveyed, but not fieldwalked, as no peat was recognised in the area, and the Shirdley Hill Sand which forms the underlying geology in the former moss area has now been extensively mined. A little fieldwalking took place in 1981 along the western slopes adjacent to the old boundary of Knowsley Park, but produced no artefactual evidence.

Central group: Rainford Moss, Reeds Moss, Windle Moss

This part of the moss complex has suffered less from direct modern disturbance, although sand extraction for the glass industry has occurred fairly extensively on much of the surrounding dryland. The mossland occupies *c*480ha and in total *c*520ha of mossland/dryland was fieldwalked. The peat across this area is, in general, deeper than that to the west, being over 5m deep in the north around Camholes Wood, while the southern Windle Moss is on average *c*2m deep. Virtually the whole of the western central mossland area lies within the Derby Estate, which has created many small scattered plantations over the last few hundred years in the farmland on and around the mosslands. These often have shelter belts of scrub grassland around them which preserve the original pre-drainage peat stratigraphy, often a little under 1m high.

East-central group: Holiday Moss, Kings Moss

The Holiday Moss/Kings Moss complex occupies an area of *c*160ha. The core of Holiday Moss (to which access was not granted totalling *c*30ha) has been heavily affected by extraction of the peat, so that only a surviving band *c*150m wide around the margins can be identified. In several areas there is a *c*30m wide baulk of upstanding peat up to *c*1m high, which is occupied by scrub woodland, between the landfill operations and the surrounding arable. At the southern end this widens out into a domed area of *c*10ha of woodland and scrub, where the peat survives in a better condition. The peat in the arable areas is generally from *c*1.3m to over 2m deep. The fieldwalking survey was extended beyond the immediate edges of the moss to test the archaeological potential of the slopes of the encircling basin, except on the open south-eastern side, where access was limited through time considerations and restrictions imposed by a farmer. Much of the dryland slopes which were walked on the north-eastern sides of the depression have been affected in large areas by the extraction of sand, as are small areas immediately adjacent to the wetland on the eastern and north-eastern side. The sandy margins of the south-western side and the upper slopes of the north-eastern side have not been affected by sand extraction.

East-central group: Holland Moss, White Moss, Bickerstaffe (east) Moss

The fieldwalking survey of this group of mosses was more limited than in other areas in its treatment of the dryland margins, which would topographically seem to have been the most attractive for early settlement. The eastern and northern sides of this group are obscured by urban development. The M58, which was taken as the northern limit to the survey area, also approximately marks the limit of the peat. Insufficient time precluded comprehensive survey of the slopes of the eastern ridge of the valley which falls down to Rainford Brook. A further restriction on the recovery of archaeology was the extensive stripping of the sand in this area, so that even where the margins were walked there is uncertainty as to how full an archaeological pattern has been retrieved.

The peat was generally found to be deep and in a good state of preservation. The best was in the uncut area of Holland Moss, where across an area of *c*30ha, the peat surface presents a broad dome-shaped topography, mirroring undulations in the underlying mineral ground. The general depth of the peat in the farmed areas is between *c*1.5-2m, although it deepens at the northern end of White Moss to over 4m in places.

Archaeology and palaeoecology

Early Mesolithic *c*8000–6800 BC

The earliest dates for the inception of organic deposits in this central area, together with the regional pollen record, suggest that prior to *c*7000 BC the landscape was dominated by a light pine, birch, and hazel woodland cover across the geographically predominant sandy areas. Knowsley Park Moss, which lies in a relatively small steep-sided basin, may have been the only wetland in the area, except perhaps for small areas alongside the major river valleys, where local waterlogging may have taken place. At this time therefore adjacent settlement may have been attracted to gentler slopes away from the Knowsley basin's immediate margins. One small site (Me 096; Fig 84) is in such a position, although it is difficult to characterise how early or late within the Mesolithic the site may be from the few blades it has produced.

Although most of the mosses probably had a core of greater age, more general wet conditions affected the basins of the central mosslands at the same time as the development of mixed deciduous forest, at approximately the beginning of the sixth millennium Cal BC. The spread of wetland may have provided more encouragement for human settlement in order to exploit it. The majority of sites and findspots around the central and eastern mosslands (which appear to be mainly of Mesolithic or perhaps Neolithic date) probably reflect this to some extent, as they are found on the sandy soils of the Shirdley Hill Sand series. The onset of mixed deciduous forest may have been delayed in these sandy areas, for up to an estimated half a millennium, in contrast to the adjoining clay areas (Innes and Tomlinson 1991).

The archaeology of the central mosses as a whole, however, is limited (Figs 80, 84). To some extent this is possibly an artifice of the sand extraction around the mossland fringes, but alternatively it may be reflecting real differences in exploitation of these areas in comparison with the coastal region (*see Chapter 7*). The evidence consists of a few flint concentrations and a small number of widely scattered individual finds. While the latter group is mostly difficult to date, the small concentrations of material recovered, which are all found on the western side of the complex, are probably datable to the Mesolithic period, on the basis of the incidence of blades. One piece, from the three that make up the concentration (A 27), a large blade 70mm long, may suggest that the activity at this location is of early Mesolithic date, while the other sites are assumed to be late Mesolithic on the basis of comparison with excavated sites at Ditton and Croxteth Park (Cowell 1991a, in prep b).

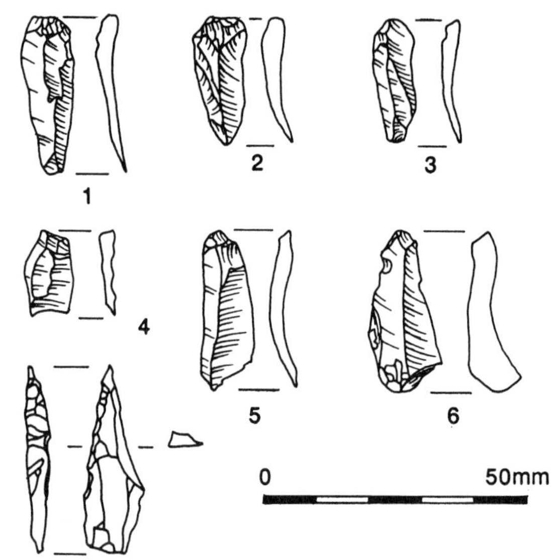

Fig 32 Central mosslands: Mesolithic flintwork from fieldwalking, site A23 Kirkby (1-6); site Me 106 (7), Rainford

Material of probable early Mesolithic date is rare on the eastern side of the Mersey, the only other probable location coming from the eastern mosses at Holiday Moss. One findspot (Me 106; Fig 32.7), from the summit of the sandy ridge between the Holland and Holiday Mosses, suggests Mesolithic activity was taking place in the area, and the piece seems very likely to be early Mesolithic. Careful rewalking of the site on several occasions has not produced more than the one artefact, so it seems doubtful that the site can be of an early Mesolithic type found on the Wirral (Cowell 1992c), or possibly those on the Sefton coast (*see Chapter 4*),which have been interpreted as potential base camps. Although it may be a casual loss, it does point to the fact that these inland aquatic environments were being exploited to some degree for hunting and gathering at these early dates. A possible tranchet axe has also been found recently in Bickerstaffe (A Hallam pers comm) which may further reinforce the pattern. It seems unlikely that this exploitation was on more than a casual basis, as larger sites potentially of this period tend to be concentrated in the major river valleys further to the west (*see Chapter 7*).

In the eastern mosses, at Holiday Moss, a band of clay is found at a level in the peat profile beneath the pine maximum of the late Boreal, *ie c*6300 BC. This has been interpreted as a possible woodland clearance phase resulting in increased erosion and runoff of the topsoil (Baxter and Taylor 1981). The scarcity of archaeological material in the area, however, is at odds with this notion, and it is questionable whether clearance as such was likely in the lightly wooded environments at the time.

Early Mesolithic sites are found in the Pennines 25km to the east, and are probably located in the Alt Valley *c*18km to the west. There is no evidence from the flint raw material that, as on the eastern side of the Pennines (Jacobi 1978), the same communities are involved in the exploitation of both the uplands and the lowland environments of southern Lancashire. In the early Mesolithic, hunting strategies may have been organised around specific favourable locations which were used to 'intercept' prey, as opposed to the more wide ranging, small-task, group-based strategy put forward for the late Mesolithic (Myers 1987, 145).

There is little archaeological evidence in the central mossland area to support the idea of such activity in the relatively light pine, birch, and hazel woodland cover of the area during this early period. The major landuse may therefore have been concentrated on specialist tasks associated with a few people only. Although the total survey sample from these areas of the central mosslands is small, it does appear that the major valleys and the low sandy ridges in the north of the county were the main corridors for any links with the uplands to the east, rather than through the more varied landscape further south. The main upland wetlands, flanking the Carboniferous ridge, appear not to have been a major focus for hunting activity during this period.

Later Mesolithic *c*6800 BC–4000 Cal BC

If the pattern of early Mesolithic activity appears highly dispersed, then that of the later Mesolithic seems only slightly more concentrated. By the middle part of the later Mesolithic period (*ie* beginning sixth millennium Cal BC), radiocarbon dating and palaeoecological evidence show that the central and eastern mosses were established as wetlands. Landscapes of fen carr woodland with reedswamp and occasional areas of open water were therefore widespread across the north-central and north-eastern parts of the county. By this time the surrounding landscape was more densely forested than in the early Mesolithic. The mosslands would therefore have provided welcome and extensive breaks in the thick forest. They would also be places where wildfowl and fish might be exploited as well as the woodland animals that would use these areas to browse and for water.

The archaeological evidence, though quite sparse, reveals two possible trends (Figs 80, 84). Along the fringes of the west-central mosslands two main concentrations of struck flint have been identified, at Kirkby (A 23) and Knowsley Park (A 96), with a further probable site on the eastern edge of Knowsley Park Moss (Me 100). Both the former sites are heavily dominated by blades, although they only comprise 17 and 4 pieces respectively. The comparisons with the excavated surface site in the Ditton Valley, near its confluence with the Mersey in the south of the county (Cowell 1991a), suggests that small blade scatters may point to well defined task sites. Although the flint material from the other site in Knowsley Park (Me 100) is small and less homogeneous in make-up, it includes a blade and a few waste scraps and flakes, which may reflect a similar type of occupation site, albeit less visible on the surface.

The little archaeological material that is found around the western edges of the west-central mosslands dies out almost completely around the central and eastern mosses. The pattern here is one of a few, single, struck flint pieces without any evidence of flint concentration sites. This pattern is also found around the west-central mosses in addition to the flint clusters. This lack of material may partly be a result of sand quarrying, which is particularly noticeable in the east, such as along either side of the Rainford bypass which runs along the valley of the Rainford Brook. The quarrying, however, is evident around the whole mossland complex, so a more homogeneous pattern might be expected if it was the sole reason for the bias. A few areas around the northern limits of the central mossland complex were not walked, mainly due to problems of access, but it seems unlikely that these could drastically change the current sparse pattern of archaeological material.

In contrast to the eastern and central mosses, typical locations for flint material in the west-central area are low sand ridges. In the former region (where such ridges either define the wetland or are visible in the peat surface margins) careful walking of unquarried areas produced little evidence of prehistoric material, either dispersed or clustered. The nature of the sparse and highly scattered

Table 11 Central mosslands: Simonswood Moss B, palaeoecological summary

cms	Date (Cal BC)	Local Pollen Zones	Ecology of Moss	Regeneration/Disturbance Phases	
0			Drainage of moss in historic times.	**SMB-11**	AP 20% TP increasing clearance. Cereal type pollen *c*5% with arable and grassland weeds.
35		Tree pollen 50% total pollen.	Very wet acidic raised mire with cotton grass and *Shagnum*.	**SMB-10**	Oak/ alder woodland with ash and birch partly restored. Some open ground with possible evidence of arable activity at one level.
85		Zone characterised by oak, alder, birch, and hazel.		**SMB-9**	Major evidence of forest clearance. Open vegetation. Probable mixed farming. Cereals, *Cannabis,* and both arable and ruderal weeds recorded.
115	790-257			**SMB-8**	Forest regeneration, mainly oak. No human activity.
145		Elm, ash, and lime low throughout zone.	Birch and heather increases as fen carr dries out.	**SMB-7**	More intensive land use. AP falls sharply. Cereal type pollen recorded possible arable and pastoral activity.
165	1154-780	Heather fluctuates although important for major part of zone. **SMB-a**		**SMB-6**	Sustained evidence of low intensity clearance. *P. lanceolata* consistently recorded. Herb pollen suggests pastoral landuse with fire used to clear forest.
225			Alder carr	**SMB-5**	Dense damp forest. Heliophyte shrubs disappear. Herb pollen wetland types.
265	1950-1450	Grasses and sedge pollen low.		**SMB-4**	Limited forest clearance as in Phase 2. Oak falls, ash, and lime rise.
295	2600-1829	Herbaceous pollen increases towards top of zone.		**SMB-3**	Tree regeneration. Heather and birch on mire surface.
325				**SMB-2**	Some evidence of small, brief clearings. Alder falls, heather rises with a little weed pollen.
345	3023-2280	*Sphagnum* present throughout.	Reedswamp	**SMB-1**	Dense oak / alder woodland. Some light-responsive shrubs recorded.
390					

distribution around the eastern side of the central mosses may result from the geographical nature of the area. To the south-east of the main Simonswood/Kirkby Moss complex, the surrounding dryland consists of boulder clay, either at the surface or with only a thin capping of coversand. This mainly manifests itself in low, discontinuous ridges around the edge of the mosslands. The exception is the Mossborough Hall ridge, which is a dominant feature of the central part of the whole moss complex, approximately dividing the central mosses into two main areas. Access to the former wetlands might have been hindered where patches of heavier woodland lay on the clayier subsoils. To the north-east the wetland basins appear from the ditch sections to be deeper than anywhere else, often 5-7m. Such deep basins may, in the early stages of wetland formation, have provided relatively steep sided hollows, which, even where adjacent to sandy subsoils, may have been less easily exploited than the shallower western basins with their hummocky pre-Flandrian surfaces. Other attractions of the western group of mosslands may have lain in a mosaic of lighter woodland, sandy soils, and river valleys running to the coast. These western mosses on the whole also have shallower gradients between the changing dryland/wetland interface, so facilitating wetland edge exploitation.

Of the later Mesolithic occupation sites, only Me 100, in Knowsley Park, is positioned next to the present edge of the moss. A flint implement (Me 101), of later prehistoric date, found *c*125m to the south, lay *c*20m within the peat. The other sites in the west-central mossland area are located some way off the present wetland edge, on average at distances of *c*400-500m. They are positioned on small, almost imperceptible eminences which form the highest localities in their immediate vicinity. The lateral extent of these peat complexes formerly extended much further (the present extent of the west-central mossland area is known through documentary evidence to be reduced in size (LRO, DDM 14\3, 1696 Survey) and has contracted in modern times due to the wasting of the peat after drainage, turf cutting, and agricultural use. This means that the original relationship of the archaeology with its contemporary wetland environments is obscured. It is possible that sites were once closer to the edge of the former wetland. These Mesolithic sites may therefore have been positioned a little way from the contemporary wetland edge in order to exploit it. If so, this would seem to have been at the expense of ready-made clearings along the wetland edge, but this might have been compensated for by the human presence in the woods not preventing access to the water's edge for animals, thus allowing a reasonably predictable source of game for hunting. Other sites, positioned nearer the contemporary Mesolithic wetland edge, may be hidden by the subsequent spread of the wetland, although none have been found in the several hundred metre band which separates the presently known sites from the current mossland edge.

The sites are too small and few in number to provide much insight into the type of exploitation that was taking place in these environments, but the evidence from similar, excavated sites elsewhere in the county may act as a guide. One, only *c*3km to the west at Croxteth Park (Fig 81), is located in a river valley whose floor, from map evidence, contained areas of peat. It was identifiable by a very small, highly dispersed concentration of struck flint (Cowell in prep b). This evidence would suggest that the mossland sites could be temporary locations where industrial activity (*ie* raw material procurement and initial preparation) was not an important factor, but rather the maintenance of tools and blanks as well as limited processing activities (scrapers, cutting flakes) were more dominant. Croxteth Park also suggests that even single figure surface flint distributions can represent several visits to a site.

When trying to interpret the distribution of flint material in this area, the biases produced by the normal problem of site visibility and the local, more special, case of bias caused by sand extraction must be taken into account. However, the absence of widespread concentrations of flint material in the landscape around the central mosslands might suggest that small, specialist groups making repeated visits to a restricted series of locations may have been the norm. The occasional site, such as the blade concentrations around the west-central mosses (A 23, Me 096), could thus represent slightly more intensive occupations.

The palaeoecological evidence provides some possible clues to the type of activity that might be represented by this evidence. Around Knowsley Park Moss, the sandy soils carried a vegetation of mixed deciduous forest of oak, elm, and hazel, with pine and birch frequent. During the sixth millennium Cal BC the local woodlands started to be invaded by alder, so that oak/alder woodland became dominant (Table 13). At this time the woodland cover around the site was quite dense, but in the centuries before *c*4460-4044 Cal BC (5440±80 BP; GU-5239) two small disturbances to the forest are noted. One is suggestive of open ground (Table 13, phase 2); in the other, there is no real effect on the overall woodland cover, but indicator weed types suggesting disturbed ground are present (Table 13, phase 4)). At Simonswood Moss C, at a similar date,

Table 12 Central mosslands: Simonswood Moss C, palaeoecological summary

cms	Date (Cal BC)	Local Pollen Zones	Ecology of Moss	Regeneration/Disturbance Phases
383				
		Oak, elm, and alder major tree taxa with pine and birch.	Raised mire with *Sphagnum*.	**SMC-3** Woodland returns to former levels.
386				
	4669-3970	**SMC-a** Hazel and heather high.	Charcoal band	**SMC-2** AP (mainly oak) falls to 25% TP. Pollen from plants indicative of open conditions and fire recorded with increase in grasses.
388				
			Fen carr	
		Herbaceous pollen of all types represented.	Shallow lake	**SMC-1** AP c 40% TP. Open woodland may be caused by earlier disturbance or natural processes.
390			Reedswamp	

*c*4669-3970 Cal BC (5440±160 BP; Birm-1222), oak is reduced sharply and weed pollen is high. In this case it is associated with a 20mm thick band of charcoal within the peat (Table 12, phase 2). A charcoal band is also recorded in a lowland context outside the survey area in West Lancashire, at Hoscar Moss (Cundill 1981), 10km north-east of Simonswood.

Around the eastern mosses there is similar palaeoecological evidence to suggest that changes occurred to the vegetation, although these are dated crudely by pollen assemblages to the late Mesolithic. At Holland Moss a phase of high grass pollen values, with pastoral weeds and *Artemisia* and *Rumex* also common, is suggested as being of mid-Atlantic date, *c*5000 Cal BC by the pollen assemblage (Baxter and Taylor 1981). This type of activity may have been taking place within the context of quite widespread burning across the whole central and eastern moss complex. The NWWS stratigraphic survey identified the presence of carbonised material within the peat, which was particularly strong near its base, in White Moss, Holland Moss, Simonswood Moss, and Knowsley Park Moss. This may be interpreted as representing similar occurrences of burning of the surrounding woodland, as is found at Simonswood Moss C. This burning of woodland is a recognised feature in some of the Pennine uplands, where it occurs in the same locations as Mesolithic flint scatters. It is interpreted as clearance at the woodland margins to encourage the growth of new shoots in the burnt-out woodland, to improve woodland browse and so enhance subsequent years' hunting opportunities (Mellars 1974). There is perhaps an intriguing possibility, therefore, that the evidence at Simonswood could represent the same phenomenon in the lowlands.

However, there is little material archaeological indication of this activity other than the few potential occupation sites. Most of the archaeological material from around both the western and eastern side of the mosses comprises scattered pieces of struck flint, and though it is tempting to link this with the phases recognised in the palynological record from Knowsley and Simonswood Mosses (where woodland cover remains constant but weeds of open or disturbed ground are present), this artefactual evidence is unlikely to be associated with human

Table 13 Central mosslands: Knowsley Park Moss, palaeoecological summary

cms	Date (Cal BC)	Local Pollen Zones	Ecology of Moss	Regeneration/Disturbance Phases
		KP-n Pine and birch characteristic with high oak, alder, and hazel. NAP values rise.		**KP-21** Very intensive clearance. High values of cereal, ruderal, and arable weed pollen recorded. Modern plantation on top?
30				
	240-440	Oak and alder main pollen types.		**KP-20** Major clearance, increasing pastoral and arable weeds. Cereal-type pollen and *Cannabis*.
50			Omobrotrophic mire communities dominated by cotton grass developed after mid-Flandrian II.	
65				**KP-19** Regeneration to open woodland with ash and hazel.
				KP-18 Significant clearance of limited extent.
70	930-790	Fluctuating values of birch and heather.	Some evidence for variations in the wetness of the mire surface.	**KP-17** Woodland regeneration.
95	1258-900	Low elm, lime, and ash.		**KP-16** Brief, intense reduction of woodland (oak). Peaks of bracken and ruderals.
				KP-15 Regeneration to more open woodland.
120	2028-1670	Ruderal herb pollen increases with fluctuating peaks.		**KP-14** Temporary, large-scale reduction in alder. Ruderal weeds recorded and light demanding trees/shrubs increase.
130				**KP-13** Regeneration of closed canopy woodland. May be incomplete.
140		**KP-m**		
				KP-12 Significant forest recession (oak, alder). Peak of heather pollen and ruderal weeds recorded.
145		Bracken, sedges, and *Sphagnum* prominent.		**KP-11** Woodland regeneration. Heather increases on mire surface. Possible drier phase.
180	3325-2788			**KP-10** Minor disturbance with some weeds and bracken. No real increase in light demanding trees/shrubs.
185				
				KP-9 Sucessful regeneration of closed woodland. Elm does not recover.
230	3510-3100	Rise of birch and pine marks upper boundary of zone.		**KP-8** Small-scale clearance (oak and elm). Birch rises. Peaks of ruderal weeds and bracken. Secondary Elm Decline.
240				
				KP-7 Regeneration of secondary woodland. Elm recovers. Probably some open ground persisting.
300				**KP-6** Significant disturbance. Ruderal weeds, bracken and ash recorded. Major expansion of heather.
310	4340-3970	**KP-l** Oak, alder, and hazel major pollen types. Elm and lime lower. Birch and heather low. *Sphagnum* significant. Elm Decline marks upper boundary.		**KP-5** Unbroken mixed woodland.
340	4460-4044			**KP-4** Limited small disturbance. Some ruderal weeds but little effect on woodland.
		KP-k Oak, alder, and hazel major components. Low herb pollen, but peaks of *Sphagnum*. Top of zone marked by start of continuous curve of lime pollen.		**KP-3** Woodland with possibly less dense canopy.
410	4998-4300			**KP-2** Brief period of more open conditions with some plants suggesting limited clearance.
445	5322-4906	**KP-j** Oak and alder present with pine, hazel, birch, and elm. Oak rises at top of zone. Herb pollen low. Heather well represented.	Fen carr	**KP-1** Natural woodland succession. Very little open ground.
480			Reedswamp	
520	6690-6410	**KP-i** Main pollen types oak and pine with hazel, elm, birch, alder, and heather. Alder expansion marks upper boundary.	Lake mud	

activity of such intensity to warrant woodland clearance.

Neolithic/early Bronze Age *c*4000–1650 Cal BC

In this period the evidence becomes a little more restricted in scope, and difficult to date, so only general trends across this large cultural span can be identified. Palaeoecological evidence suggests that in the north-east relatively dry conditions obtained in the wetlands during the Neolithic or early Bronze Age. The archaeological evidence from around these areas is totally restricted to single findspots, either from fieldwalking or from chance finds prior to it, and only comprises about five artefacts in all. The flint retrieval pattern for this period is slight in the extreme, while it is very difficult generally to attribute single flint pieces to a particular period unless they are distinctive implement types. For the present purpose all material that does not display Mesolithic traits has been assigned to this period, while only a Neolithic flint axe (A 29) can be adequately dated. This find came to light during sand stripping in the 1970s, suggesting that other finds from the topsoil could still be recovered from walking after its reconstitution. The area was walked in detail at the time but nothing else was found, and any cultural levels undisturbed by ploughing have been quarried away. A second Neolithic axe, from the slopes of the Carboniferous ridge near Billinge, 2km south of Holiday Moss (just outside the area of Fig 84), suggests that the wooded slopes of the major upland ridge in the area were settled to some extent during the fourth or early third millennium Cal BC.

The distribution and character of the archaeological material echoes that of the Mesolithic, and together with the palaeoecological evidence suggests that exploitation of the mosslands changed little throughout this period. It argues for no more than very light episodic exploitation of the areas surrounding the mosses, presumably based heavily on hunting. There is a general lack of arrowheads of this period, however, with the exception of a late Neolithic example from west of Knowsley Park Moss (A 31).

After the elm decline, which at Knowsley Park is dated shortly before 4340-3970 Cal BC (5290±80 BP; Birm-1191) (Table 13, phase 5/6), the palaeoecological evidence shows little change from the earlier period for the western group of mosses. Similar short disturbance phases are recognised by the appearance of ruderal weeds and bracken as at Knowsley Park (Table 13, phases 8 and 10), although the clearance immediately after the elm decline appears to have been slightly more intensive than those before or after it (Table 13, phase 6). The increase of ash at this level may suggest burning had taken place in the vicinity. On each occasion these episodes are followed by woodland regeneration. A similar pattern is seen at Simonswood Moss B (Table 11). The evidence for this site only covers the period from the late Neolithic onwards, but two similar phases are seen between 3023-2280 Cal BC (4090±140 BP; Birm-1217) and, towards the end of the early Bronze Age, 1950-1450 Cal BC (3390±140 BP; Birm-1219) (Table 11, phases 2 and 4). The latter disturbances make little impression on the woodland and are mainly recognised by the appearance of grasses and bracken.

The sparse archaeological evidence is insufficient to support the notion of artificial woodland clearance at this time. One possible interpretation is that Mesolithic clearance techniques may still have been in operation. Other disturbance phases, such as phase 2 at Simonswood Moss B or phase 10 at Knowsley Park, fit the archaeological evidence better, in that they seem to represent little if any change in the woodland cover. If this does reflect human interference it may reflect an 'averaging out' of small, time-constricted episodes in different places over a long period of time. If the various types of clearance and archaeological evidence are associated with farming, then it would seem to reflect small-scale pastoral activity within and around the central mosslands. Woodland clearance (at least of a sustained nature) and cereal farming did not appear to form an important landuse within the area. The one exception from the interior comes from Prescot, 4km to the south (*see Chapter 6*). This lies in a different topographical area, where relatively small mossland basins are located on the edge of the sandstone where it adjoins the boulder clay.

Archaeological evidence for settlements of the late Neolithic/early Bronze Age is mainly confined to coastal areas on the Wirral and Sefton (*see Chapter 7*), and there is some pollen evidence for cereal farming at this time both here and to the south of the central mosslands (*see Chapter 6*). The lack of archaeological sites of the later Neolithic/early Bronze Age around the central mosslands might imply mobile and or seasonal use of the margins of the wetlands in these areas. The wet grasslands around the fens and alder carr prevalent on the west central mosslands through this period, would have been ideal for summer grazing or fodder.

Such grazing areas might be expected to be important for Bronze Age burial grounds (Fleming 1971; Green 1974), and as witnessed in similar environments in the Pennines (Barnes 1982), but despite this no record has ever been made of such sites around these central mossland areas. The nearest such burials are found in the lowlands overlooking the lower Mersey, around Winwick (Chitty 1981).

Late prehistoric period *c*1650 Cal BC – AD 43

The scarcity of archaeological evidence around the central mosslands continues beyond the early Bronze Age, since there are no indications that any of the small amount of non-blade flintwork should date later than this. Pottery, other than Very Coarse Pottery found on several Cheshire sites and dating to after 600 Cal BC (Keen and Hough 1993, 78), is rare on sites in the North West and would not survive long in the ploughsoil. The main evidence for the period is metalwork, which comes only from the dryland areas in the south of the county and in North Wirral (Davey 1976). The recent discovery, through aerial photography, of farmstead enclosures possibly of this date (Philpott 1991), has provided another source of archaeological evidence. In Merseyside these are found away from the wetlands, being located mainly in valleys running through sandstone areas. One site to the east of the central mosslands, however, lies on the edge of Holcroft Moss/Chat Moss in Cheshire (Nevell 1989).

The palynological evidence is at some odds with this archaeological void, since at Simonswood Moss B there are two episodes, one dated 1154-780 Cal BC (2730±100 BP; Birm-1220) (Table 11, phase 7) the other dated 790-257 Cal BC (2380±80 BP; Birm-1221) (Table 11, phase 9), that show tree cover being reduced quite significantly. At the same time weeds seen in earlier disturbance episodes occur, but here in association with cereal pollen and arable weeds. At Knowsley Park, a disturbance episode (Table 13, phase 18) dated to 930-790 Cal BC (2670±50 BP; Birm-1176) shows a major woodland reduction, but with no evidence of the arable cultivation taking place slightly to the north. As with earlier episodes, at both sites regeneration of woodland is complete between the disturbances.

This evidence presents something of an archaeological problem as it seems that, whatever the doubt about some of the earlier disturbance episodes having had a human cause, the presence of arable cultivation in the later episodes points conclusively to associated human activity in the area. The similarities in detail between these late Bronze Age/Iron Age clearances and most of the earlier episodes, save for the presence of cereals and arable weeds, could be used to argue a stronger case for a human association with the earlier episodes. In the late prehistoric period the mossland areas would have been very wet, *Sphagnum*-dominated mires implying that the farming activity would have been restricted to the surrounding dryland.

Without convincing artefactual evidence by which these sites can be identified, the main source for their identification must rest in the first instance on aerial photography. Existing coverage, however, reveals no convincing settlement and/or field systems around the mosses that might be dated to this period. To the south-east, however, the Great Woolden Hall site on the edge of Holcroft/Chat Moss (Nevell 1989) suggests such sites might still remain to be identified in the northern sand areas associated with the wetlands. The preponderance of metalwork and sites actually within the wetlands in the Fens (Pryor 1991) and the South West (Coles and Coles 1986), might suggest similar activity in the central mosslands, but the survey has provided no such evidence.

Romano-British period *c*AD 43–410

The archaeological evidence for this period is slightly stronger in the county, and is continually growing through a programme of aerial survey (Philpott 1994). It consists in the main of pottery scatters and cropmark enclosures, which are found across the sandstone areas in the south of the county and along some of the main valleys in the boulder clay areas. This is particularly so in the east, where the main Roman road runs from Wilderspool to Wigan, and from there to the settlement at Walton-le-Dale on the Ribble. A number of these sites have produced evidence of pre-Roman settlement (Philpott *et al*, 1993; Cowell and Philpott 1994) as have sites in Cheshire (Nevell 1989, 1991). This suggests that the late prehistoric pattern of settlement may be mirroring that of the Roman period to a large extent.

No archaeological evidence for Roman settlement has come from the NWWS field survey or the aerial photography from around the central mosslands. A chance find of a bronze bracelet from close to the edge of the western Bickerstaffe Moss, 1km to the north of Simonswood Moss, has recently been identified at the Liverpool Museum. That this might reflect more than a casual loss in antiquity is suggested by the pollen diagram from Knowsley Park. Here, a similar clearance phase to those seen in the late prehistoric period, associated with arable weeds and cereals, is dated to Cal AD 240-440 (Table 13, phase 20). A possible location for the farmstead associated with this activity, as suggested by place name evidence (Ekwall 1922), might be represented by the small sandstone hill to the west of the moss known as Bury Hill (*burh*, a fortified farm or settlement) (Fig 81). It is covered by thick pine woodland, and the resulting litter surface makes identification of archaeological sites difficult. The case for the identification of a site of this period on such place name evidence might be strengthened by the recent discovery of a rectangular cropmark in arable land, near a settlement, Arbury, with the same *burh* element in the place name, just over the south-eastern boundary of the county in Cheshire (R Philpott pers comm). The Bury Hill site lies almost on the junction of the sand and the boulder clay close to the major stream in this area, while the mossland is only 800m to the north-east. Such a location would seem to be ideal for a farmstead site, utilising a range of resources based on arable land, woodland, pasture, and meadowland, for some of which the moss would have been of value, although not seemingly of primary importance for location purposes.

Another site, the medieval moated farm at Mossborough Hall (Fig 81), on the large boulder clay ridge at the south edge of the west-central mosslands, may be another candidate for a Roman site, from its place name and location. Two farms with *burh* elements, Little Longborough and Big Longborough (Ekwall 1922), are located close together *c*2km away on the southern end of the ridge, midway between Bury Hill and Mossborough, suggesting that more Roman sites may yet be located around the mosslands.

Early medieval period AD 410–1066

While knowledge of the archaeology of the Roman period is slowly being expanded by survey programmes across the county and the immediate surrounding areas, that of the pre-Norman Conquest period remains locked within a few already identified sources, which in the area of the central mosses are chiefly place names. Following the withdrawal of the Roman army from Britain, Ekwall (1922, 231) suggests on linguistic grounds that the area of Lancashire between the Ribble and the Mersey was colonised from the Anglian kingdom of Northumbria, mainly after the battle of Chester in AD 614, when this area became part of the kingdom of Northumbria. In the second quarter of the seventh century it was added to the Mercian kingdom of the West Midlands, to be rapidly returned to Northumbrian control before the mid eighth century. The mid ninth century influx of population of Danish origin was mainly confined to eastern England, and seems not to have penetrated the North West, although the early tenth century Norse influx, by way of Ireland, is recorded as being particularly strong in the North and West (*see Chapter 4*).

New settlements in this part of South Lancashire are therefore likely to have been largely founded during the second phase of Anglian or English settlement, between the seventh and tenth centuries, in a fairly unpopulated landscape of isolated farmsteads. They can be traced from their place names, largely concentrated on the central sandstone and boulder clay areas to the west of the main surviving mosses (Ekwall 1922). The probable increase of population from external sources, and improvements in agricultural technology, probably allowed the first widespread clearance of the woodlands in these areas, although the pattern of landuse indicated in Domesday Book and late medieval documents many centuries later (Farrer and Brownbill 1907) implies that changes were still relatively limited in this early phase of settlement (*see below*).

Settlement during this period must have been sparse, probably based on large estates, with the dense woodland recorded in the pollen diagrams of the later prehistoric period still much in evidence until after the twelfth century (*see below*). The pre-Conquest Norse place name of Kirkby (*Kirkiu-byr*, 'church settlement'; Ekwall 1922) suggests an early nucleated settlement on the eastern side of the late medieval township, where sandstone outcrops within the Shirdley Hill Sand. The church is in fact a chapel dedicated to St Chad, a seventh century saint and Bishop of Lichfield (in Mercia). The dedication and the Norse place name suggest a church was founded here, possibly as early as the seventh century, which subsequently became the focus for Norse influence in the area. The positioning of the village and the presumed arable around it, is some way off the peat areas, suggesting that these were looked upon as marginal to the location of the main settlement.

The large majority of place names in the eastern part of the county, including the mossland areas, are of English origin, a very small proportion are British, and only two can be ascribed with certainty to a Scandinavian source (Cowell 1982; Chitty 1981). The British place names, with the exception of Eccleston (*ecclesia*, church, *tūn*, settlement) (Ekwall 1922), which is likely to have been one of the earliest settlements, lie just to the east of Winwick parish (Fig 33), and form the fringe of a group of British names in and around Newton Hundred (*eg* Ince, Wigan, Bryn, Kenyon, Penketh, Culceth) (*see below*). Ekwall (1922, 225) has suggested that 'in this district a British population was left in undisturbed possession for a comparatively long time. As the district is not separated by natural boundaries from the surrounding areas, the inference is plausible that it was in the old days chiefly an inaccessible forest'. All the evidence points to the fact that elsewhere the area was still very scantily populated, providing ample land for incoming settlers without the need to penetrate the thickly wooded areas to the east.

Later medieval period AD 1066–1500

The area of the central mossland complex has been the subject of an in-depth documentary study of landscape development (Chitty 1981; Cowell 1981, 1982) which allows this period to be quite well understood for the area. At Domesday, within the area between the Ribble and Mersey there were six Hundreds (the units of pre-Norman and medieval administration), which appear to have been treated as large royal manors for the purposes of assessment. The western part of the central mosses was included in West Derby Hundred, where the manors of Knowsley and Kirkby (including Simonswood), held by Uctred, were included in an area with eight vills with 'woodland two leagues long and broad with two eyries of hawks'. In the whole Hundred there was land for four ploughs (Farrer 1906). The area seems to have been fairly evenly settled by the eleventh century, although the actual density of population was probably not high (Farrer 1899).

The east-central mossland areas fell within the Hundred of Warrington, with the probable division between it and Newton Hundred to the east lying along the Black Brook and Sankey Brook, which later marked the boundary between Prescot parish to the west and the parishes of Wigan and Winwick to the east (Fig 33). None of the berewicks held as subsidiary manors is named, nor any details given except for the land held in demesne, that is for the use of the king or lord himself (Farrer 1899), so the earliest historic settlements are not as easy to identify as in West Derby Hundred.

In the modern district of St Helens, covering mainly parts of Warrington and Newton Hundreds, there are eleven ancient townships (Fig 33) which seem to approximate to their medieval predecessors, the vills (Chitty 1981). These units of land are very large when compared with the Wirral, for example, where there are 39 townships in a similar sized area. Wirral was relatively densely populated during the earlier medieval period, and there can never have been the same pressure on the land in St Helens. However, in the course of the medieval period many small estates developed within the vills, sometimes styled 'vill' in their own right, some of which had achieved separate manorial status (*eg* Hardshaw in Windle), but most of which had no special status within the vill or manor. Some of these occur so early that it seems probable that they represent pre-Norman estates (Chitty 1981).

The entries concerning plough teams are insufficiently consistent to assess the extent or distribution of arable land; however, some attempt to calculate the total number of plough teams of the area has been made (Farrer 1899), and while such calculations must be considered with caution, it is clear that by 1086 the area was settled, if only sparsely, and farmed. Farrer's calculated population 'implies a density of about two people per square mile. This density..... must be multiplied by a factor of say four or five, for the actual population' (Terret 1962, 407).

The Domesday Book also indicates that clearance of the woodland during the pre-Conquest period had been limited. In Warrington Hundred no woodland is recorded, although this seems likely to have been an oversight or, more probably, was included with the assessment for Newton, where between 130 and 140 square miles, almost double the area of the Hundred, is entered. It is probable that this was a global figure including uncleared forest extending into the adjacent Hundreds of West Derby, Warrington, and Salford. Nonetheless it emphasises the nature of the land in this area. According to Farrer's very approximate calculations, something less than 7% of the land in Newton Hundred was under cultivation and less than 10% in Warrington Hundred (Farrer 1899, 22-26).

Around the central mosslands it seems that the large areas of moss, heathland, and scrub woodland were a disincentive to supporting a large population dependent on intensive arable farming. The extent of the woodland recorded at Domesday suggests that, apart from pre-Conquest nucleated settlements such as Kirkby or Eccleston, much of the rest of the land remained largely uncleared forest in the eleventh century. This feature is reflected in the development of the landscape in the late medieval period, as seen for example in the place names around the eastern part of the area, most of which are found fairly late in written form in the thirteenth century. Almost half the early names consist of descriptive elements relating to woodland or forest clearance, for instance Windleshaw, Hardshaw, Cranshaw (OE *scaga*, shaw, thicket, grove), Shirdley, Birchley (OE *leah*, tract of open land, clearing), Barrow, and Lunt, deriving from the OE and ON for grove; and more obviously Ashton, Garswood, and Woodfall (Ekwall 1922). With the exception of the first of these, all are located off the mosslands.

In the western half of the central mosslands, the woodland extent probably heavily influenced the distribution of parks and enclosed fields in the late medieval period, which to some extent are likely to be a reflection of the widespread extent of the

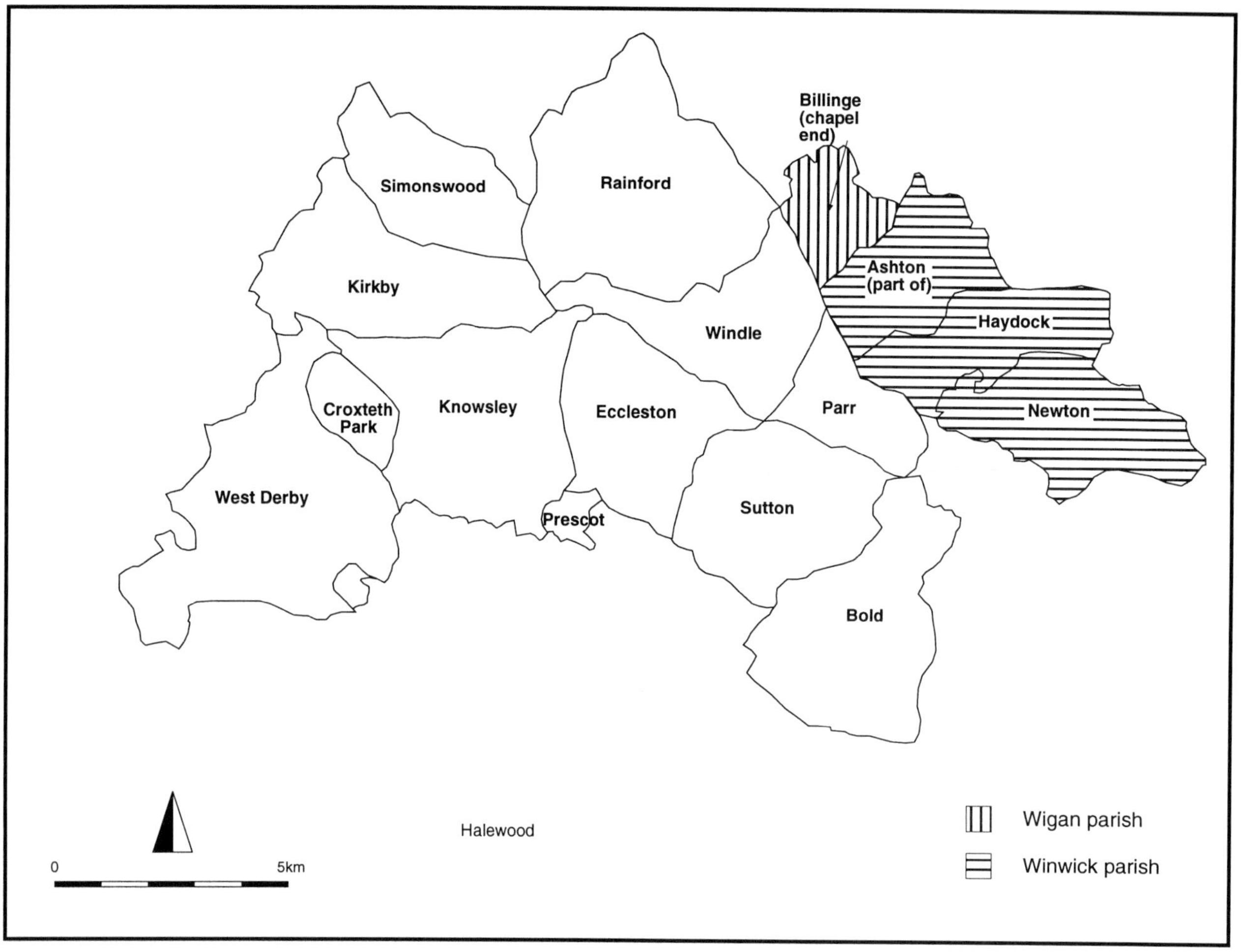

Fig 33 Central mosslands: Townships and parishes

mosslands and the lack of their direct exploitation. One township, West Derby (Fig 33), slightly to the west of the present distribution of mosses, at Domesday formed the chief manor of the Hundred of the same name, and with its six berewicks stretching to Halewood in the south and Liverpool in the west, formed the royal demesne within the Hundred (Farrer 1899, 19), with the site of the castle at the borough of West Derby. Later in the medieval period this demesne land formed the heart of the Royal Forest of Lancaster in this area, and was the main hunting area of the forest, about half of which was still wooded (Farrer 1906, 284). In medieval England the term 'forest' implied an area of land subject to a special legal code and under the direct control of the king through his appointed officers. The object of 'forest law' was the preservation and protection of animals for hunting purposes, and of the woodlands in which they lived (Cunliffe Shaw 1956). Much of the woodland used for hunting in the vill of West Derby appears to have lain in the east of the area, where place names such as Gillmoss and Blackmoor attest to former areas of now disappeared mossland, and the woodland probably also extended into the adjoining Croxteth Park, where the 1st edition OS map shows localised areas of mossland which have now disappeared. In these areas therefore the wet woodlands were prized as hunting areas rather than for farming and settlement.

Little of the area of the surviving mosslands to the east and north of West Derby fell within the bounds of the hunting forest proper, except for Simonswood, which in the early thirteenth century was taken from Richard son of Roger, lord of the vill of Kirkby, by Henry II and emparked to enlarge the Royal Forest (Cunliffe Shaw 1956). Territorial grants of land could be made within these demesne areas with reservation of the hunting rights, although no medieval references have yet been found to suggest medieval clearance in the parks in this area.

Other evidence relating to the wooded nature of the medieval landscape in this area (outlined in

Cowell 1982) comes from a number of private hunting parks, which tend to be most concentrated in the western part of the study area, although areas of grassland were found within the bounds of medieval hunting parks (Cantor 1982). The earliest reference to Knowsley Park is in 1292, when Robert de Lathom claimed to have a 'certain wood that his father enclosed with paling' (Cunliffe Shaw 1956, 122). The boundaries of this wood do not appear to have survived, possibly because of later post-medieval landscaping, but historic map analysis suggests that the mossland areas lay to the north of the medieval park but had become included within the park by the sixteenth century (Cowell 1981, 38-9). In 1337, John of Kirkby was granted a licence to 'impark 500 acres of land, wood and moor within the manor' (PRO, Cal Charter Rolls, 4, 390). Landscape analysis of later maps suggests that this may have lain away from the mossland and nearer the hall to the south-east of the village (Cowell 1982). In general, however, these parks appear to have used the mossland areas for part of their terrain, as has already been seen in the royal demesne parks within the Forest of Lancaster at Simonswood, West Derby, and Croxteth Park. In some townships, which contained actual areas reserved for hunting, the effect on the landscape was restrictive and the medieval enclosures remained well into the post-medieval period until thrown open to cultivation after disemparking. In others, such as West Derby, open fields existed alongside hunting forest in the thirteenth century (Cowell 1981). The townships adjoining the royal hunting forest were included within the area known as the purlieu, which became subject to forest law, without being in the actual hunting area. The Forest Eyre of 1334-6 indicates that the purlieu stretched from Croxteth Park eastwards as far as the vills of Eccleston, and to the south of the main mossland belt, Bold, Rainhill, Sutton, and Parr (Cunliffe Shaw 1956, 13).

Of the mossland townships in the purlieu of the forest only Kirkby, Rainford, and Eccleston appear to have developed an agricultural landscape based on common arable fields, as witnessed by the 'Grant from Hugh of Eccleston of 3 selions of land in Eccleston, to wit, the headland of the furlong which is called Bothum and the furthest selions in the same furlong abutting upon the said headland...' (Farrer 1900, 600). Kirkby was divided into two manors, Kirkby Beetham and Kirkby Gerard, which may have equated with the medieval nucleated settlements at Kirkby and probably at Little Briton. The contrast in field shapes around each of these two settlements may be seen on the 1769 Molyneux estate map (Cowell 1981), with several almost circular fields (delineated by roads) to the west and north of Little Briton and the much larger, more open, area of former common arable to the south and north of Kirkby village. The latter, however, is distant from the surviving peat, although there is some evidence that the peat was more extensive in the medieval period (*see below*). In general this type of open landscape seems to have been restricted, and piecemeal woodland enclosure seems to have been more the norm.

Woodland was one of the most important resources to the medieval community, providing timber for building and fencing, firewood, and grazing at certain times of the year, with each township possessing its own area of wood. The use of woodland provided a source of income to the lord of the manor, and was accordingly subject to strict laws, and woodland was often enclosed, generally by a bank and a ditch (Rackham 1980). The steady increase of population throughout the twelfth to fourteenth centuries in England led to the ever increasing need for more arable land, with woodland and waste cleared and enclosed. A document of 1298 shows that the tenants of the vill of West Derby held over 680 acres of land improved from the moss and moorland waste of the vill. By 1316 the wood of West Derby covered only 300 acres when granted with the manor to Robert de Holand by Thomas of Lancaster (Farrer and Brownbill 1907, 14). In some cases the expansion of arable was achieved by increasing the area of the common arable fields at the expense of the woodland and waste which lay around them (Hoskins 1955, 90). Throughout the thirteenth century the Assize Rolls attest the encroachment onto the common pasture for arable in the area: 'Richard de Wulcrofthead (in Eccleston) accused Robert de Eccleston and others of razing his dyke, so that their cattle entered and destroyed his corn. The defendant alleged that he wished to improve to himself a part of the common pasture of the vill of Woldscroft; whereupon Robert de Eccleston caused this encroachment to be removed' (Farrer and Brownbill 1907, 363). Another process typical of this area, however, was that known as assarting, whereby woodland and waste lying on the more marginal land, generally towards the edges of the township, were cleared to form small fields, enclosed with banks and hedges by individuals who farmed the land in severalty. The individual assarts were generally quite small, often of only an acre or so (Moorhouse 1981, 664). In Kirkby in the hamlet of Ingewaith, now known as Ingoe Lane, an identifiable area of assarts next to Alders Bridge included one of '3 acres, in the same place, (which) lie together within their own bounds' (Farrer and Brownbill 1907). Assarting of woodland or waste seems to have been the common method of agricultural expansion in the mossland townships.

The earliest recorded assart, of 1199, is known as 'Outi's ridding', which the bounds of medieval field names place in the north-eastern part of Knowsley township, where the mossland is located (Cowell 1982). Medieval place names and field names also indicate areas of Knowsley around the mosslands which were formerly woodland, but where a certain amount of clearance had taken place by the twelfth century. In 1190 a place or field called Stockley (wood clearing) (Ekwall 1922) is recorded in the vicinity of Bury Hill (Farrer 1900, 606), and by the fourteenth century it supported a settlement of some kind there. Other thirteenth century field names, such as 'waterhurst' and 'watercarr', probably located in the general area of the mossland in the northern part of the township, are also recorded (Farrer and Brownbill 1907, 158 n8).

Much the same pattern seems to have developed around the townships on the eastern half of the central mosses. The nature of the landscape is well demonstrated in the descriptions of the boundaries around grants of arable land in Windle; 'to wit, Herthefellinge (Earthfelling), by these bounds, on the eastern side of Fell Clough to the deep Moss leach at Windle, below the head of Moss leach wood, through the valley by a straight line to the east again, and so to the boundary of Parr' (Farrer 1900, 608). The descriptions of many of these grants are circumscribed by mosses, riddings, and fellings (clearings), woodland, and streams. By the fourteenth century the continued expansion of agriculture was having its effect on the once apparently inexhaustible woodlands. In 1327 Eccleston wood, referred to in a mid thirteenth century charter (Farrer 1900, 603), was the subject of a dispute over manorial rights between Alan de Eccleston, lord of Eccleston, and his 'charterers and freemen', in default of the estover which they were accustomed to have in the wood of Eccleston which now begins to decay....' (Hawkes 1936, 72). This process gathered pace everywhere, so that in 1507, Simonswood, formerly a hunting park in the royal forest, was described as 'waste ground', 'overgrown with wood of little or no value and watery or moorish and mossy ground with little or no grass on it' (Cunliffe Shaw 1956). In 1508 it was disemparked.

Post-medieval period AD 1500–*c*1850

With the increased political and economic stability of the sixteenth century, agriculture in the area generally continued to expand, and until the nineteenth century the landscape remained predominantly rural. Yates' 1786 map of Lancashire (Plate 5) depicts the final expansion of the late post-medieval rural population just before the full development of industry in the St Helens and Liverpool areas. The contrast is strong between the areas to the south and west of the mosslands on the heavier soils around the pre-Conquest centres of Huyton and Prescot and around St Helens, although its identity as a town is not long established, comprising part of the old manor of Hardshaw in Windle, with parts of the townships of Parr, Sutton, and Eccleston. The scattered nature of settlement is a feature of the remaining undeveloped rural areas today, and in some places there has been a noticeable shrinkage in settlement as the populations were drawn to the industrial towns during the nineteenth century. The road from Prescot to Warrington, for example, has a string of settlement along its length in Yates' map, in many places where none exists today.

Early clearance and nucleated settlement is represented around the outer fringes of the mosslands by Kirkby, West Derby, Eccleston, and to a lesser extent, Rainford. Within this belt lies a pattern of sparse and scattered settlement, which, with the exception of the moated site of Mossborough Hall, appears to be mainly post-medieval in date. Medieval settlements associated with assarting around the mosslands have not been located through fieldwalking. Medieval nucleated settlements never grew up in Simonswood (Cowell 1981) or Croxteth Park, and only in the late eighteenth century in Knowsley (Cowell 1982), while most of the large township of Rainford lay far from its nucleus.

The economic role of the mosslands within the townships into the nineteenth century was as common land which lay on the margins. This had been an important feature of the medieval agricultural system, with common meadow and grazing each forming part of an integrated method of managing the livestock of the local community. Each township possessed its own area of common grazing, generally located on the poorest quality land, often on the edge of the township, on which each inhabitant could graze his livestock during the summer months (Aston 1985). Eighteenth and nineteenth century maps (Merseyside SMR) show that all the townships in the central mosslands used these areas of moss as common grazing land (Fig 34). There are no medieval references to these areas being common, but it seems likely that the commons shown in the post-medieval period on the mosses continue the pattern from the earlier period. In Knowsley a mid eighteenth century estate map (Merseyside SMR) shows that the mossland here, although in the park, was divided up into strips which were the basis of common land. There are several references to the moss being used for fuel as well as grazing during the post-medieval period, from bailiffs' accounts of the late sixteenth century (Farrer and Brownbill 1907).

Plate 5 South Lancashire: Yates' Map of Lancashire, 1786 (Photo R Cowell)

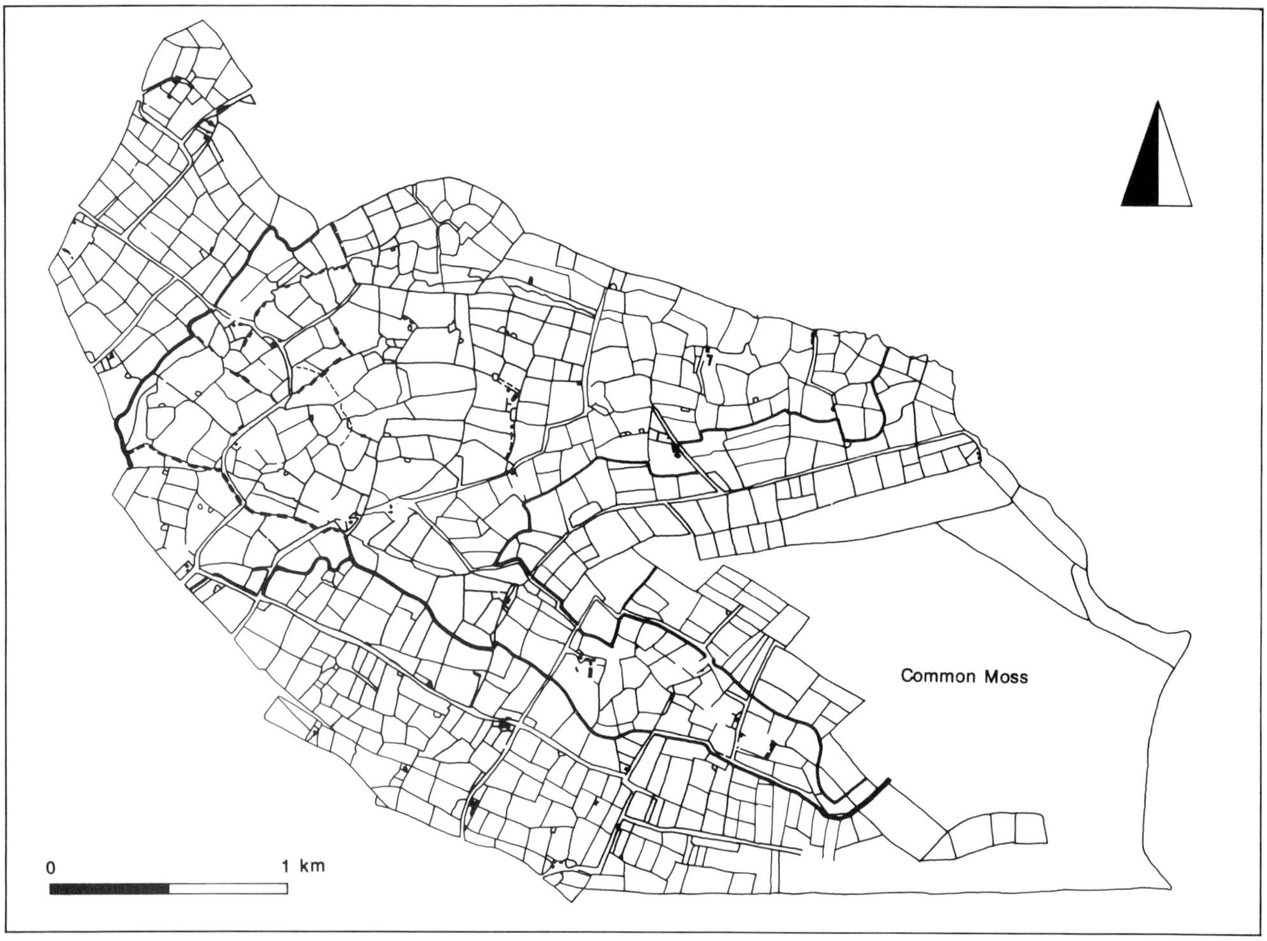

Fig 34 Central mosslands: Simonswood township, Molyneux estate map, 1769

During the fifteenth century and throughout the sixteenth century, there is considerable evidence in south-west Lancashire for the increasing importance of grassland in the agricultural economy. Chitty (1981) shows that in summaries of landholdings the proportion of pasture and meadow increases significantly, which does not appear to result from the enclosure of arable fields, since enclosure appears to have been thoroughly established by this date, nor can it be attributed to the conversion of arable land to pasture, since the acreages of arable land remain constant or even increase. It thus represents the expansion of agricultural activity onto the hitherto unexploited marginal heath and probably to a lesser extent the mossland. Quoting Walker (1939, 39-40, 55-57) she illustrates this convincingly by means of the global figures for arable, pasture, meadow, wood, and mossland given in the final concords for the county. Large areas of underused land of poorly drained or inferior quality were available for grazing and led to the rise of a mixed agricultural economy in which cattle played a prominent role, bringing a new prosperity to the area.

The movement of population onto the moss and heathland is to a large extent responsible for the pattern of isolated farmsteads which characterises the area today. The rents for this type of land were low, but it offered the opportunity to the less prosperous labourers and husbandmen to establish themselves in an independent holding. For example, from the Scarisbrick leases in Eccleston in 1663 'a messuage ... in Eccleston being newly erected ... with two acres of ground of large measure not yet enclosed but lying abroade upon the common in Eccleston called catchall moss' was rented at 2s per year, while in 1659 a tenancy of better arable land, a messuage and a further, indecipherable number of acres, averaged a yearly rent of 12s 6d (LRO, DDSc 12/19 and 26). In a survey of 1769 (LRO, DDM 14/48), 14 ancient tenements are listed in Simonswood which, on the whole, occupy the central belt of the township. The fields associated with these tenements are largely contained within long continuous curving boundaries, indicated by a thicker line on Figure 34. This area of fields corresponds to a belt of sand and boulder clay which lies to the north and east of the mossland and represents the earliest phase of colonisation of the waste, most of which probably dates after Simonswood was disemparked in 1508 (Cunliffe Shaw 1956). In 1568 rents were due from nine tenements, although only five family names are represented. The enclosure of mossland and the development of settlement in these areas can be documented from the sixteenth century, but it appears to have developed on a large scale from the mid seventeenth century and continued into the early nineteenth century. In Kirkby the Molyneux estate map shows that enclosure around the edges of the common moss had already taken place by 1769 (LRO, DDM 14/33), recognised by the small regular enclosures being added to existing farms set back onto the dryland. In the south-east of the township along Cut Lane, near the junction of the Shirdley Hill Sand and mossland, documents show that most of these settlements originated towards the end of the seventeenth century, except for the occasional farm that is early seventeenth century or earlier (Cowell 1981). The later, probably eighteenth century enclosure can be seen inside the old established continuous border of the common moss.

Around the eastern mosses the seventeenth century saw substantial stone farms, for instance at Reeds Brow, Maggots Nook, Guild Hall, and Hydes Brow, being constructed around the mossland edges in response to the new prosperity that yeoman farmers were also enjoying (Chitty 1981). Large areas of land still remained unimproved through this period, however, and this holding on the eastern side of the mosses in 1588 is typical of the division of land use: '5 messuages, a cottage, a dovecote, 6 gardens, 200a land (arable), 40a meadow, 100a pasture, 10a wood, 200a moor, 200a furze, and heath and a mine of coals in Sutton, Windhyll and Eccleston' (Farrer 1910, 146). In areas such as Kirkby, Simonswood, Rainford, and Eccleston, large areas of unimproved wasteland remained to a relatively late date. In the Stanley estate in Rainford in 1652 there was '...al yt ye wast of comon mosse and turbary ... belonginge to ye mannor of raineford ... conteyning by estimacon in ye whole 760 acres...' (Stanning 1892, 235). In 1620 the Mossborough estates, held by the Lathom family from outside the manor, contained 380 acres of moor, moss, heath, and briar (Rylands 1887, 205), so that well over 1000 acres remained waste within the township despite the encroachments which had already been made onto the mosses (Stanning 1892, 234). In Simonswood in 1769, the Molyneux estate map (LRO, DDM 14/48) shows 498 acres of common land remaining unenclosed on the moss in the south-eastern part of the township (Fig 34). In Kirkby the same source (LRO, DDM 14/33) shows 380 acres of unenclosed mossland. By the mid nineteenth century, four new farms had established themselves within the formerly unenclosed mossland areas as the process of enclosure continued on the common (LRO, DRL 1/43, Tithe map).

A feature of a number of the small moss edge farms to the north of St Helens is the growth of cottage industries, which mainly took the form of potting and pipemaking, probably undertaken on a seasonal basis (Chitty 1981). Walker (1939), in his analysis of the growth of industry in south-east Lancashire, has suggested that the economic spur to cottage industry was the comparative poverty of agricultural resources and the availability of time within a pastoral or non-intensive agricultural community. This may suggest a reason for the development of cottage industries on many of the poorer smallholdings on the mosslands, where there was a need and opportunity to supplement their incomes from other sources. Documentary sources identify seventeenth century sites around the edges of Rainford Moss, Holiday Moss, Kings Moss, Moss Nook in Rainford, and Catchdale Moss in Eccleston (Chitty 1981). Fieldwalking produced two more probable sites, at Scythe Stone, Delph Farm in Rainford (SD 484022), and near Barrows Farm on the north side of Reeds Moss (SJ 483989), which extend the pattern into a new area. Although there are early documentary references for clay industry sites in Windle, no sites were located from the fieldwalking survey around Windle Moss.

Palaeoecological survey

Knowsley Park Moss (KP) (SJ 455961)

Previous pollen investigations in the Shirdley Hill Sand areas of south-west Lancashire and Merseyside (Tooley 1978a; Cundill 1981; Innes 1986; Innes *et al* 1989) have shown that vegetation development and the history of human impact upon the environment have followed rather different patterns there than in the boulder clay plain of southern Merseyside. The early Flandrian development of vegetation cover in the sand areas seems to have been less rapid and less complete than in areas of glacial till, and human disturbance also seems to have occurred earlier. It is likely that such differences derive from the contrasting soil types formed upon these two parent materials. Such inferences have been drawn, however, from pollen diagrams which cover only short periods of the Flandrian, many of which are poorly dated, or not at all. The extensively radiocarbon dated pollen profile from Red Moss, near Bolton in Lancashire (Hibbert *et al* 1971), provides

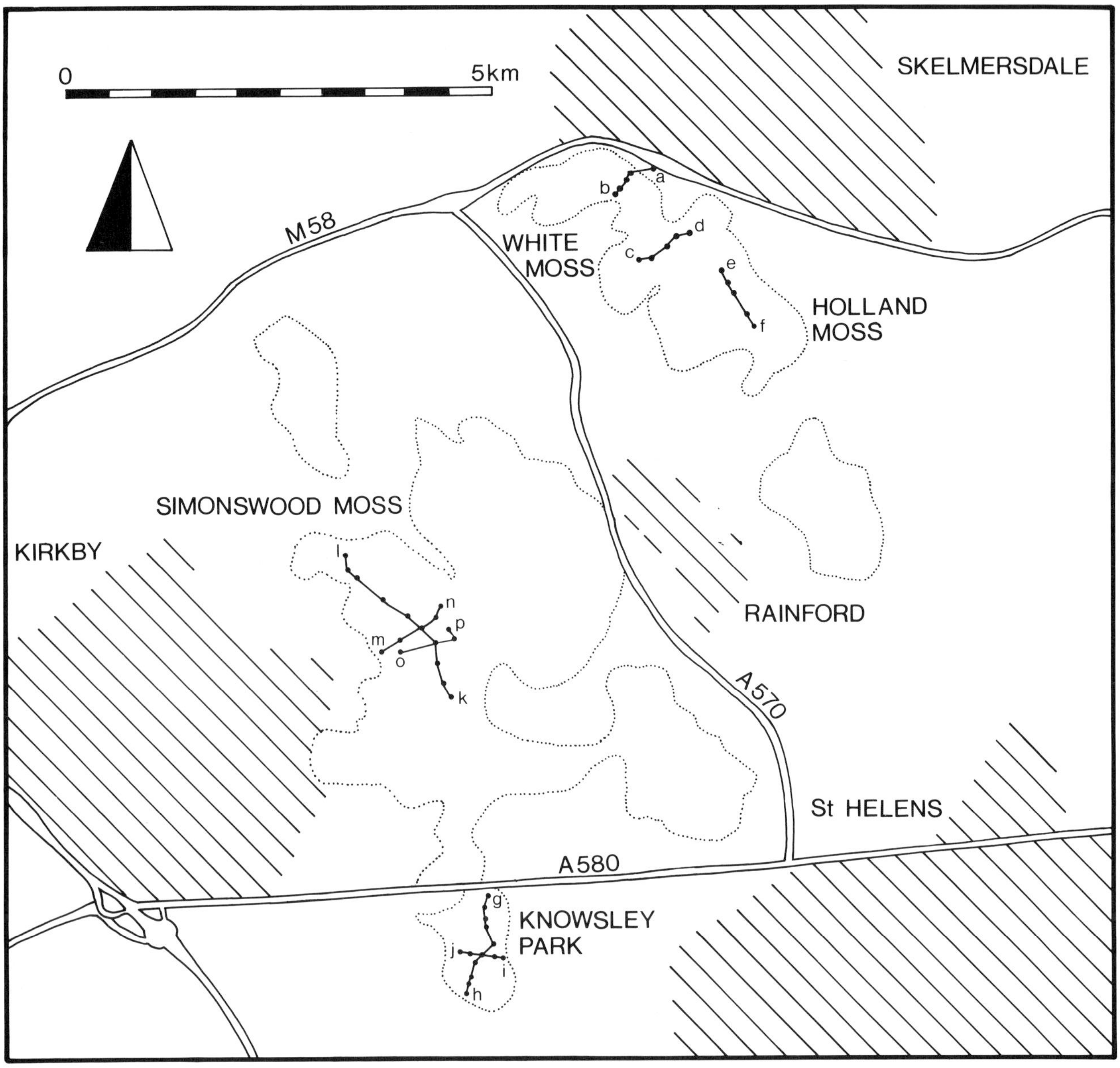

Fig 35 Central mosslands: Location of stratigraphic transects a-b to o-p

a North West regional type site with which the current pollen data may be compared. It is probable, however, that the post-glacial vegetation changes in the Shirdley Hill Sand areas may have been significantly different in timing and character to those at the control site of Red Moss, and to other dated diagrams in the region. A dated long pollen profile from the Shirdley Hill Sand area was therefore considered essential to act as a regional standard for other pollen diagrams from sites on sandy soils.

Knowsley Park Moss was chosen as a possible site for this standard diagram as it has a central location in Merseyside and lies within Knowsley Park. The park is an extensive area of arable land, parkland, and forestry situated on deep peat to the south-east of Kirkby, and is owned by Lord Derby. It has been partly protected to some degree from excessive truncation of the stratigraphy by agriculture and cutting. Little was known of the general stratigraphy apart from the fact that considerable depths of peat (up to 7m) still remained on the site.

General stratigraphy

A north-south transect of thirteen cores, and an east-west transect of five cores, was used to investigate the stratigraphy of the site (Figs 36, 38). The results indicate the presence of two shallow interconnected and elongated basins, with the more northerly one situated at a lower level. This basin contains a more truncated stratigraphy than the southern one and there are subtle differences in the peat types present.

The southern section contains the deepest peats (>6m) and also provided the site for the core used for detailed palynological investigation and radiocarbon dating. The deepest deposits overlie the centre of the southern basin and consist of a uniform succession of *Eriophorum* and *Calluna*-dominated peats overlying basal wood and/or monocotyledon peats. *Polytrichum* peats are also present in some of the basal deposits. The southern edges of the basin contain peats dominated by *Sphagnum* (mainly *S imbricatum*), particularly in the upper section of the profiles. Carbonised plant material is found in many of the profiles and tends to be more common in the lower deposits (Fig 37).

Although generally following a similar sequence to that found in the southern part of the mire, the peats of the northern section tend to have a higher *Sphagnum* content. The other main difference is in the peat depth remaining, the northern basin containing much shallower (*ie* c3m) peat profiles than the southern one. Charcoal is again encountered, particularly at lower stratigraphical levels.

Site-specific stratigraphy

A 7.2m core was retrieved from a point at SJ 455961. The surface of the core was levelled to 62m OD. A coarse brown sand lay below the first organic sediments at 7.2m. On top of this a 0.15m layer of sandy gyttja intervened before a herbaceous peat with *Phragmites* remains succeeded. At 6.8m wood fragments began to occur in the same peat type which continued up to 3.9m. After this point a change to humified *Eriophorum*-dominated peats occurred which were interrupted from 3-3.9m by fresher *Sphagnum* peats. Above 3m *Eriophorum* peats resumed dominance up to 2.45m when they were replaced once more by *Sphagnum* peats. These continued up to 1.9m. Above this *Eriophorum* and monocotyledon peats occurred until replaced at 1.4m by a mixed *Sphagnum/Eriophorum* peat, which continued up to 0.40m when highly humified amorphous peat completed the profile.

Pollen analysis

Samples were taken for pollen analysis at 50mm intervals throughout the organic sediments. Pollen preservation was good and the results are shown as Figures 39 to 42. The pollen stratigraphy has been subdivided into 14 local pollen assemblage zones as follows.

KP-a 7.20-7.06m The basal zone is characterised by *Juniperus*, *Pinus*, *Betula*, and *Salix*. Also important are a wide range of aquatic taxa, particularly *Myriophyllum alterniflorum*, and ruderal herbs such as *Artemisia* and *Rumex*. *Empetrum* also reached significant values. The upper boundary of the zone is defined by a fall in *Juniperus* pollen percentages and is approximately radiocarbon dated (7.06-7.12m) to 7355±65 BC (9305±65 BP; SRR-4515).

KP-b 7.06-6.75m This zone is characterised by *Betula* with lesser frequencies of *Salix*. *Pinus*, *Juniperus*, and *Empetrum* are present in low amounts. Few aquatic or ruderal herb types are recorded. Gramineae and Cyperaceae are present in high frequencies. The end of the zone is defined by a fall in *Salix* pollen, the end of the curves for *Juniperus* and *Empetrum* pollen, and the start of the curve for Coryloid. It is approximately radiocarbon dated (6.76-6.84m) to 7330±80 BC (9280±80 BP; GU-5246).

KP-c 6. 75-6. 60m This zone is characterised by *Betula* with lesser frequencies of Coryloid and *Pinus*. *Salix*, *Juniperus*, and *Empetrum* are almost absent, as are aquatics. Gramineae and Cyperaceae remain high. The end of the zone is defined by a sharp rise in Coryloid frequencies and is approximately radiocarbon dated (6.51-6.59m) to 7210±80 BC (9160±80 BP; GU-5245).

KP-d 6. 60-6.45m This zone is characterised by Coryloid and *Betula*, with minor values for *Pinus* and a little *Empetrum*. *Menyanthes* and *Sphagnum* peaks occur. Gramineae and Cyperaceae frequencies are low in this zone. The end of the zone is defined by the start of the *Ulmus* curve, the first *Quercus* pollen, and a decline in *Betula*. The end of this zone is approximately dated (6.36-6.44m) to 7180±80 BC (9030±80 BP; GU-5238).

KP-e 6.45-6.25m This zone is characterised by Coryloid, *Ulmus*, and *Betula*, with significant frequencies for *Pinus* and *Empetrum*. *Sphagnum* values are high. All other types are very poorly represented. The end of the zone is defined by a sharp rise in *Pinus* and decline in *Betula*, and is approximately radiocarbon dated (6.16-6.24m) to 6930±90 BC (8880±90 BP; GU-5243).

KP-f 6.25-5.90m This zone is characterised by *Pinus*, Coryloid, and *Ulmus*. Continuous curves for *Quercus* and *Calluna* begin. *Betula* falls to moderate values, while Coryloid declines through the zone. Some *Empetrum* occurs and *Sphagnum* values are high. The end of the zone is defined by a rise in *Quercus* frequencies and is approximately radiocarbon dated (5.81-5.89m) to 6700±80 BC (8650±80 BP; GU-5242).

KP-g 5.90-5.60m This zone is characterised by *Pinus*, *Quercus*, and Coryloid. *Betula* and

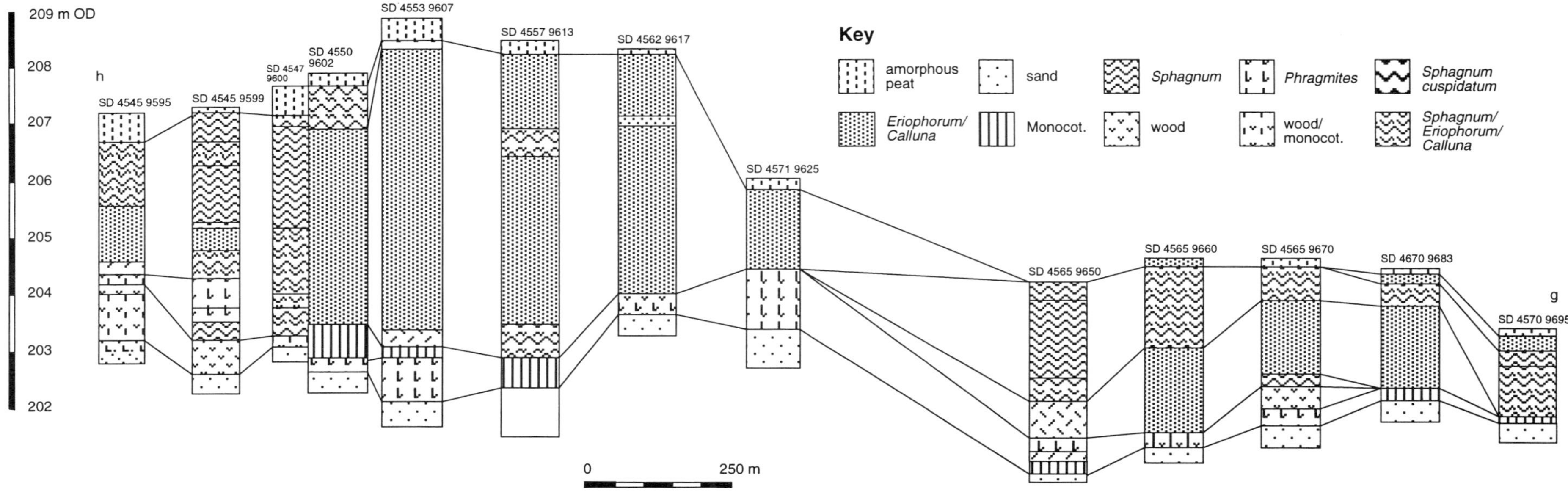

Fig 36 Central mosslands: Knowsley Park south-north stratigraphic transect n-g

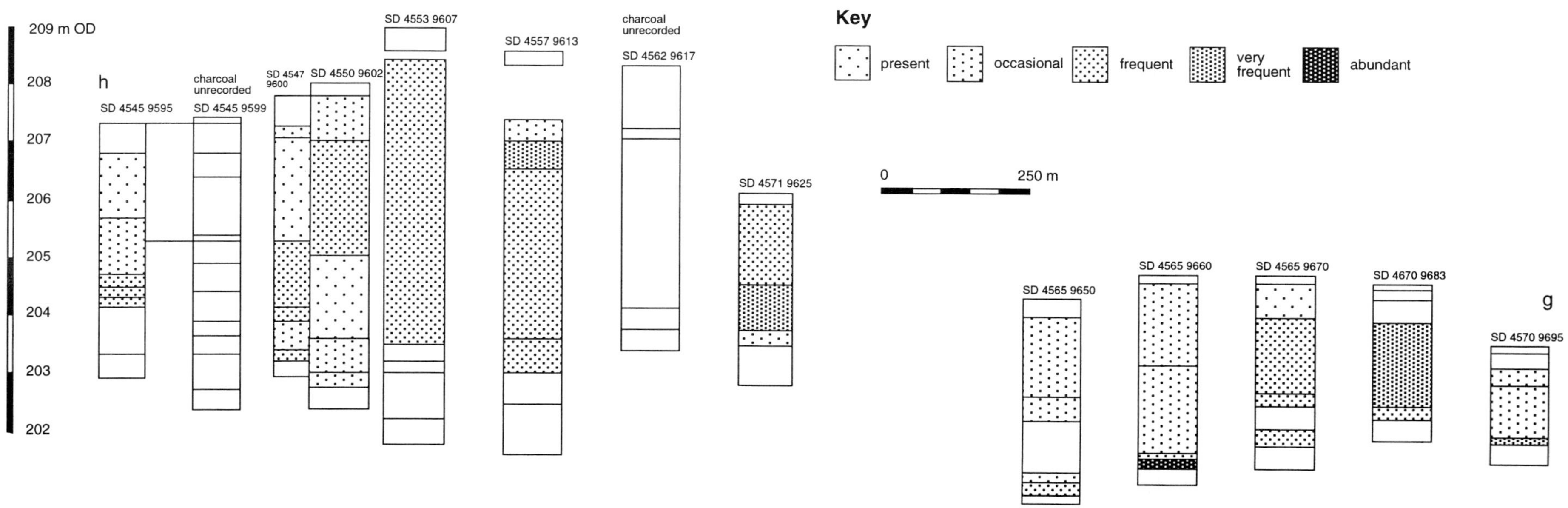

Fig 37 Central mosslands:Knowsley Park south-north stratigraphic transect, carbonised material n-g

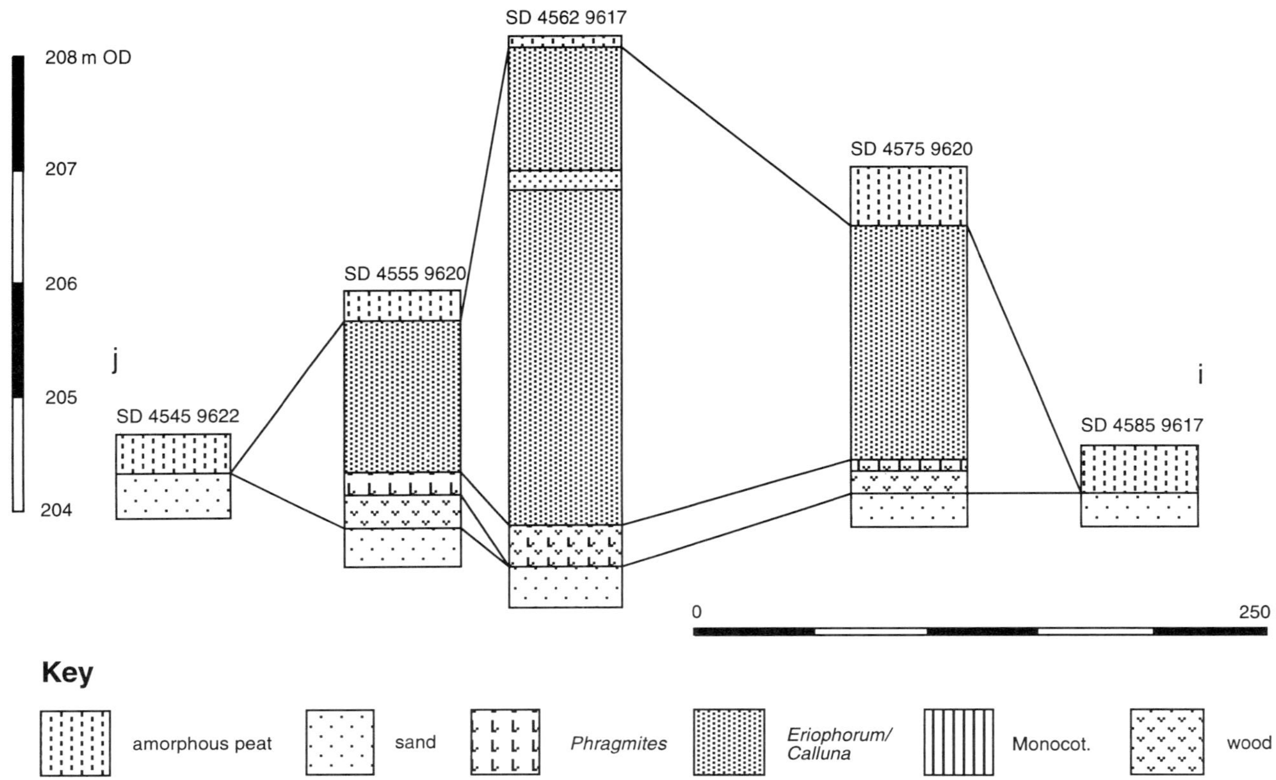

Fig 38 Central mosslands: Knowsley Park west-east stratigraphic transect j-i

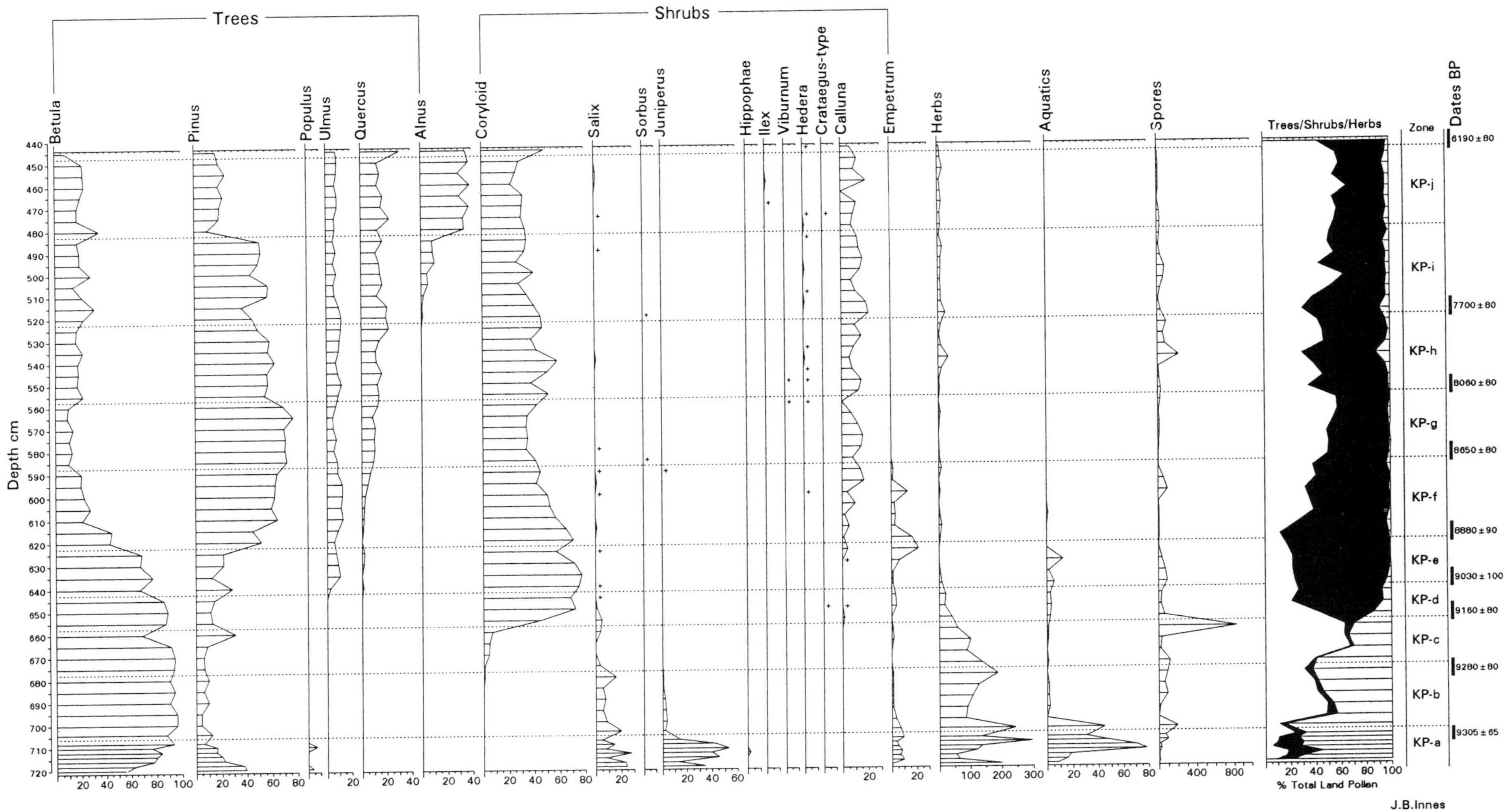

Fig 39 Central mosslands: Knowsley Park Moss, early Flandrian percentage arboreal pollen diagram (trees plus group)

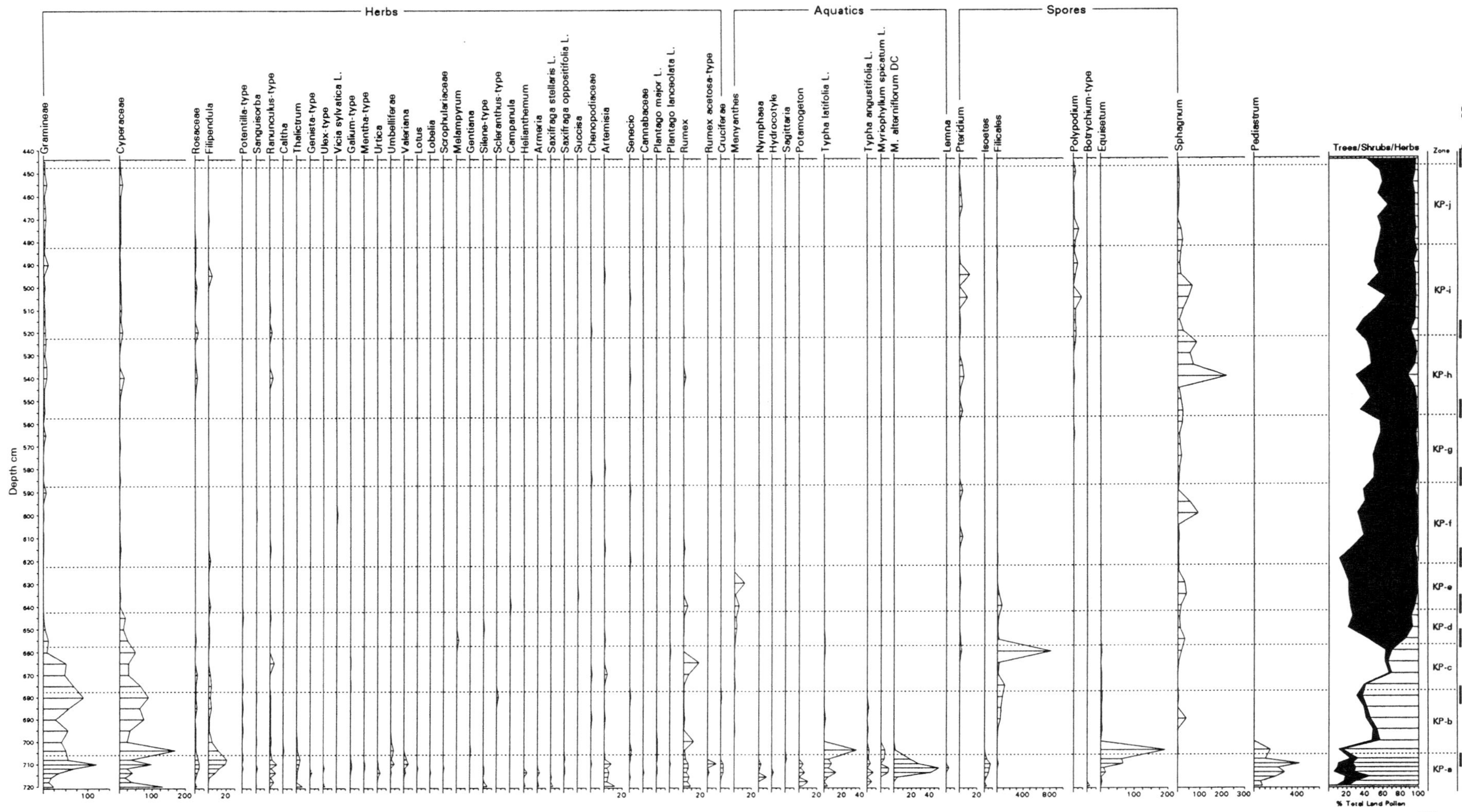

Fig 40 Central mosslands: Knowsley Park Moss, early Flandrian percentage herb pollen diagram (trees plus group)

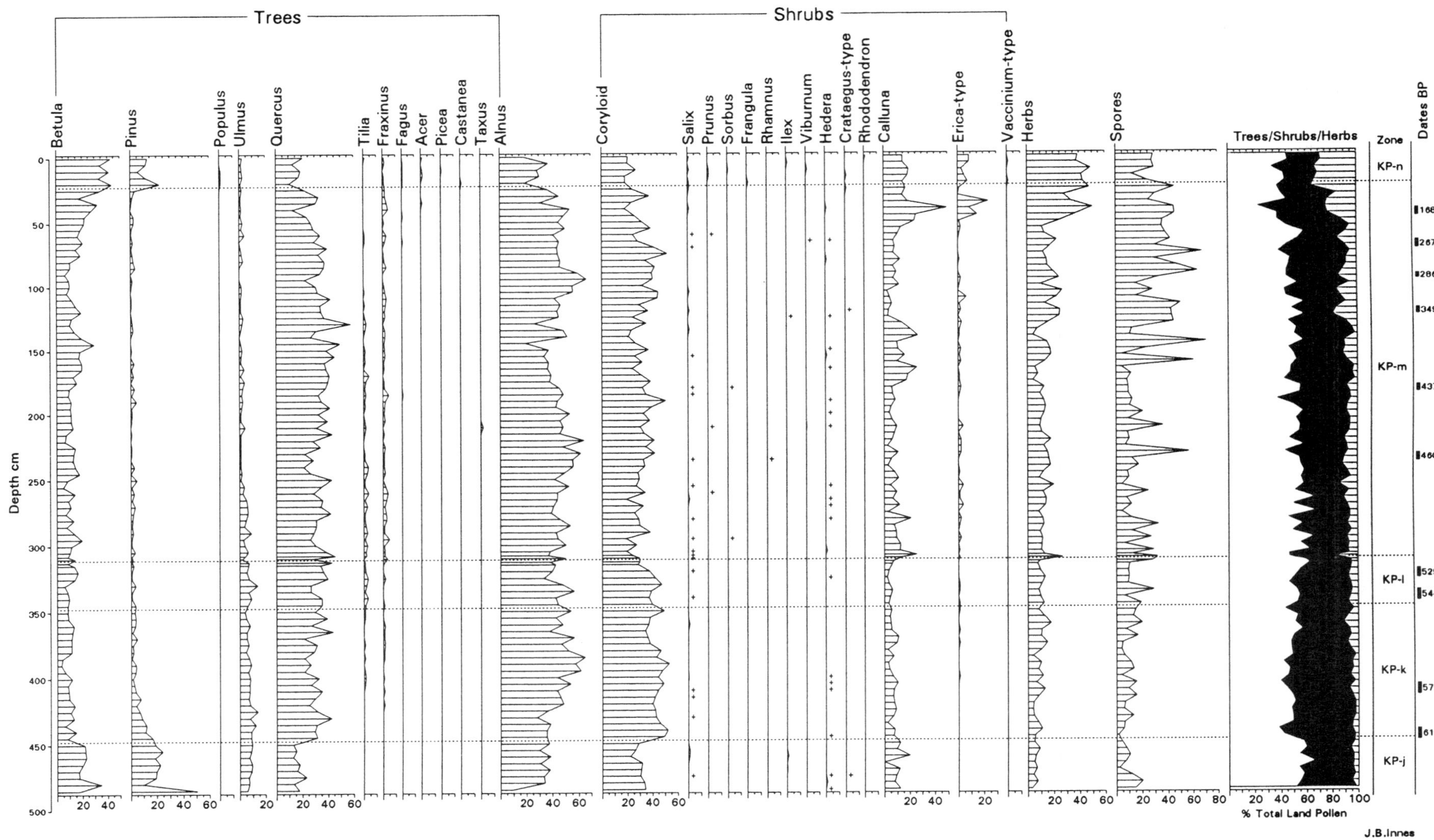

Fig 41 Central mosslands: Knowsley Park Moss later Flandrian percentage arboreal pollen diagram (trees plus group)

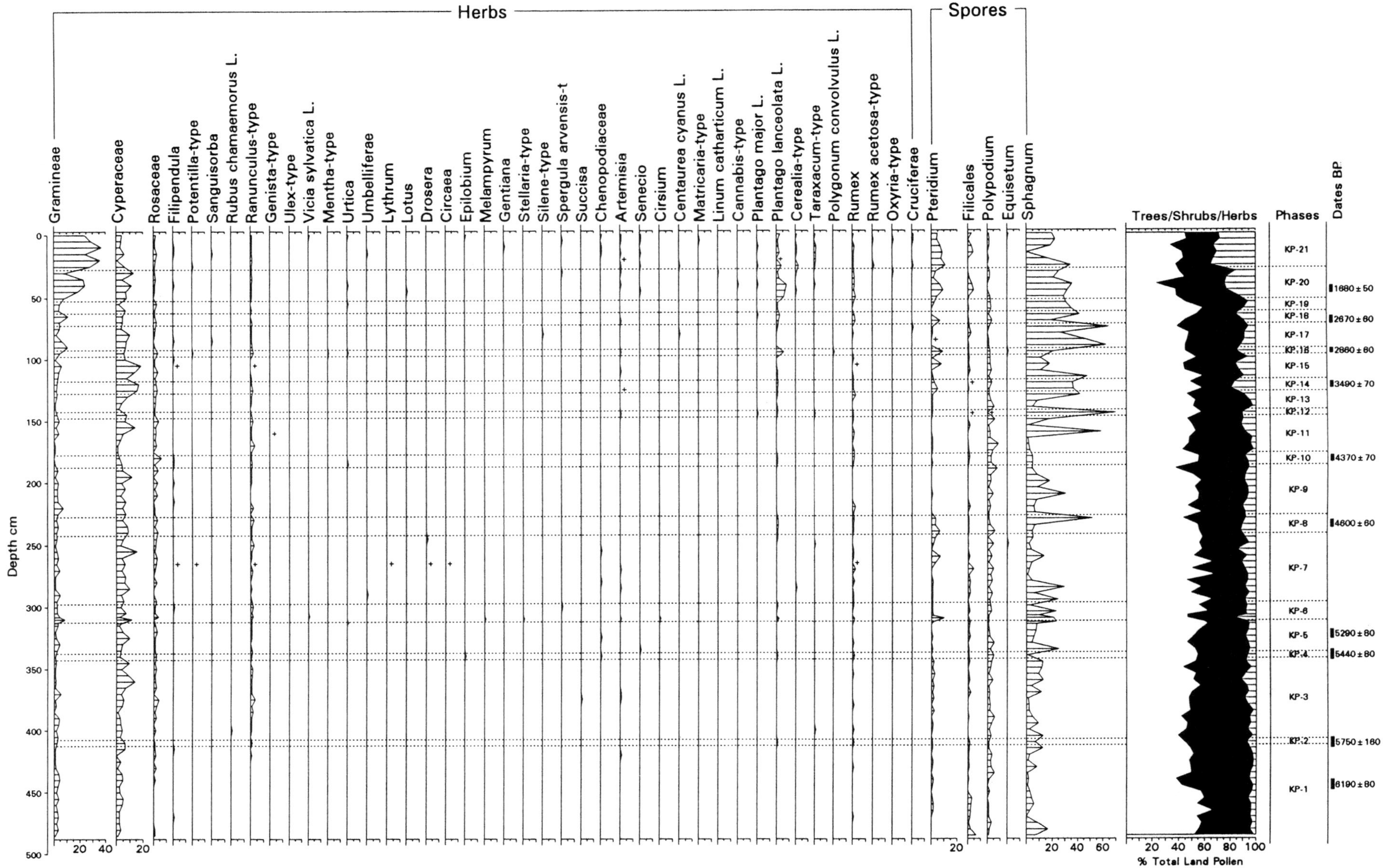

Fig 42 Central mosslands: Knowsley Park Moss later Flandrian percentage herb pollen diagram (trees plus group)

Ulmus are reduced in frequency. *Calluna* is the only other significant pollen type, although *Sphagnum* is still present. The end of the zone is defined by a decline in *Pinus* frequencies and is approximately radiocarbon dated (5.51-5.59m) to 6110±80 BC (8060±80 BP; GU-5241).

KP-h 5.60-5.25m This zone is characterised by *Pinus*, Coryloid, and *Quercus*. *Pinus* frequencies decline while Coryloid, *Quercus*, and *Betula* all show small but significant increases. *Ulmus* remains important. *Sphagnum* and *Calluna* remain the only other major taxa. The end of the zone is defined by the start of a low but continuous *Alnus* curve and is approximately radiocarbon dated (5.16-5.24m) to 6690-6410 Cal BC (7700±80 BP; GU-5240).

KP-i 5.25-4.85m This zone is characterised by *Pinus* and *Quercus*, with major contributions from Coryloid, *Ulmus*, and *Betula*. *Alnus* and *Calluna* are the only other important taxa. The end of the zone is defined by a sharp rise in *Alnus* and decline of *Pinus*. It has not yet been radiocarbon dated.

KP-j 4.85-4.50m This zone is characterised by *Alnus* and *Quercus*, but with major frequencies of *Pinus*, Coryloid, *Betula*, and *Ulmus*. *Calluna* remains well represented. All herb pollen types are in low frequencies only. The end of the zone is defined by a rise in *Quercus* frequencies and is approximately radiocarbon dated (4.41-4.49m) to 5322-4906 Cal BC (6190±80 BP; GU-5237).

KP-k 4.50-3.45m This zone is characterised by *Quercus*, *Alnus*, and Coryloid, all of which are markedly increased in frequency. *Ulmus* remains important but *Pinus* and *Betula* fall to low levels. *Fraxinus* and *Tilia* are recorded occasionally. Herb pollen values are low, but peaks of *Sphagnum* occur. The end of the zone is defined by the start of a low but continuous *Tilia* curve and is approximately radiocarbon dated (3.36-3.44m) to 4460-4044 Cal BC (5440±80 BP; GU-5239).

KP-l 3.45-3.13m This zone is characterised by *Quercus*, *Alnus*, and Coryloid, with lesser amounts of *Ulmus* and *Tilia*. *Betula* is very low and *Pinus* almost absent. *Calluna* falls to low values and *Sphagnum* remains significant. Other pollen types are very low. The end of the zone is defined by a fall in *Ulmus* pollen frequencies and is approximately radiocarbon dated (3.2-3.27m) to 4340-3970 Cal BC (5290±80 BP; Birm-1191).

KP-m 3.13-0.25m This zone is characterised by *Quercus*, *Alnus*, and Coryloid. *Betula* and *Calluna* increase during the zone and show sporadic peaks but remain in only moderate frequencies. *Ulmus*, *Tilia*, and *Fraxinus* values are low. Non-tree pollen types are more abundant and *Pteridium*, Cyperaceae, and *Sphagnum* are prominent. Ruderal herb pollen records increase and show occasional peaks. The end of the zone is defined by a rise in *Pinus* and *Betula*, and a fall in *Quercus*, *Alnus*, and Coryloid pollen frequencies. The end of the zone occurs shortly after Cal AD 240-440 (1680±50 BP; Birm-1177).

KP-n 0.25-0m The uppermost zone is characterised by *Pinus* and *Betula*, but with high frequencies of *Quercus*, *Alnus*, and Coryloid. *Calluna* values remain important. *Pteridium* and Gramineae values are increased. Open ground herbs such as *Artemisia*, *Plantago lanceolata*, and *Taraxacum*-type become more common. *Fagus* and *Acer* are consistently present. Total non-tree pollen percentages are greatly increased.

Ecological history

Peat formation began at Knowsley with the colonisation by aquatic vegetation of wet hollows in the recently stabilised Shirdley Hill Sand forming the present deepest sections of the basins.

The basal organic deposit at Knowsley Park is a gyttja formed under lacustrine conditions. Aquatic pollen types are abundant during this early phase, with *Pediastrum*, *Myriophyllum alterniflorum*, *M spicatum*, *Potamogeton*, *Nymphaea*, *Isoetes*, and *Typha latifolia* all in high frequencies and reflecting open water conditions. In this phase the ecological conditions existing on the drier ground around the site appear to have favoured transitional herbaceous communities, probably on shallow unstable soils, as a range of ruderal type weeds are present, including *Artemisia*, *Rumex*, *Valeriana*, *Armeria*, *Saxifraga*, Cruciferae, and several others. Such a pollen assemblage would usually be present in late glacial or earliest Holocene environments around 8000 BC. The late radiocarbon date of 7355±65 BC (9305±65 BP; SRR-4515) for the end of this tundra-type weed phase, however, suggests that the soil instability of the Shirdley Hill Sand areas, at least in places, sustained some open ground vegetation well into the Holocene. This agrees with the findings of Innes *et al* (1989) in the area of Holiday Moss. The peak values for the early Holocene successional shrub taxa *Empetrum*, *Juniperus*, and *Salix* at this stage persist much later than at Red Moss, where *Betula* woodland quickly became established in the earliest Holocene. This supports the view of late soil development and establishment of complete shrub and tree cover around this site. The good sequence of dates in this lower part of the profile means that there is no reason to suspect that the basal date is in error.

Surrounding these hollows, a carr woodland developed and the site is likely to have consisted of a mosaic of wetland habitats including extensive reed-swamp communities. High Gramineae, Cyperaceae, *Equisetum*, and *Menyanthes* percentages reflect these environments, and *Phragmites* macrofossils are important in the monocotyledon peats of the lower stratigraphy. The stabilisation of the local sandy soils allowed rapid movement through the early post-glacial successional vegetation communities, until first birch and then mixed birch/hazel woodland became established. This was clearly dense, as herb pollen, provided almost entirely by wetland taxa, is reduced to very low levels indeed. Diversification of the forest occurred, with firstly *Quercus* and *Ulmus* entering the community to be followed at 6930±90 BC (8880±90 BP; GU-5243) by the rapid rise to prolonged dominance of *Pinus*. The sandy soils around Knowsley Park will have favoured the establishment and maintenance of pine/hazel Boreal forest. The presence of detrital wood remains in the peat which formed during later Flandrian I shows that wet carr woodland developed around the site as the environment became less aquatic. Few fen and carr herbs are present in the assemblage. The record of *Sphagnum* throughout the lower profile probably reflects the general acidity of the environment in these sandy soil areas.

Burning affected the vegetation over wide areas of the mire complex during this phase, and it is conceivable that this may have favoured pine around the site, and also influenced the subsequent establishment of ombrotrophic peats dominated by *Eriophorum* which developed in mid-Flandrian II. These accumulated to form an extensive raised mire system which was periodically affected by fire (Fig 37). The tendency for *Sphagnum*-dominated peats to occur towards the outer edges of the system, while *Eriophorum*/*Calluna* peats dominate the deeper central areas, suggests that wetter conditions prevailed on the fringes of the mire for much of its development.

Forest disturbance

There are indications of forest disturbance at several intervals in the Knowsley Park diagram which may have been the result of human activity. Twenty-one phases in the history of forest clearance were recognised, and are described below.

KP-1 7.20-4.05m From the base of the diagram to mid-zone KP-k in chronozone Flandrian II, there are no indications of vegetation disturbance in the pollen record which may reliably be ascribed to human activity. While major vegetation changes took place in that time, these represent seral plant communities which developed following the start of the Flandrian post-glacial period. Climatic, edaphic, and tree immigration rate factors influenced the successions which took place during the development and diversification of mixed woodland during Flandrian I and early Flandrian II. From zone KP-d onwards tree and shrub pollen dominate the assemblages totally, and despite a few peaks of *Pteridium* spores there are no taxa which demonstrate breaks in the forest canopy once it became established.

KP-2 4.10m A brief phase of disturbance occurs at this level in which *Plantago lanceolata, Rumex, Artemisia,* and *Fraxinus* suggest the creation of some open ground and thus the increased provision of light into the woodland. The clearing was not large enough to increase the ratio of non-tree pollen to total pollen. The ruderal weed pollen suggests human activity may have been responsible for this limited break in the forest. This phase is approximately radiocarbon dated (4.07-4.15m) to 4998-4330 Cal BC (5750±160 BP; Birm-1223).

KP-3 4. 05-3.45m Although the sporadic presence of *Fraxinus* and *Pteridium*, and some increase in *Betula* and Coryloid frequencies, suggests that the forest in this phase may have been of a lighter type than hitherto, there is no convincing evidence of renewed disturbance of the woodland environment. While perhaps less dense, the tree canopy appears to have remained unbroken.

KP-4 3.40m A second small-scale phase of woodland opening occurs at this level when ruderal weed pollen return to the assemblage. *Rumex, Plantago lanceolata*, Chenopodiaceae, and *Epilobium* all occur. There was no appreciable impact upon the total tree pollen percentage, however, and so once again the effects of the creation of open ground must have been limited. This small disturbance of the woodland is approximately radiocarbon dated (3.36-3.44m) to 4460-4044 Cal BC (5440±80 BP; GU-5239).

KP-5 3.35-3.13m Although occasional weed pollen occurs during this phase, it is isolated and cannot be interpreted as renewed disturbance. Unbroken mixed woodland tree canopies prevailed. The increase in *Tilia* and *Fraxinus* could be regarded as being caused by a change in forest composition after the previous episodes of disturbance, but this is a phase of regeneration and stability in the forest.

KP-6 3.13-3.00m This is a significant phase of forest disturbance, with several grains of *Plantago lanceolata* recorded as well as *Rumex, Urtica, Artemisia, Stellaria,* and *Cirsium.* A very high peak of *Pteridium* and a major increase in *Fraxinus* are also associated with this episode of woodland recession. *Ulmus* is the tree most affected and Coryloid seems also to have declined. A big increase in *Calluna* follows this disturbance, so that perhaps some heath communities on disturbed sandy soils were created temporarily after the opening of the woodland. This disturbance is radiocarbon dated approximately (3.2-3.27m) to soon after 4340-3970 Cal BC (5290±80 BP; Birm-1191) and so represents the elm decline of the Flandrian II/III transition.

KP-7 2.95-2.45m This is a phase of woodland recovery in which no distinct evidence of forest disturbance occurs, although some weed pollen is recorded at intervals which suggest that open ground persisted in places. The increased *Fraxinus* presence points to the woodland being of a more secondary nature at this time, although still very dense. *Ulmus* recovers to a level almost equal to its abundance prior to the elm decline.

KP-8 2.45-2.30m Significant disturbance occurs in this phase with *Plantago lanceolata, Rumex,* and *Pteridium* all showing peaks in values. Few other ruderal types occur, however, and the opening of the woodland canopy appears to have remained small in scale. *Betula* was favoured by this forest disturbance. *Quercus* and *Ulmus* appear to have been the trees most affected, with the latter almost becoming absent. This phase may be regarded as a secondary elm decline and is radiocarbon dated approximately (2.32-2.38m) to 3510-3100 Cal BC (4600±60 BP; Birm-1172).

KP-9 2.25-1.90m Apart from a few isolated records of weed pollen there are no indications of disturbance in this phase. Tree regeneration appears to have been complete with closed canopy conditions present in the area and characteristic secondary woodland trees such as *Betula* and *Fraxinus* not prominent. *Ulmus,* however, does not recover.

KP-10 1.90-1.80m This is a minor phase of forest disturbance, in which *Plantago lanceolata, Rumex, Urtica,* and *Pteridium* occur but tree pollen values are not affected, and light-demanding trees and shrubs are not greatly encouraged. Some diversification of the woodland did take place, however, as *Fraxinus* and *Sorbus* do occur and *Fagus* is recorded for the first time. This phase is radiocarbon dated approximately (1.8-1.85m) to 3325-2788 Cal BC (4370±70 BP; Birm-1173).

KP-11 1.80-1.50m This phase is one of undisturbed woodland conditions with almost no indications of breaks in the tree cover. Little herb pollen occurs and the assemblage is dominated by major forest trees. *Calluna* is increased and could represent heath vegetation, but in the absence of any other open ground plants could equally be associated with communities on the mire surface, as the stratigraphy indicates a relatively dry surface at this time.

KP-12 1.45-1.45m This is a phase of significant forest recession, with the appearance of ruderals including *Plantago lanceolata, P major-media, Rumex, Artemisia, Taraxacum*-type, and *Pteridium. Quercus* and *Alnus* were significantly affected by this disturbance and *Tilia* is not recorded. A major peak of *Calluna* occurs which is likely to be linked to the opening of the forest and creation of open ground. No radiocarbon date is available for this phase but by extrapolation from dates above and below this point it is probably of about 2000-3000 Cal BC in age.

KP-13 1.40-1. 25m During this phase there is a return to closed canopy woodland with few indications of disturbance, although some isolated ruderal records and initially high *Calluna* frequencies suggest that regeneration may have been incomplete after the previous episode of disturbance. Increased *Fraxinus* values also support this view.

KP-14 1.25-1.15m This is a major phase of forest disturbance in which many ruderal types are recorded, including *Plantago lanceolata, Rumex, Artemisia,* and *Pteridium. Alnus* appears to have been adversely affected at the start of this phase although it later recovers. Trees and shrubs favoured by increased light become more common, including *Fraxinus, Crataegus*-type, and *Ilex.* This phase is radiocarbon dated approximately (1.2-1.25m) to 2028-1670 Cal BC (3490±70 BP; Birm-1174).

KP-15 1.15-1.00m This phase is designated as one of woodland regeneration without significant renewed disturbance. Although some ruderal pollen types do occur at intervals, these are considered indicative of incomplete regeneration after the previous phase of woodland opening, rather than new activity. *Pteridium* is particularly high but in the absence of other major indicators of clearance could represent areas of persisting open ground. A more open type of woodland is probable by this time.

KP-16 0.95m This is a brief but intensive phase of woodland recession in which *Quercus* is much reduced and is replaced by *Alnus* and

Coryloid. Very high percentages of *Plantago lanceolata* and *Pteridium* occur as well as lesser values of *Artemisia, Urtica,* and *Polygonum*. A significant removal of tree cover and creation of open ground seems to have taken place during this phase, which is radiocarbon dated approximately (0.93-0.97m) to 1258-900 Cal BC (2860±60 BP; Birm-1175).

KP-17 0.95-0.70m This is a phase of woodland recovery during which there is little real evidence of renewed disturbance. Weed pollen is particularly low in the earlier part of the phase, and the forest cover appears to have been quite dense. Sporadic occurrences of ruderal pollen occur towards the end of the phase, but not in association or in sufficient numbers to be interpreted as evidence of human activity.

KP-18 0.70-0.65m This phase is one of major disturbance of the woodland vegetation, with high percentages of *Plantago lanceolata* and *Pteridium*, and also records for *Rumex* and *Artemisia*. Tree values remain high, however, with no particular type particularly reduced. This disturbance, although substantial, must have been of rather limited extent, and the landscape in the area remained well wooded at this time. The phase is radiocarbon dated approximately (0.67-0.73m) to 930-790 Cal BC (2670±60 BP; Birm-1176).

KP-19 0.60-0.55m In this phase, weed pollen is present in small quantities only and so is regarded as representing continued open ground rather than the results of renewed clearance. While tree pollen values remain high, it is apparent that regeneration may have been only partial and a more open woodland with much *Fraxinus* and Coryloid existed. This phase is, however, not regarded as one of active disturbance.

KP-20 0.55-0.25m This is a major phase of forest clearance during which total tree pollen values are much reduced. *Quercus* was the most badly affected but *Alnus* and Coryloid were also markedly reduced in abundance. Secondary trees *Betula* and *Fraxinus* were encouraged by this diversification of the woodland, and *Fagus* and *Acer* are also recorded. In the early stages *Plantago lanceolata, Rumex, Pteridium,* and *Calluna* occur in high frequencies, and are joined in mid phase by *Senecio*-type, *Artemisia, Plantago major-media,* and *Taraxacum*-type. Most significantly, cereal-type pollen grains of *Triticum*-type are also recognised at this time, and point to human agricultural activity as being the cause of disturbance of the woodland. A grain of *Spergula arvensis* late in the phase supports this view. *Cannabis*-type grains also are recorded in this assemblage, and these may also be associated with cultivation. The early part of this major phase of disturbance is radiocarbon dated approximately (0.42-0.48m) to Cal AD 240-440 (1680±50 BP; Birm-1177).

KP-21 0.25-0m This phase is one of very intensive forest clearance, and the consistent presence of cereal-type pollen grains of *Triticum*-type shows that agricultural activity was the main reason for woodland removal. Cereal percentages are particularly high at the start of the phase but continue throughout. *Plantago lanceolata* and *Pteridium* are very high, but several other clearance indicators are also present, such as Cruciferae, *Rumex, Plantago major-media, Artemisia, Taraxacum*-type, *Senecio*-type, Chenopodiaceae, *Urtica,* and *Epilobium*. Types associated with arable cultivation include *Centaurea cyanus* and *Spergula arvensis*. Changes also occur in the composition of the forest, with *Quercus, Alnus,* and Coryloid all declining, especially in the latter half of the phase where *Betula* and *Pinus* become much more abundant and *Acer* and *Fagus* increase in frequency. This final change in woodland composition may be associated with the establishment of plantation woodland in more recent times, and the relative decline in farming intensity from the peak of the first half of the phase may reflect this change in landuse. No radiocarbon date is available for this phase.

Simonswood Moss

At the time of sampling for the palynological investigation described here (early 1980s), Simonswood Moss was one of the few areas in the central mosslands of Merseyside in which parts of the raised bog deposits had remained only slightly disturbed. The moss was drained in the last century to be used for game shooting, and large areas of the central part of the moss have had at least 2-3m of peat cut from the surface and removed for commercial sale. This process has accelerated in recent years under more systematic extraction techniques, and much of the site's stratigraphy has recently been severely truncated, but at present peat up to 4m deep still exists in several places, and *c*2-3m was frequently encountered over much of the site.

Two areas which had originally escaped severe damage, although still affected by drainage, were situated at the northern and southern extremities of the moss. At the southern end a woodland plantation, mainly of pine, covered a large area. At the northern remnant, however, the vegetation was composed of semi-natural mire communities typical of situations where drainage has severely restricted new peat growth, and consisted of a mixture of *Molinia*, Ericaceae (including some invading *Rhododendron*), *Betula*, and *Pteridium*, with some isolated patches of *Sphagnum* in wetter areas. In general, however, the surface of the bog was dry.

General stratigraphy

The stratigraphic transects are too limited in their scope to enable the accurate mapping of basin topography, but they hint at the presence of two shallow basins: a southern one situated at a higher altitude than a northern one, with a higher ridge of sands intervening (Figs 43, 45, 47).

The organic deposits are underlain by an undulating sand over which lie basal deposits comprising a mixture of wood or wood/monocotyledon peats. These levels almost uniformly contain significant amounts of carbonised plant material. In the northern section of the transect these deposits are succeeded by *Sphagnum*-dominated peats, while in the middle section of the transect *Eriophorum/Calluna*-dominated peats occur. The southern section contains a more complicated stratigraphy with monocotyledon and *Phragmites* basal peats overlain by wood peats and succeeded by *Polytrichum* and *Aulocomnium* peats, before a switch to *Sphagnum*-dominated deposits. The most abundant occurrence of carbonised material is largely confined to the lower wood/monocotyledon layers.

The stratigraphy indicates an initial peat-producing system dominated by reedswamp/monocotyledon communities, which was succeeded by carr. The vegetation of the southern part of the system then changed to a mosaic of *Eriophorum* and *Polytrichum*-dominated vegetation, reminiscent of heathland, while the northern section of the site appears to have experienced wetter conditions resulting in *Sphagnum*-dominated vegetation. Later, this community was to extend to cover the whole mire surface.

A burning episode, almost certainly identical with that identified by Innes and Tomlinson (1983a; *and see below*), and associated with limited evidence for forest clearance, can be seen to be of a widespread nature rather than merely confined to a localised area within the Simonswood site (Figs 44, 46, 48).

Pollen analysis

Cores for pollen analysis were recovered from three locations at Simonswood Moss: two in the area of cut peat in the centre of the moss to retrieve the earlier sediments near the centre of the basin (core A at SJ 452996 and core C at SD 448010), and one from the northerly uncut area to recover more recent peat deposits (core B at SJ 445996). Of these, cores B and C were used for analyses, the latter as it proved to be the deepest (*ie* >4m), and contained freshwater lake muds at the base.

Simonswood Moss B (SMB) (SJ 445996)

The altitude of core SMB was levelled to 42.67m OD.

A coarse dark sand lay beneath the deepest organic deposits below 3.98m. The basal organic deposit comprised humified *Phragmites* peat with occasional wood fragments, and this extended to 3.76m. Highly humified wood peat occurred above this until 3.3m, when it was succeeded by an *Eriophorum* peat with *Betula* wood fragments. Above 3.15m *Eriophorum*-dominated peat became established until *c*1.92m, when a fresher *Sphagnum* peat with some wood fragments and *Eriophorum* replaced it. A fresh *Sphagnum* peat occurred from 1.92m up to 0.71m when it became more humified. Above this lay humified amorphous peat.

Pollen analysis

Samples were taken for pollen analysis at intervals of 10cm through the core, and the resulting pollen diagrams are shown as Figures 49 and 50. Although changes in the pollen frequencies of the major taxa occur, these are not sufficiently great to warrant the formal division of the diagram into separate zones. Accordingly the whole of the diagram is assigned to a single local pollen assemblage zone SMB-a, which may be described as follows.

SMB-a 3.95-0m The zone is dominated by *Quercus*, *Alnus*, *Betula*, and Coryloid. *Ulmus*, *Fraxinus*, and *Tilia* are present in low frequencies throughout. Tree pollen consistently accounts for about 50% of total pollen. *Calluna* fluctuates, but is of major importance except at the start of the zone. Gramineae and Cyperaceae pollen values are low, although the former increases near the end of the zone. Other herb pollen frequencies are uniformly low, although the variety of types recorded increases towards the end of the zone. Ruderal weed pollen and cultural indicators such as cereal-type pollen become more common in the later part of the zone, whereas wetland herb pollen is present in low frequencies in all parts of the diagram. *Sphagnum* is present throughout the zone.

Ecological history

The stratigraphic succession at Simonswood Moss B shows that peat formation began in early Flandrian III as a result of waterlogging of the sand

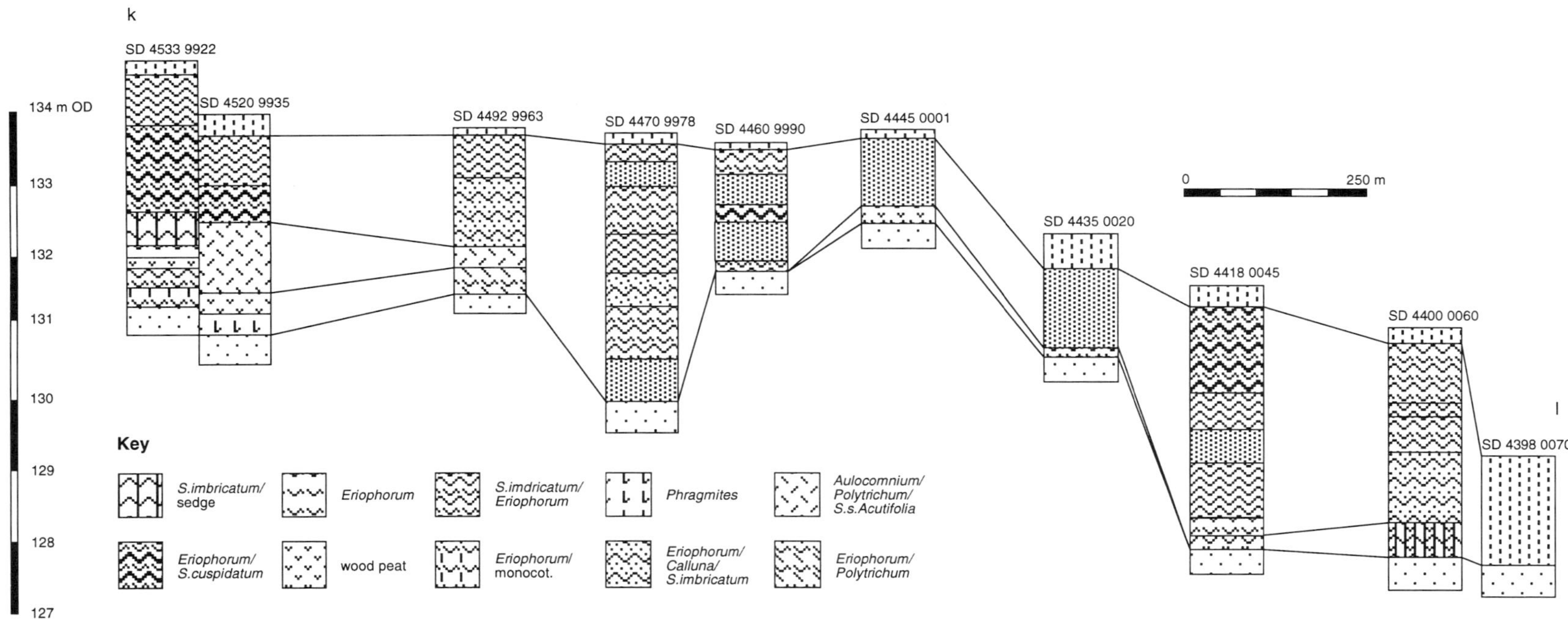

Fig 43 Central mosslands: Simonswood Moss south-east to north-west stratigraphic transect k-l

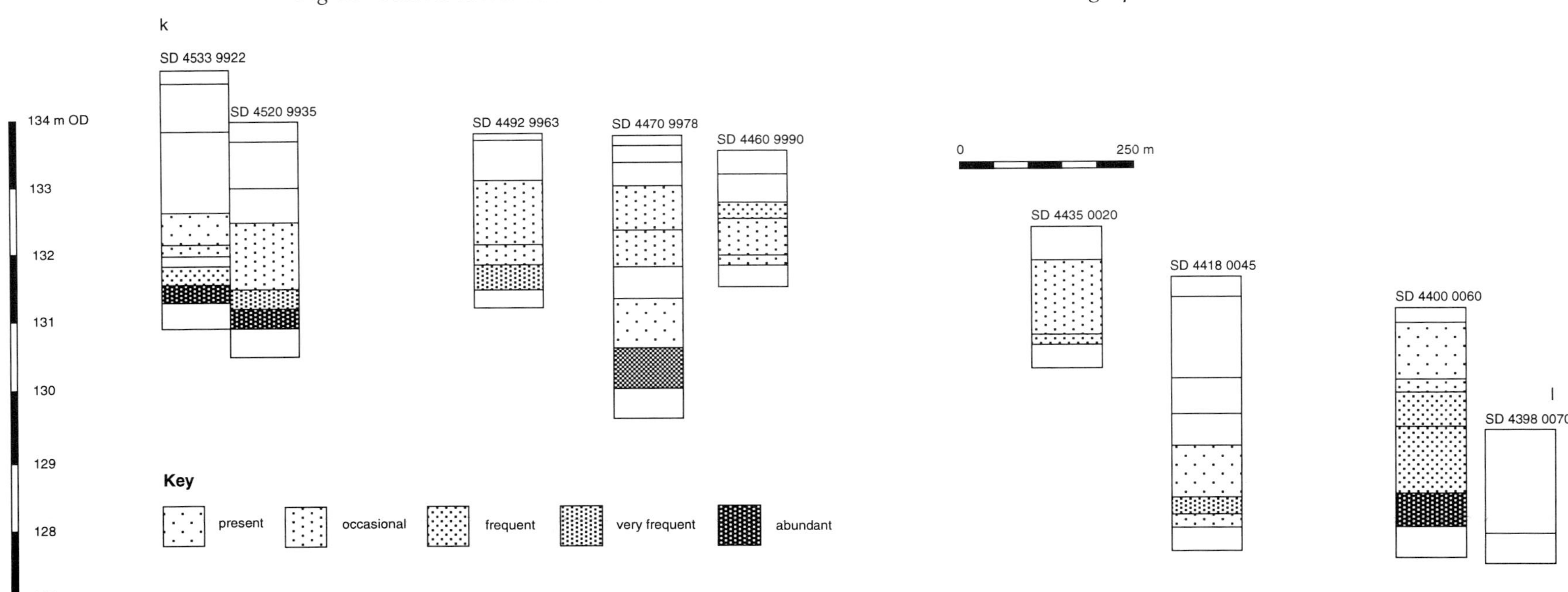

Fig 44 Central mosslands: Simonswood Moss south-east to north-west stratigraphic transect, carbonised material k-l

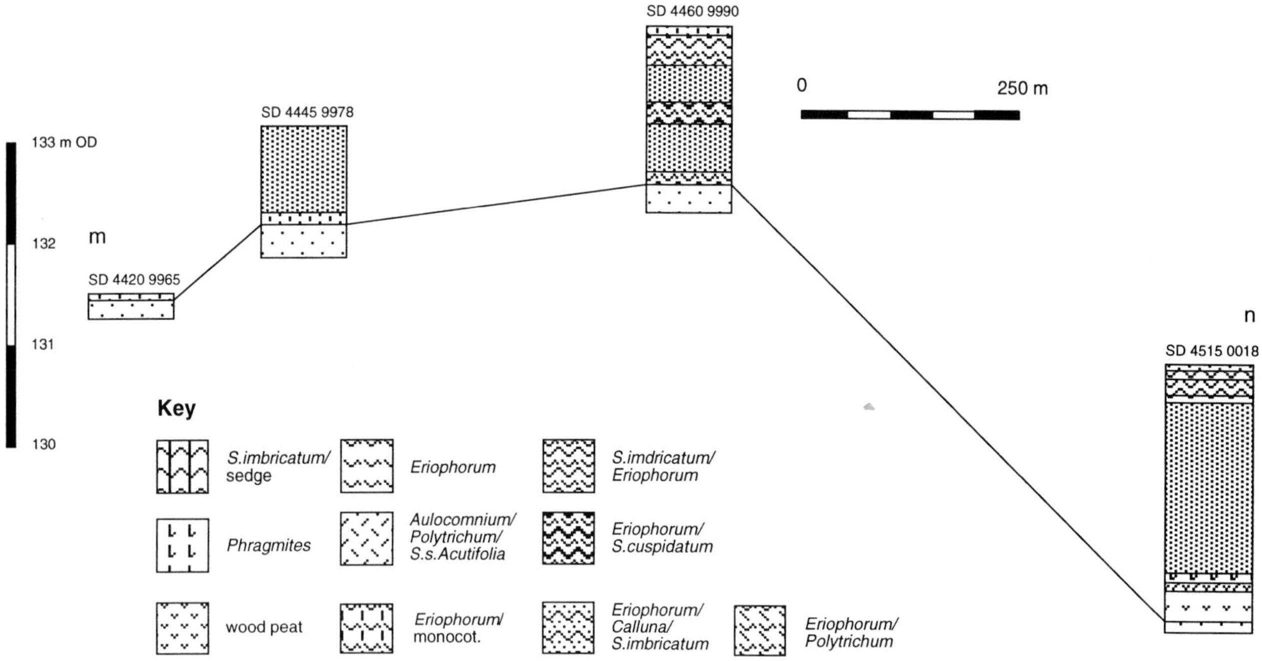

Fig 45 Central mosslands: Simonswood Moss northern south-west to north-east stratigraphic transect m-n

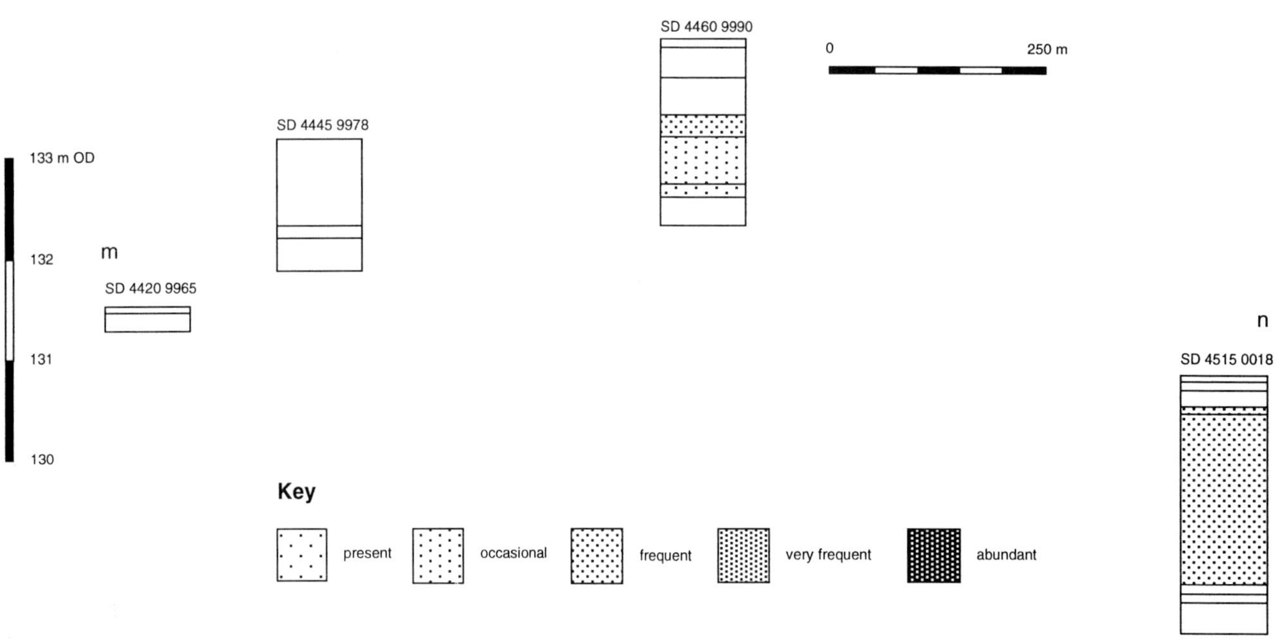

Fig 46 Central mosslands: Simonswood Moss northern south-west to north-east stratigraphic transect, carbonised material m-n

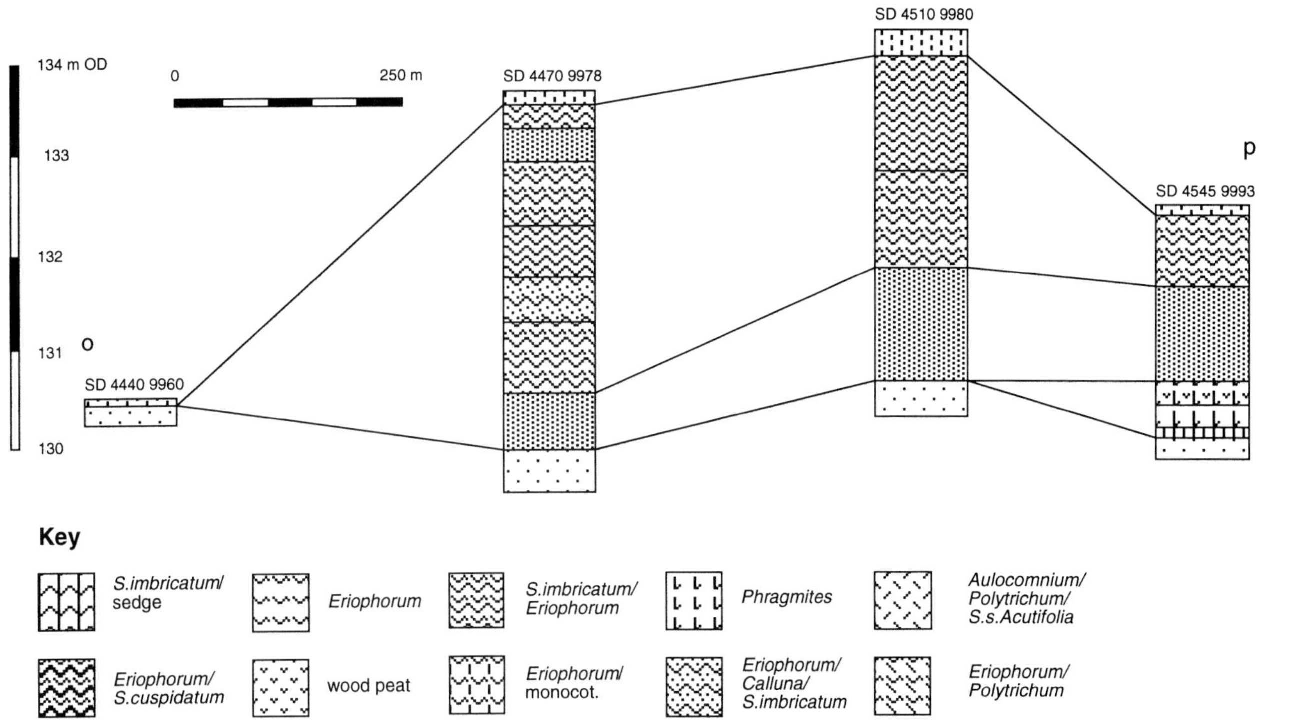

Fig 47 Central mosslands: Simonswood Moss southern south-west to north-east stratigraphic transect o-p

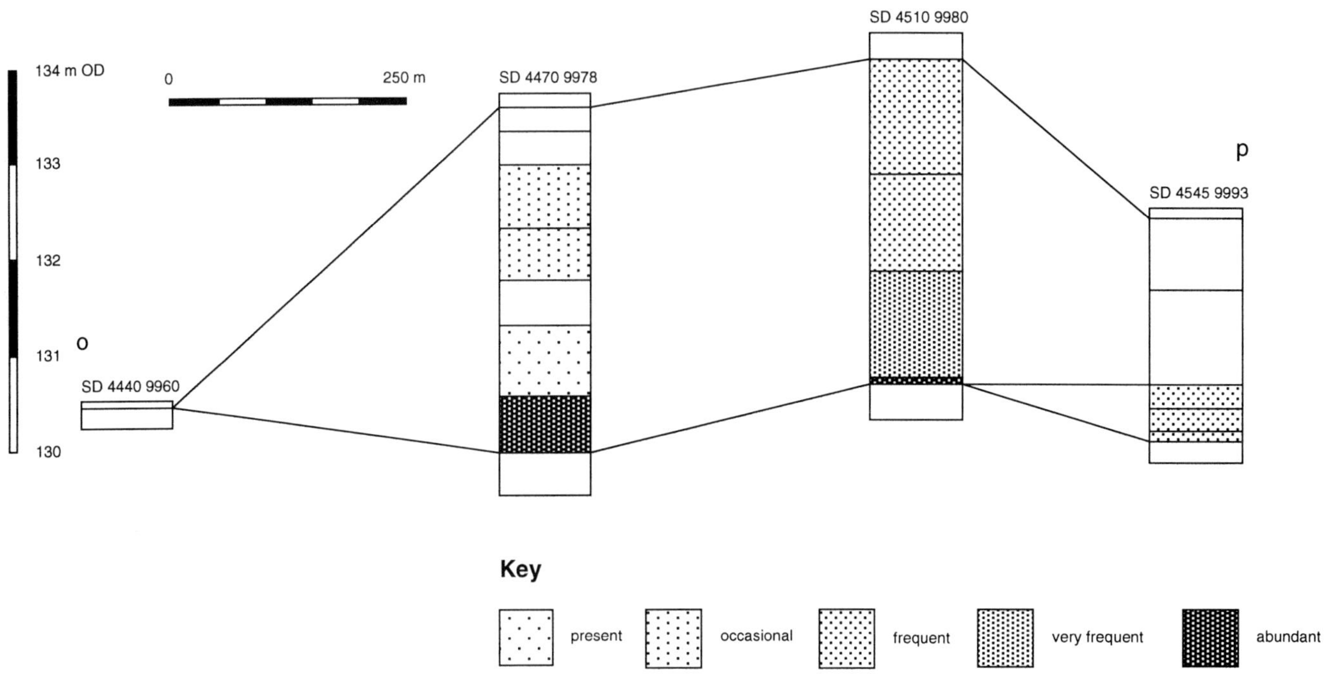

Fig 48 Central mosslands: Simonswood Moss southern south-west to north-east stratigraphic transect o-p, carbonised material

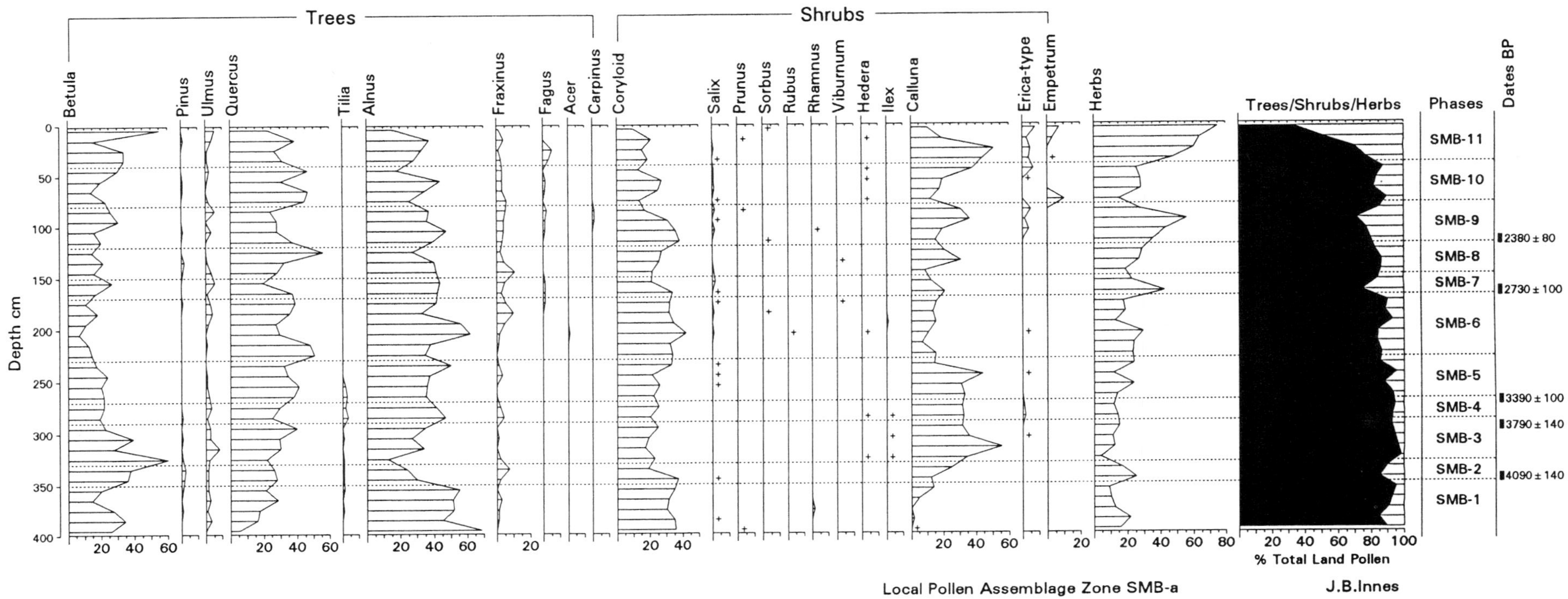

Fig 49 Central mosslands: Simonswood Moss B, percentage arboreal pollen diagram (trees plus group)

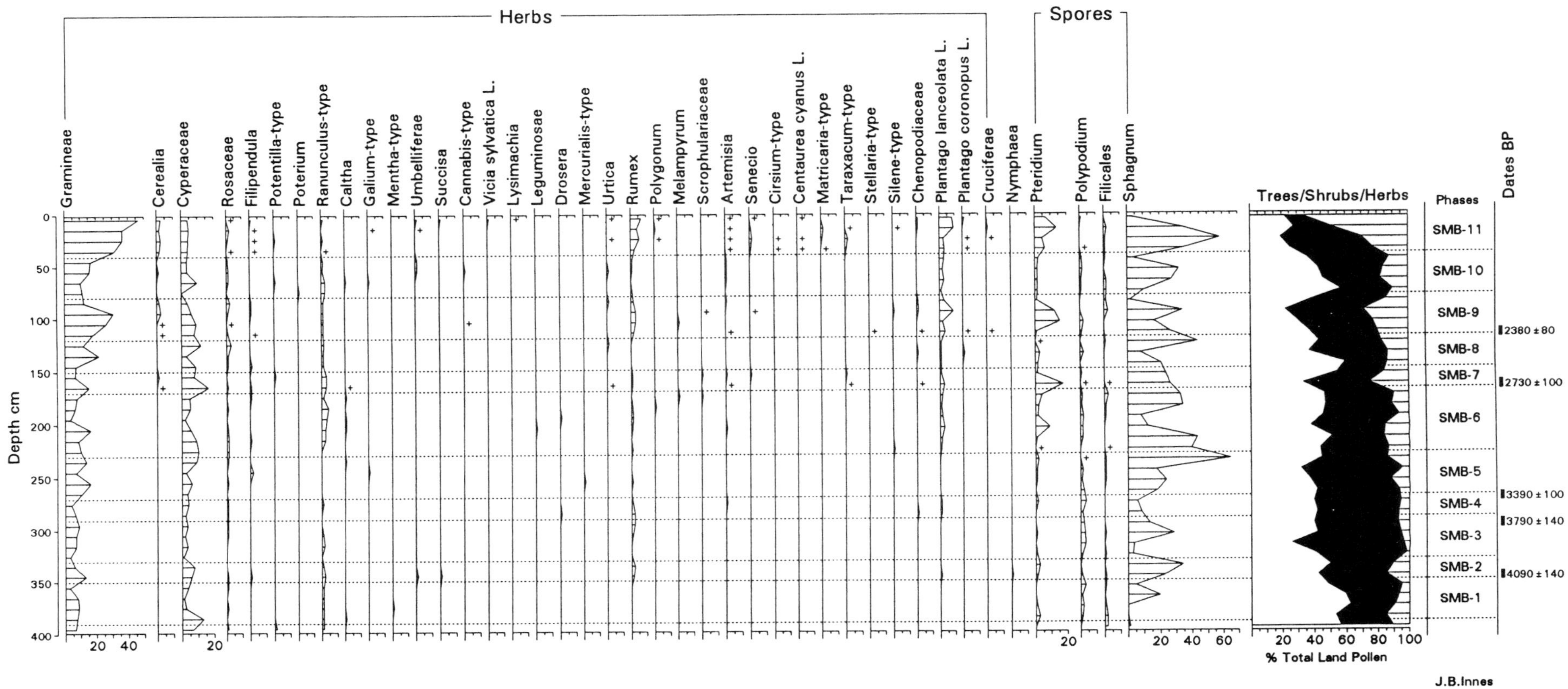

Fig 50 Central mosslands: Simonswood Moss B, percentage herb pollen diagram (trees plus group)

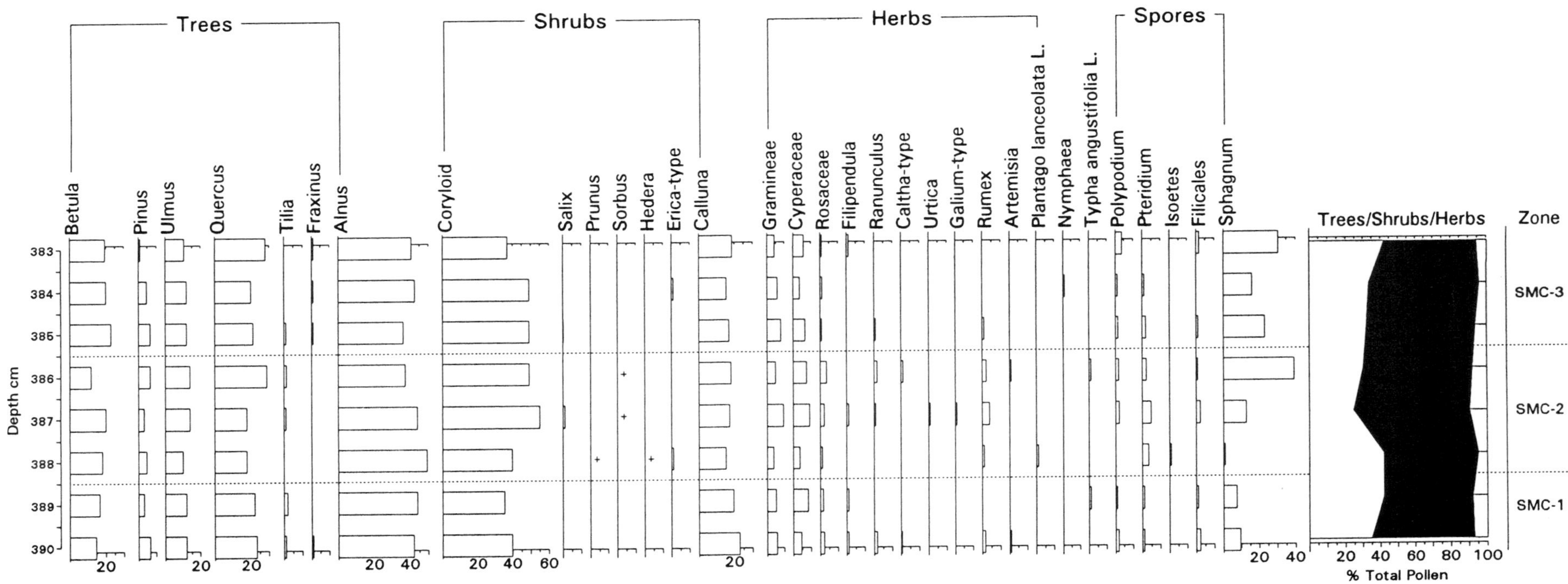

Fig 51 Central mosslands: Simonswood Moss C, percentage arboreal and herb pollen diagram (trees plus group)

surface, as peat spread from the central basins of the Simonswood complex. The absence of aquatic pollen at the base of the diagram suggests that lacustrine conditions never occurred at this site, but that a gradual paludification took place, followed by the colonisation of the ground surface by reedswamp vegetation, with *Phragmites* followed by *Alnus*-dominated fen carr. *Potentilla, Caltha, Filipendula,* and *Rhamnus* occurred in these carr habitats. As the nutrient status of the mire declined, *Betula* assumed dominance of the drying fenwood, until acid raised bog with Cyperaceae, mainly *Eriophorum*, and *Sphagnum* prevailed. Around 1100 Cal BC *Sphagnum* became the peat-forming vegetation in this part of the mire system, and continued to dominate the bog surface until recent times. This area of the mire clearly remained very wet throughout that time, until the effects of modern drainage took hold.

Landscape history and forest disturbance

There are indications of forest disturbance in several parts of the diagram which may have resulted from human activity. Eleven phases in forest clearance history are recognised, as described below.

SMB-1 3.90-3.45m There are no real indications of woodland disturbance during this phase, and the vegetation comprised mixed oak forest in which *Alnus* was particularly important. Some light-responsive shrub taxa occur, like *Prunus* and *Salix*, and isolated *Pteridium* spores are present. There are no open ground weeds to suggest fresh breaks in the canopy, however, and non-tree pollen values are very low and confined to local wetland types.

SMB-2 3.50-3.30m There is some evidence in this phase of the creation of small openings in the woodland cover. Small peaks in the frequencies of Gramineae, *Rumex*, *Pteridium*, and *Fraxinus* occur, while a grain of *Plantago lanceolata* is recorded. Arboreal pollen values fall, but not sufficiently to suggest that the breaks in the tree canopy were more than very limited and temporary. *Alnus* is the tree most affected by the change, while *Calluna* was encouraged as its pollen curve rises from this point. The beginning of this phase is dated approximately (3.43-3.51m) to 3023-2280 Cal BC (4090±140 BP; Birm-1217).

SMB-3 3.30-2.90m This is a phase characterised by tree regeneration, during which signs of forest disturbance are no longer apparent. A fall in *Alnus* pollen and a rise in *Calluna* and *Betula* occur, but these are best explained by changes due to succession after the previous zone and in the mire system itself rather than by renewed disturbance. No ruderal indicators are present and full woodland conditions had returned to the area around the wetland.

SMB-4 2.90-2.70m A further phase of forest clearance occurs, in which total dryland tree pollen values fall and *Rumex* and *Plantago lanceolata* are again recorded, as well as single grains of *Artemisia* and Chenopodiaceae. The nature of this phase resembles that of phase 2, and was also apparently of quite limited extent. *Tilia* and *Fraxinus* increase at the expense of *Quercus* and some change in woodland composition may have occurred. The total extent of tree cover was not greatly decreased. The beginning of this phase is dated approximately (2.93-3.01m) to 2600-1829 Cal BC (3790±140 BP; Birm-1218) and the end is dated approximately (2.68-2.76m) to 1950-1450 Cal BC (3390±100 BP; Birm-1219).

SMB-5 2.70-2.30m There is little evidence of forest clearance in this phase, which is characterised by high *Alnus* and *Quercus* frequencies. Secondary tree taxa which may indicate a more open type of woodland, such as *Betula* and *Fraxinus*, are not important. Shrub pollen of heliophyte types also fades from the record, and a relatively dense forest canopy seems to have existed. Only *Salix* values increase and these may be associated with the wetland rather than the forest communities, as may the continued high *Calluna* curve. Herb pollen frequencies are low and non-mire taxa are confined to probable wetland types. While some *Pteridium* spores occur, these may be derived from the shaded woodland environments as easily as from areas which remained open.

SMB-6 2.30-1.70m This phase contains evidence of major disturbance of the vegetation. A continuous *Plantago lanceolata* curve is established and remains present for the rest of the profile. *Pteridium* and *Rumex* values are also significant. These records, together with high frequencies of Gramineae and *Ranunculus*, would suggest a predominantly pastoral nature of the phase if caused by human activity. Isolated grains of *Artemisia, Polygonum,* and *Melampyrum* support the other evidence for more open conditions, the latter perhaps suggesting fire played a part in disturbance. This episode is not of very high intensity except at individual levels such as 2.05m, where a clear recession of tree values occurs. It is sustained, however, and seems to have caused changes to the woodland character, *Fraxinus* becoming much more prominent from this time onwards. Other indicators of more open woodland like *Fagus, Acer, Ilex, Salix, Sorbus,* and *Viburnum* confirm this change. The culmination of this phase is radiocarbon dated approximately (1.63-1.72m) to Cal BC 1154-780 (2730± 100 BP; Birm-1220).

SMB-7 1. 70-1.50m This phase represents a period of highly intensive landuse. A marked recession of tree cover occurred, which can be attributed to human activity as grains of cereal pollen occur, with *Artemisia*, Chenopodiaceae, *Taraxacum*-type, and *Senecio*-type. *Pteridium* rises very sharply and *Plantago lanceolata*, *Rumex*, *Urtica*, and Gramineae all increase. *Quercus* is the tree most affected. Mixed agriculture seems to have made a considerable difference to the vegetation in this phase and caused a real decline in local woodland cover.

SMB-8 1.50-1.20m Although the *Pteridium* and *Plantago lanceolata* curves continue at a low level, there are no convincing indicators of new disturbance and the woodland appears to have become established again. Tree pollen dominates the total sum and *Quercus* recovers from its decline of the previous phase. This phase is characterised by forest regeneration and an apparent absence of human activity.

SMB-9 1.20-0.80m This phase marks a major episode of forest clearance and represents a greater impact on the vegetation than anything hitherto recorded at the site. Tree pollen values fall sharply and a large-scale replacement of woodland by open vegetation occurred. Indicators of arable agriculture are present in large numbers, with cereal pollen continually present, and *Cannabis*-type and Cruciferae also recorded. Possible arable weeds present are *Artemisia*, Caryophyllaceae, Chenopodiaceae, and *Senecio*-type. *Plantago coronopus* and *Melampyrum* also occur. Familiar ruderal types like *Rumex*, *Plantago lanceolata*, and *Pteridium* occur in very high frequencies indeed, and a considerable area of open ground must have existed. Coryloid, *Calluna*, and Gramineae all increase in response to the removal of the tree cover, and *Fagus*, *Acer*, and *Carpinus* occur at the culmination of the phase, showing the change in forest composition of the regenerating woodland. This phase is evidently a very severe episode of forest clearance for mixed farming. The start of this phase is radiocarbon dated approximately (1.14-1.22m) to 790-257 Cal BC (2380±80 BP; Birm-1221).

SMB-10 0.80-0. 40m This phase marks a reversal of the trend towards more open conditions, with the recovery of tree pollen values and evidence for open ground or active landuse much reduced. *Quercus* and *Alnus* woodland was partly restored and taxa such as *Betula* and *Fraxinus* are of subsidiary importance. The continued *Pteridium*, *Plantago lanceolata*, and *Rumex* curves show that some open areas remained, but a general re-advance of tree cover characterises the pollen record. At one level a single cereal grain, with *Cannabis*-type and *Artemisia*, occurs but this is not typical of the phase. Low Coryloid and *Calluna* curves suggest that as woodland was re-established it became rather dense.

SMB-11 0.35-0m This final phase is one in which deforestation takes place with increasing intensity. Tree pollen in the upper levels is reduced to only 20% of total pollen, and agricultural and waste ground indicators occur in high frequencies. Although poorly transported, cereal pollen grains reach almost 5% of total pollen, and clearly significant cereal cultivation must have been going on nearby. Probable arable crops or weeds include *Artemisia*, *Matricaria*-type, *Centaurea cyanus*, *Vicia*, *Cirsium*, *Taraxacum*-type, *Polygonum aviculare*, and Chenopodiaceae. Much open ground or grassland existed, as Gramineae frequencies are very high, and *Rumex*, *Plantago lanceolata*, and *Pteridium* all reach 30% of tree pollen. *Succisa*, *Plantago coronopus*, and *Urtica* also occur. *Quercus*, *Alnus*, and Coryloid are all greatly reduced in this phase, and only *Betula* seems to have been increased, probably as scrub on waste ground or on the bog surface.

Simonswood Moss C (SMC) (SJ 448010)

The Simonswood Moss C core was situated nearer to the centre of the moss than core B, and was levelled to a surface altitude of 45.78m OD.

Below 4.02m coarse brown sand occurred which was overlain by a 0.04m band of sandy gyttja with *Phragmites* remains. Between 3.63-3.92m a *Phragmites* peat with occasional wood fragments was present, but with a prominent charcoal lens occurring at 3.86m. This peat type continued in a more humified state up to 3.25m, where it was replaced by a well humified *Eriophorum* peat which extended to 2.05m, becoming less humified between 2.75-2.85m. Between 2.05-1.57m fresh *Eriophorum* peat with some *Sphagnum* occurred. A *Sphagnum*-dominated peat band was present between 1.50-1.57m, after which there was a return to *Eriophorum/Sphagnum* peat. Above 0.50m the stratigraphy comprised highly humified amorphous peat.

Pollen analysis

Pollen analysis at this core was restricted to the peat containing the lens of charcoal near the base of the profile, as the upper part of the core was considered likely to duplicate the information recorded in core B. Samples were analysed at 10mm intervals and the pollen diagram through this charcoal lens is shown as Figure 51. A single pollen assemblage zone is recognised.

SMC-a 3.90-3.83m Characterised by high frequencies of *Ulmus, Quercus,* and *Alnus,* with *Pinus* and *Betula* also significant. Non-tree pollen values are high with Coryloid and *Calluna* prominent. Herbaceous pollen is well represented, including aquatic, wetland, and ruderal types. Fern and moss spores are present in high frequencies. Tree pollen accounts for less than 40% of total pollen. This zone is radiocarbon dated to 4669-3970 Cal BC (5440±60 BP; Birm-1222).

Ecological history

The pollen assemblage agrees with the radiocarbon date in placing this charcoal-rich peat within the pre-*Ulmus* decline, Flandrian II chronozone. The black gyttja beneath the sampled sediment was deposited in a shallow lacustrine environment. This shallow lake was slowly colonised by reedswamp, with taxa such as *Typha angustifolia, Nymphaea, Filipendula,* and *Caltha.* A substantial amount of charcoal was deposited into these wetland environments following fire. This fire or fire sequence certainly affected the mire communities themselves (as demonstrated by the gross stratigraphical investigations), but may also have affected the mire catchment. A sharp increase in *Sphagnum* occurs above the charcoal lens and it is possible that one of the effects of the nearby fire was to increase the input of water into the wetland because of increased runoff from the disturbed slopes. This runoff from the sandy soils around the wetland could have led to a more acid environment and promoted growth of *Sphagnum.* Succession through fen and carr vegetation to ombrotrophic mire then took place.

It is conceivable, therefore, that the establishment of ombrotrophic communities in these areas may have been, if not caused, at least encouraged by the repeated burning of the initial carr and minerotrophic communities occupying these areas. This effect would be enhanced if combined with clearance of woodland in the surrounding catchment area. Further research is required to explore this possibility.

Landscape history and forest disturbance

Three phases in forest clearance history are recognised at Simonswood Moss C, as described below.

SMC-1 3.90-3.89m In the peat below the charcoal layer tree pollen values are at their highest, but still only reach 40% of total pollen, and an open wooded landscape is suggested beyond the bog itself. *Ulmus, Quercus,* and *Alnus* formed the woodland but *Pinus, Betula,* Coryloid, and *Calluna* are all prominent and may have grown as local shrub communities. The woodland may have been naturally open or could have been subject to previous disturbance. Occasional grains of *Artemisia* and *Rumex* indicate some open ground nearby.

SMC-2 3.88-3.86m This corresponds to the charcoal layer in the stratigraphy, and the pollen assemblage strongly supports the evidence of fire disturbance of the vegetation. Tree pollen values fall to 25% of total pollen, *Quercus* being most affected, and a real recession of tree cover seems to have occurred. *Plantago lanceolata* and *Artemisia* are recorded, while *Rumex* reaches a peak of almost 5% of tree pollen. *Pteridium,* an early coloniser of burned clearings, is present in high frequencies. *Urtica* and *Galium*-type occur and Gramineae also increases. The opening of the woodland canopy encouraged the growth and flowering of light-demanding shrubs during the regeneration phase after the fire, with Coryloid in particular showing greatly enhanced frequencies. *Salix, Prunus, Sorbus, Calluna,* and Rosaceae are other taxa which respond favourably.

SMC-3 3.85-3.83m This phase is one of successful regeneration of woodland with tree pollen values returning to pre-disturbance levels. Although Coryloid and *Calluna* remain high for a time they are declining towards the top of the sampled profile as closed canopy conditions return. *Fraxinus* is introduced into the assemblage after the disturbance, and *Betula* frequencies remain higher than before. *Pinus* returns to high values and was clearly important in the local woodland. Ruderal herb pollen is no longer recorded and any areas which did not regenerate trees remained occupied by scrub and heathland associations. There is no evidence of renewed disturbance. Although the mixed oak forest trees characterise the assemblage again, the open nature of the woodland remains its most significant feature.

White Moss (SD 482046)

White Moss and Holland Moss (*see below*), which lie almost entirely within Lancashire, probably once formed a discrete mire macrotope, now truncated and fragmented. Both sites, however, retain extensive sections of deep (*ie* >5m), well preserved peat.

At White Moss transects were placed across two sections of the deepest peats. On the northern fringe of the mire a short north-east to south-west transect of five cores (Fig 52) was undertaken, in order to

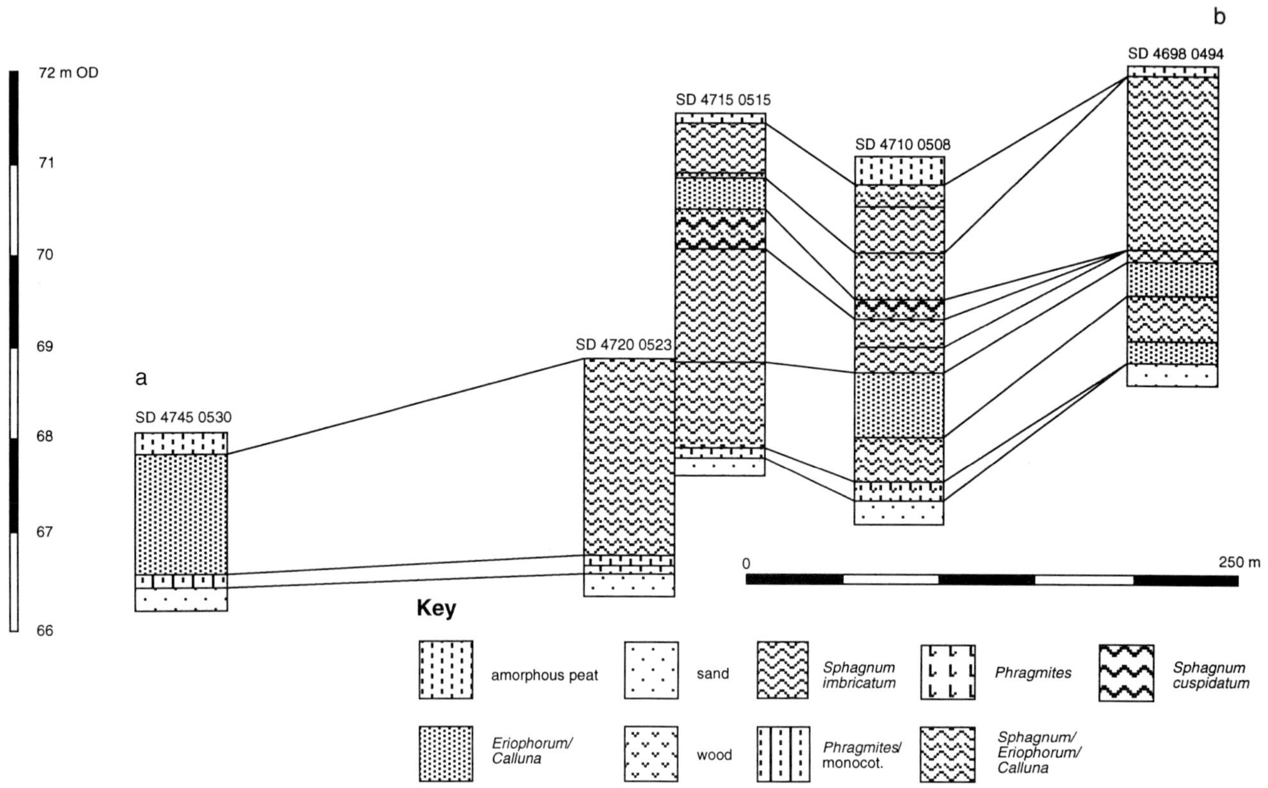

Fig 52 East-central mosslands: White Moss (north) north-east to south-west stratigraphic transect a-b

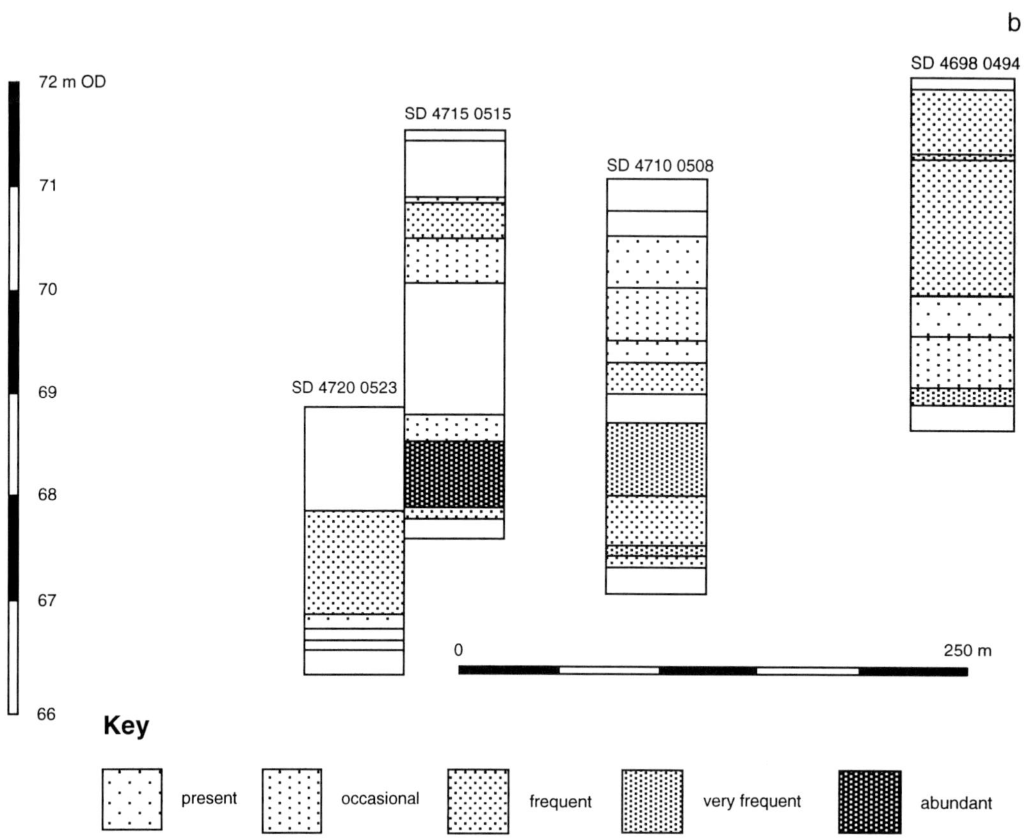

Fig 53 East-central mosslands: White Moss (north) north-east to south-west stratigraphic transect a-b, carbonised material

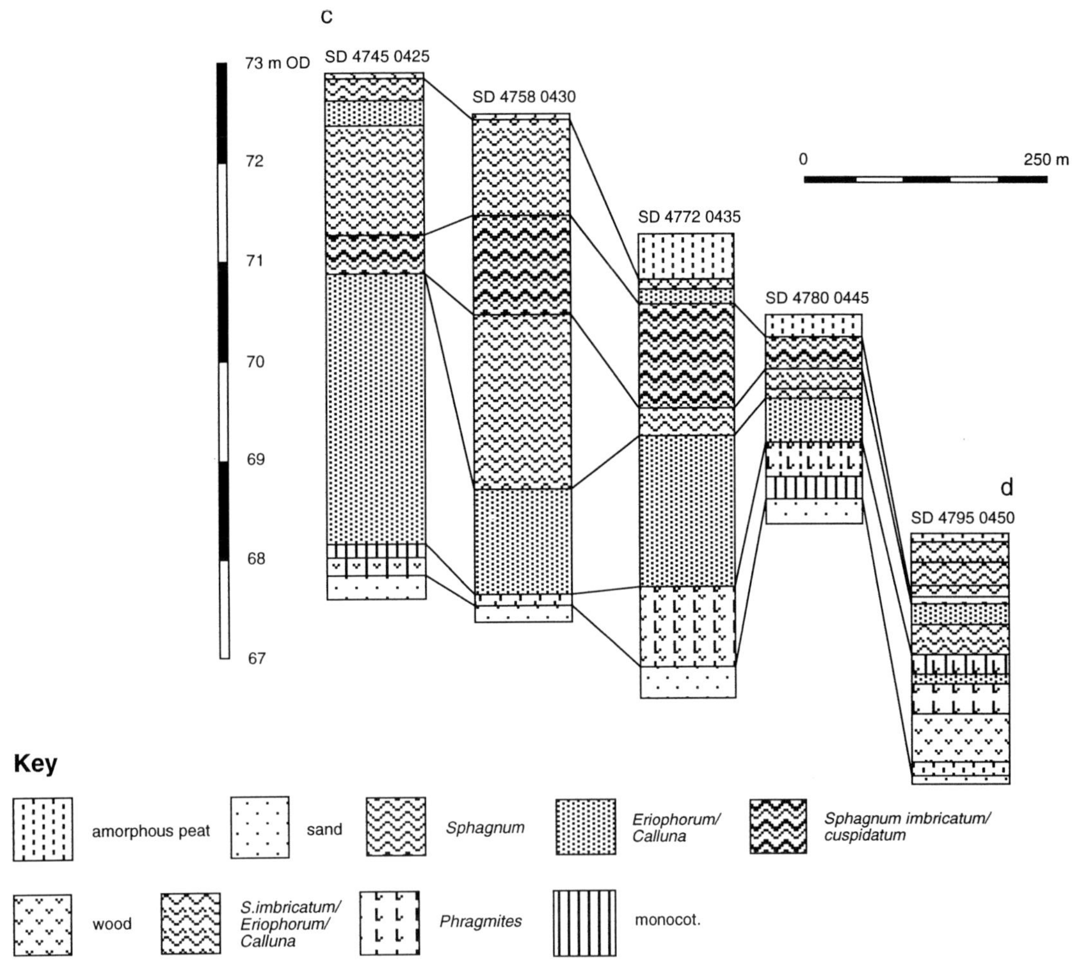

Fig 54 East-central mosslands: White Moss south-west to north-east stratigraphic transect c-d

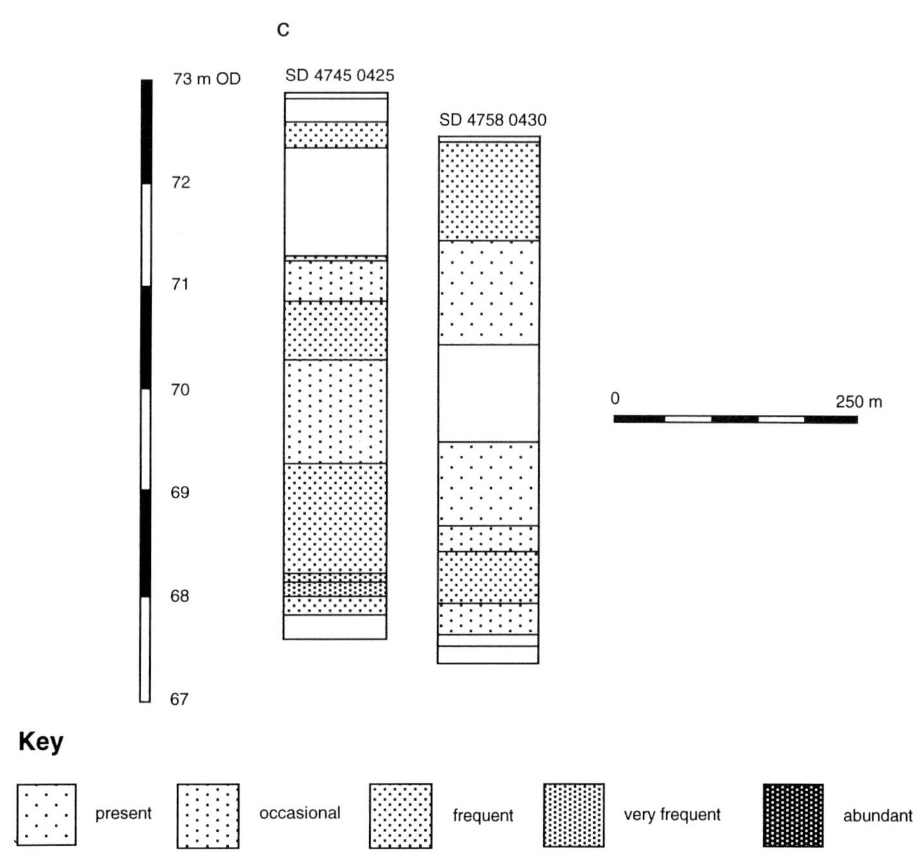

Fig 55 East-central mosslands: White Moss south-west to north-east stratigraphic transect c-d, carbonised material

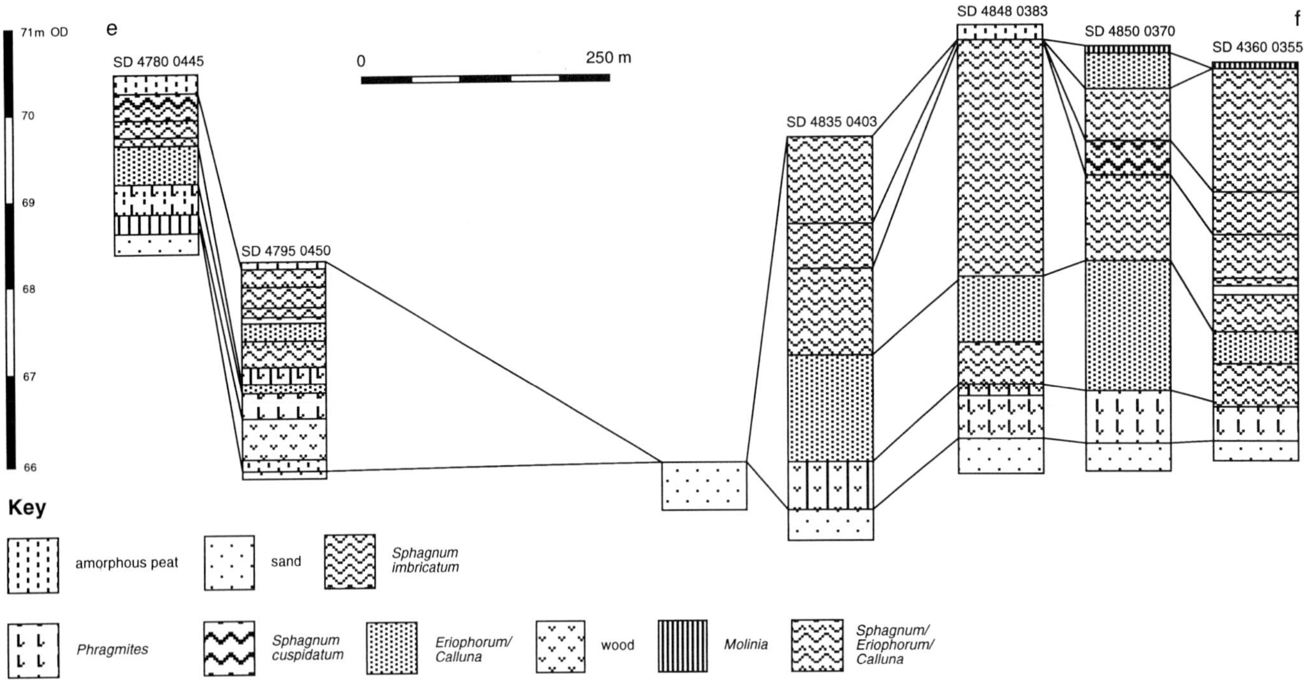

Fig 56 East-central mosslands: Holland Moss north-west to south-east stratigraphic transect e-f

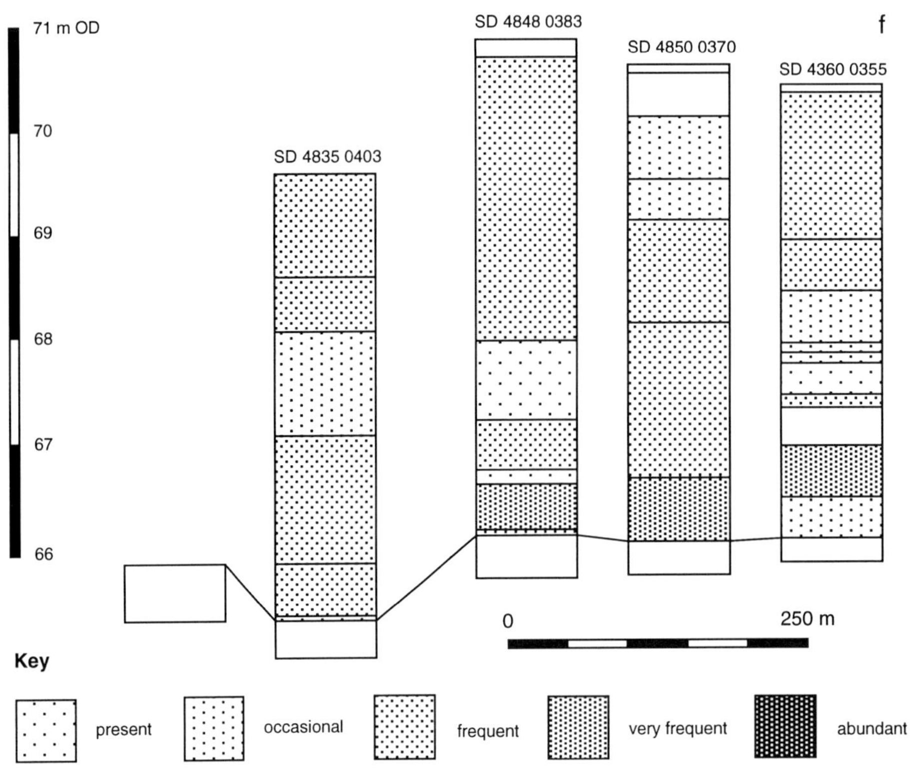

Fig 57 East-central mosslands: Holland Moss north-west to south-east stratigraphic transect e-f, carbonised material

investigate deep peats which are scheduled to be removed by industrial development in the near future. Further to the south a south-west to north-east transect of five cores (Fig 54) traversed deep peats under semi-natural vegetation and the cuttings of the White Moss Peat Works.

The northern transect follows a gradual rise in the underlying sands to the south-west. Overlying the sands peat reached up to 4m in depth, and was truncated by cuttings in the north-east. Basal deposits consisted mainly of monocotyledon Peats, but these are replaced by *Eriophorum/Calluna* peats in the south-westernmost core (SJ 46980494). Succeeding the basal deposits are sequences of peat mainly dominated by *Sphagnum*, *Eriophorum*, and *Calluna*. Carbonised material is notable for its concentration in the lower deposits of all of the cores but is also present in upper parts of the profiles, particularly core SJ 46980494 (Fig 53).

The southern transect indicates a downward slope in the underlying sands to the north-east; the apparent exception to this, seen in core SJ 47800445, is due to this core being offset further to the north. A general succession from basal wood peats to monocotyledon peats can be discerned, which are in turn followed by *Eriophorum/Calluna*-dominated peats and finally *Sphagnum*-dominated peats. Pool development is indicated in these upper peats by the presence of lenses of *Sphagnum cuspidatum*, an aquatic species. Samples for laboratory examination were only taken from the two south-westernmost cores, but they indicate that carbonised material, although only present in small quantities, tends once more to be concentrated in the lower reaches of the stratigraphy (Fig 55).

Ecological history

The northern transect represents a cross-section through the southern edge of a shallow basin, whose bottom is probably situated further to the north-east. An original wet hollow was colonised by aquatic vegetation represented by the monocotyledon peats, which was subsequently replaced by ombrotrophic communities represented by the *Sphagnum/Eriophorum/Calluna* peats. The lack of a monocotyledon basal deposit on the higher elevation, occupied by core SJ 46980494, indicates paludification of higher ground as the mire developed. To the south the second White Moss transect suggests that the underlying mineral surface dips again to form another shallow basin which initially experienced drier conditions than areas further north, allowing brushwood peats to develop. Flooding of this woodland is indicated by the overlying monocotyledon peats, which are a precursor to the establishment of ombrotrophic communities initially dominated by a 'dry' *Eriophorum/Calluna* facies, before developing into a wetter *Sphagnum*-dominated variant of raised mire. The widespread occurrence of significant thicknesses of *S sect Cuspidata* peats in the upper sections of this transect indicates a widespread phase of wetter conditions which affected the mire surface for some time. It seems likely that this represents a recurrence surface, although it is impossible to decide with accuracy which period it may belong to.

Holland Moss (SD 488038)

Holland Moss is situated immediately to the south of White Moss and it is probable that both mires formerly comprised one discrete macrotope. However, this is now fragmented and the mosses form visually distinct units. Holland Moss is easily the most impressive of the surviving deep peat areas in the region, its yellow whale-backed swell of *Molinia* and *Calluna* being a prominent landmark seen to the south of the minor road from Skelmersdale to Rainford, which separates it from White Moss. The moss is not completely undamaged, however, as attempts to remove peat cover on the eastern side of the mire in the 1970s apparently led to slumping of the peat body (Mr. Smith, owner, pers comm). This is manifested by a sharp break of slope visible on the site's eastern flank. The stratigraphy of the western portion of the mire appears to have been unaffected by this event.

A north-south transect of four cores indicated a domed profile in the underlying sands, which is reflected in the surface topography. Monocotyledon/*Phragmites* peats overlie the basal mineral deposits and are succeeded by monocotyledon/wood peats. Predominantly *Eriophorum*-dominated peats overlie this sequence, while *Sphagnum*-dominated peats occupy the upper halves of the profiles. Carbonised material tends to be present through much of the profiles, although there is a distinct trend to greater concentrations in the lower deposits (Fig 57).

Ecological history

The transect probably represents a cross-section through the northern edge of a shallow basin extending to the south, which is likely to have provided the locus for the original greater Holland Moss mesotope. The sand is seen to be dipping downwards again to the north, in core SJ 48350403, towards the White Moss system, with which peat the present Holland Moss stratigraphy was probably contiguous before reclamation.

The profiles indicate that aquatic and reedswamp conditions initially became established in the original shallow basin, which became terrestrialised by hydroseral succession by carr woodland. Ombrotrophic communities, initially of a 'dry' facies dominated by *Eriophorum / Calluna*, replaced these carr communities, until in the latter stages of the mire's development 'wetter' *Sphagnum*-dominated vegetation became established. Again, charcoal, although present sporadically through many of the profiles, exhibits a tendency to be concentrated in the lower reaches of the stratigraphy.

Dating

No absolute chronology is available for the White/ Holland Moss complexes, but by comparing stratigraphy with that recorded from Cundill's pollen diagram from White Moss (1981) some broad estimates can be attempted. The basal woody fen peats can be confidently assigned to the Mesolithic period. This suggests that the landscape of this time was characterised by extensive stands of fen carr and reedbeds, with occasional areas of open water. The overlying ombrotrophic peatland communities, mainly dominated by *Eriophorum / Calluna*, probably began to develop in the Neolithic, with *Sphagnum* becoming the dominant species sometime in the Bronze or Iron Age. The possible increase in surface wetness may have occurred in the Iron Age or Romano-British period, although this is highly conjectural. However, this would correspond to evidence from another area in the region, Over Wyre, where stratigraphy and macrofossil analysis indicate climatic perturbation around this time.

Conclusions

The archaeology of this area contrasts strongly with that of the coastal mossland areas studied in Chapters 3 and 4. The little material evident before the survey commenced, found particularly around the western side of the moss complex, has not been augmented significantly. The evidence for prehistoric archaeology in the central mosslands is still very scattered and would seem to represent apparent light 'background' material on the whole. The widespread extraction of the Shirdley Hill Sand from much of this area may have biased the recovery rate, but even where small sand islands were identified as having appeared through peat wastage since the sand extraction phase of the 1970s, no archaeological material was recovered, even with more intensive fieldwalking. It seems very likely therefore that there is a real difference in the prehistoric period between the settlement around the coastal mosslands and those of the interior mosslands.

Although the archaeological evidence around the mosslands is scattered and light, the palaeoecological record is both extensive and detailed at individual sites. It covers the period from the end of the last ice age to the late Roman period at Knowsley Park Moss and Holiday Moss, and provides much evidence for potential human interference throughout the second half of this period. This allows a contrast with the apparently more intensively settled coastal areas.

All the mires in the central mosslands had begun forming by the late Mesolithic period. From then into the Neolithic and early Bronze Age, small-scale human activity was taking place in the forests around the central and west-central mosslands. This was apparently unconnected with arable farming, although there is significant evidence for fires in the wetland ecosystems which may have affected catchment areas too. This may be associated with hunting episodes and utilisation of the wetlands, initially for fish and wildfowl, which probably formed the basis for the economic exploitation of the area into the first millennium Cal BC. The archaeological evidence suggests that this activity was more sustained around the western edge of the mosses, particularly in the early prehistoric period, where the few flint concentrations are found.

Later activity is mainly suggested by the pollen record. Relatively major deforestation episodes are signified in the palynological record relating to the late Bronze Age, where they appear to be associated with mixed arable farming. This suggests that settlement may have become more sedentary and may have been particularly concentrated around the west-central and central mosses. However, archaeological evidence for sites relating to this activity, derived from fieldwalking, chance finds, and aerial photography, is absent from the area of the mosslands and surrounding dryland areas.

This pattern of palynological evidence, suggesting mixed farming coupled to a lack of identifiable farmsteads around the mosses, continues into the Iron Age and Romano-British periods, and there is some evidence that at least parts of the mosses may still have been important in the Roman period, particularly around the southern edges. This may be a reflection of the foundation of the first settled farmsteads, apparently in the late prehistoric and Roman periods, on the sandstone areas in the south of the county. The mires were probably more suited to poor grazing by this time. By the medieval period, a good range of documentary evidence shows that the mosslands had become peripheral for settlement and mixed farming, forming open mire environments interspersed with areas of woodland, and formed the core of a landscape dotted with hunting parks and forest. In the sixteenth century, the first records occur for settlement and limited enclosure of the mosses. It is not until the mid seventeenth century that this process became widespread around the central and eastern mosses. The resulting pattern of dispersed farmsteads forms the basis for an agricultural landscape that has largely avoided industrial and modern urban development.

Significance of the central mossland evidence

The main artefactual evidence to come from the NWWS, as with many of the dryland areas to the south of the county, is biased to the Mesolithic, or at least to blade-orientated flint assemblages, which may include some early Neolithic material. Its significance is increased by association with a range of palaeoecological evidence which provides an environmental context for the human exploitation of the area.

One of the most important pollen sites in the region is recorded from Knowsley Park Moss. This extensively radiocarbon dated pollen diagram provides a record of the natural vegetational succession, from the earliest post-glacial until shortly after the late Roman period. This can now be seen as a companion to the Red Moss diagram, which has been regarded as the regional Flandrian palynological type site for the North West (Hibbert *et al* 1971). Unlike the latter site, however, the Knowsley Park evidence provides evidence of human activity in the area, from the late Mesolithic to the Roman period, that fits into a context of similar changes seen in other pollen diagrams from the west-central and central mosslands.

Mesolithic woodland burning episodes

A feature common to all the sites examined during the gross stratigraphical part of the survey is the presence of carbonised plant material. This is developed to a greater or lesser degree between mires, but one striking aspect is the tendency for burnt material to be concentrated in lower stratigraphical levels of the profiles, particularly in basal monocotyledon and wood peats. Although the dating evidence, currently confined to Knowsley Park and Simonswood, is limited, it would seem reasonable to place these events in the Mesolithic. If this is accepted then there are parallels for such burning episodes from other peats of this period in the region, notably at Hoscar Moss in West Lancashire (Cundill 1984), thought to date to *c*5900 Cal BC, and at Starr Hills at Lytham in the Fylde (Tooley 1978a), where a date of *c*5500 Cal BC was obtained for a burning episode.

Turning to areas of Britain outside North West England, Mesolithic burning has been recorded extensively throughout Scotland, with the exception of the Central Highlands (Edwards and Ralston 1984). In most of these cases the burning has been interpreted as of anthropogenic origin, and associated vegetation changes have been attributed to the artificial manipulation of fire.

On the basis of the current evidence it is impossible to state definitively whether the Merseyside Mesolithic burning evidence is natural or artificial in origin. Most authors, however, have tended to assign prehistoric charcoal found within peats to human influence, for several reasons. Firstly, it is thought unlikely that British woodland is prone to natural combustion, due to the damp nature of the undisturbed ecosystem and high moisture content of most native deciduous trees when alive (Rackham 1980). This effect might be still more pronounced in the carr communities which appear to have been affected by burning in the Merseyside peats. Natural fires would probably have been even less likely in the reedswamp communities which were also frequently affected.

If artificially induced fire is being recorded in the Merseyside peats it may be as a result of chance isolated fires linked to domestic activity, or through deliberate attempts at larger-scale manipulation of the landscape. The widespread occurrence of charcoal in the central mossland peats tends to support the latter suggestion. In this case the concept of the 'fragile ecosystem' proposed by Simmons *et al* (1981) becomes relevant. This promotes the notion that Mesolithic interference at relatively less stable ecotonal boundaries may have a major effect on the dominant plant species. Reedswamp and carr woodland can be accommodated within the notion of such fragile ecosystems, being seral in nature and forming an open environment set amongst the more robust closed forest of the time. There have been suggestions, based on ethnological parallels, that Mesolithic people may have attempted to manipulate reed and carr habitats by burning, in order either to drive game (Mellars 1976) or to encourage vegetation attractive to grazing ungulates (Simmons *et al* 1981).

Whatever the purpose, if any, of such widespread burning of the wetland surface, there remains the intriguing possibility that ecological succession may have been deflected or altered in some way by human activity. It is striking how often concentrations of carbonised material appear to cluster immediately below the onset of ombrotrophic conditions. A similar phenomenon has been extensively documented from numerous blanket mires (Moore *et al* 1986; Moore 1975), usually associated with palynological indications of forest clearance. Many authorities now accept the probability that human activity affected the timing and extent of mire development in the uplands (Simmons *et al* 1981; Moore 1986). The possibility exists, therefore, that a similar process may have been operating in the lowlands. The establishment of ombrotrophic communities in these areas may have been, if not caused, at least encouraged by the repeated burning of the initial carr and minerotrophic communities occupying these areas. This effect would be enhanced if combined with clearance of woodland in the surrounding catchment area. There is limited evidence for the latter activity from Simonswood Moss C and Tooley's (1978a) studies, but further research is required to explore this possibility.

Later burning episodes

The reduced frequency of carbonised material in later sections of the profiles, if of anthropogenic origin, attests to a continuing interference with the vegetation of these systems, albeit at a lower level of intensity. This is undoubtedly a reflection of the less attractive qualities which low productivity ombrotrophic systems offered to people in terms of economic exploitation. Much of the carbonised material recorded from the ombrotrophic levels is Ericaceous, suggesting light fires 'clipping' leggier heather in a fashion similar to a modern-day 'muirburn' (Muirburn Working Party 1977).

Archaeological and palaeoecological potential and threats

The peat across the central mosslands is mainly quite deep, averaging *c*2m in the west-central and eastern mosses, but in parts of the central area this reaches between *c*3-7m deep. It is well preserved in a number of locations, although wastage of peat continues in many areas as a consequence of agriculture. Small areas do, however, remain unreclaimed and carry a semi-natural vegetation. The most noteworthy are in the eastern mosses, particularly on Holland Moss, but also around the fringes of Holiday Moss, and in the west around the Simonswood peat cutting area. These surviving peat archives are therefore important in retaining palynological records for the study of past vegetational changes. Their value is enhanced by their recognised potential to reveal features of archaeological significance.

The most immediate threat to the peat comes from the need for the daily disposal of large amounts of domestic refuse from the large urban centres that surround them. This is a continuing and increasing threat and is illustrated at Kirkby Moss, where an area of *c*25ha of urban waste has already been directly tipped on to the peat surface; this has now been returned to farmland. Developers are also investigating the feasibility of extracting peat for subsequent use of the excavated area for landfill, on former farmland around Kirkby.

6

THE SOUTHERN MOSSES

The mosses studied in previous chapters have survived mainly within extensive, deep basins in the northern third of the county, where the dominant geology is the Shirdley Hill Sand. The physical background to the southern part of the county contrasts strongly with the main areas surveyed as part of the NWWS. In the boulder clay and sandstone areas in the southern half of the county, such deep basins are rarer. Medieval and post-medieval surveys, rentals, and court rolls (*see below*) make it clear that areas of peat were quite common across the landscape here, but these deposits were never as extensive as in the north. In these southern areas surface peat has very rarely survived the urban and industrial growth that characterises this part of Merseyside. This fact has restricted the work of the NWWS here. The comparative study of the peat from the southern area therefore gives opportunities to investigate how the exploitation of the wetlands might have been affected by different geographical and topographical conditions.

This part of the county is characterised by a zone where extensive boulder clay plains meet both the southern end of the Carboniferous sandstone outcrops, which provide the highest relief in this part of the region, and the south-eastern edges of a series of lower Triassic sandstone ridges and outcrops that extend westwards towards the mouth of the River Mersey.

The main area of peat lies within the Metropolitan district of St Helens (Fig 58). Here a small belt of relict mossland is located on the south-eastern edge of the town of St Helens, comprising the adjacent mosses of Sutton, Parr, Bold, and Burtonwood (hereafter grouped under Parr Moss for brevity). The second peat area lies within the Metropolitan district of Knowsley, *c*8km to the south-west. It is located on the southern edge of the township of Eccleston, adjacent to the former boundary of the medieval town of Prescot.

Landuse

More than any of the other mossland areas in the county, except Wirral, these mosses are located in areas that have made them vulnerable to urban and industrial development. The southern areas attracted many of the nucleated medieval villages in the county, and later, during the Industrial Revolution, the almost continuous belt of urban development stretching from Liverpool to St Helens that subsumed the earlier settlements. This means that hardly any surface indications of the former mosslands have survived, and so they have to be relocated from documentary sources. From time to time, however, their presence is manifested by exposures uncovered during construction work, such as was the case for the Prescot site presented in this volume.

The area of the Parr Moss complex is first recorded in 1230 (Farrer and Brownbill 1907, 358 n5) but not until the post-medieval period was it shown on maps. The earliest of these is an estate map of Burtonhead *c*1580 (Fig 60), where it is shown as a large, oval, open area known as 'Markelsse moss', surrounded by a hedge. It was still depicted as unenclosed on Yates' map of Lancashire (1786) (Plate 5), as it essentially still was in 1849 when the 1st edition OS 6" map shows it as being divided between the four townships of Sutton, Parr, Bold, and Burtonwood. About half the mossland area was drained and enclosed for agriculture by the start of the nineteenth century, relating mainly to the shallower outer parts. The south and east parts, where the peat is deeper, were reclaimed between 1848 and 1913. In the early twentieth century, although industrial development had grown up all around the mossland, Parr and Sutton mosses still remained open ground. It was not until later in the twentieth century that the potential of the mossland was severely curtailed, when it was used as a tip for the now disused adjoining Bold colliery. The tip is *c*40m high and covers *c*80% of the former area of the moss.

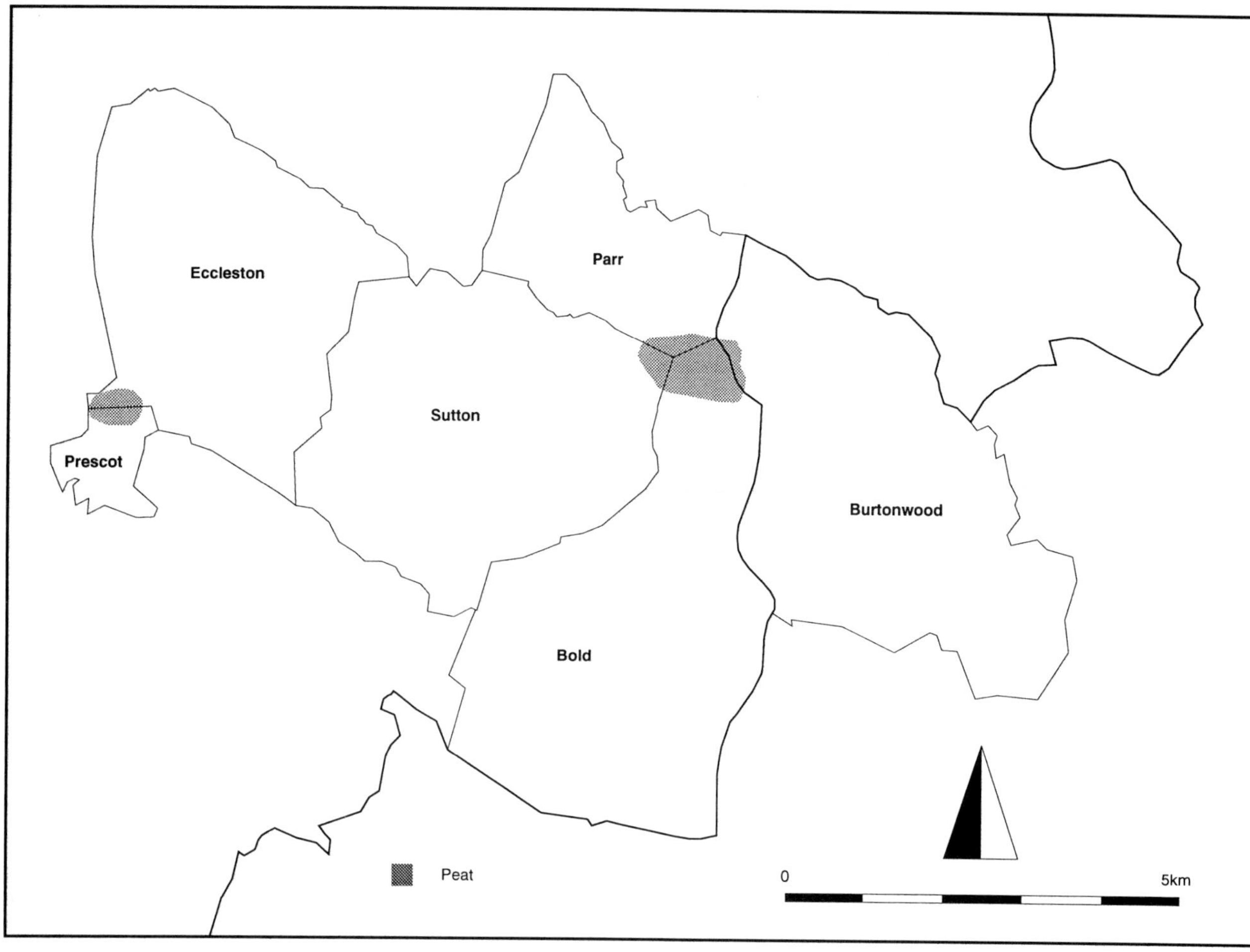

Fig 58 Southern mosslands: Location of survey area

A small area, totalling 12ha, on the east of the former moss, is all that survives as open ground, which is left as semi-natural vegetation.

Prescot Moss was granted to the town in 1508, having formerly lain within the township of Eccleston. During the sixteenth and seventeenth centuries the court leets mention the many restrictions applying to the use of the moss by the townsfolk (Bailey 1937), when it was being used mainly for grazing animals. By the late eighteenth century, however, it had been enclosed as farmland lying on the edge of the town. The urban development that now covers this area is mainly of late nineteenth and early twentieth century date, although the beginnings of this growth are represented by several buildings shown on the southern part of the former mossland on the 1848 tithe map (LRO, DRL 1/65).

Previous research

There is little evidence from the prehistoric and Roman periods in this area. The nature of the landuse has prevented systematic survey, while few chance finds have been reported from the vicinity of these mosslands. This may be partly due to the lack of a local antiquarian tradition that could have recorded any sites or material that were being destroyed as industrial and urban growth took place from the mid nineteenth century onwards. In contrast, more attention has been paid to the nature and development of this urban growth. The industrial history of St Helens has been treated in detail (Barker and Harris 1959), while a more general historical development of the growth of St Helens has also been produced, deriving from the compilation of the county SMR (Chitty 1981).

The town of Prescot is one of only three medieval boroughs in the county, and has seen the detailed publication of surveys and other documents relating to its early growth (Bailey 1937). Since the late 1970s it has also received considerable archaeological investigation. Topographical surveys (Davey 1978; Philpott 1988) identified its medieval core and post-medieval growth, in response particularly to the seventeenth century and later potting industry. Subsequently, rescue excavations have taken place (Holgate 1989; McNeil 1989; Philpott in prep) in addition to a series of small-scale, *ad hoc* salvage excavations (Cowell and Chitty 1989), and regular monitoring of developments. In all this work no material earlier than the medieval period has been recovered.

The only work on these mosses that relates to earlier periods of settlement and landuse derives from a programme of palaeoecological survey work undertaken by Liverpool Museum in the late 1970s. In 1979, when development took place in Warrington Road, on the southern edge of the former Prescot Moss, a peat core was taken and subsequently analysed for its pollen record. Two cores for palynological study were also taken on Parr Moss as part of this survey programme. The detailed results from this work are presented in the palaeoecological survey section.

Geology and topography

The mosses in this area have developed in an undulating landscape caused by a series of low hills and ridges, typically *c*50-60m OD, of Carboniferous and occasionally Triassic sandstones, around which glacial boulder clay has been deposited. The ridges of high ground, generally trending north-south, lie at the southern end of the Billinge ridge upland and continue westwards through the southern part of Knowsley Park and Prescot towards Liverpool. In the study area they form the basis for the modern growth of Prescot, Eccleston, and the northern part of St Helens, but this whole zone has historically attracted the greatest density of settlement.

The south and south-eastern parts of St Helens have been built over a basically boulder clay landscape which forms an extensive plain to the south and east, and is found in all the valleys formed within the series of sandstone ridges. Parr Moss is located on the north-west edge of this boulder clay plain at *c*30m OD, with the slightly higher ground on which the town of Sutton has grown up lying immediately to the west. The mossland is located between the major valley of the Sankey Brook, approximately 1km to the east, and the tributary of the Sutton Brook, which skirts the mossland to the west before joining the main valley 1km to the north. From Sutton, the land rises gradually to the west, to *c*75m OD, where it forms an east-west ridge.

On the western edge of this ridge, 7km away, lies the town of Prescot. There is quite a steep fall from the ridge into the valley of the Prescot Brook to the west, and more gradual falls to the north and south. The moss lay in a small basin at the foot of the gentle slopes on the northern edge of the town, where there is a boulder clay capping to the sandstone. Immediately to the north and south of the valleys either side of the Prescot ridge lie the Carboniferous sandstone outcrops of Knowsley Park and Whiston, respectively. To the north, in the park, the ground is at its highest, at 90m OD, while the Carboniferous ridge gradually narrows and lowers to *c*50m OD south of Whiston, *c*5km to the south. A substantial sandstone ridge also runs south-eastwards from Prescot, forming the focus for the medieval village and late nineteenth century town of Rainhill.

Flandrian geology

Both wetlands began as open water occupying hollows in the glacial deposits, around which reedswamp gradually developed. At Prescot this stage developed quite early, as by 5957-5560 Cal BC (6850±80 BP; GU-2723), during the late Mesolithic, an alder-dominated fen carr had colonised the site.

At Parr Moss the reedswamp may have become established a little later, as fenwood did not develop here until after *c* 3898-3100 Cal BC (4740±40 BP; Birm-1224), although the sampling site may not be typical of the mire as a whole. As the swamp dried out birch became particularly dominant, along with alder and other shrubs (Table 15). Towards the late fifth millennium Cal BC the fen woodland became replaced by ombrotrophic communities as cotton-grasses (*Eriophorum* sp) spread, along with heather and *Sphagnum* moss. The latter gradually came to dominate the surface of the bog, which remained as an active ombrotrophic mire until cutting and drainage in the post-medieval period.

Prescot Moss appears to have suffered a degree of soil erosion around the hollow, estimated to around the beginning of the fifth millennium Cal BC (Table 14), as the peat in the upper part of the core contains much clay suggesting possible inwashing of sediments. At this time pine, birch, and hazel woodland was common around the hollow. Part of the profile also appears to have been truncated at this point, while the anomalous date of 7051-6709 Cal

BC (one sigma: 7980±60 BP; GU-2724) from a birch stump further suggests intrusion into the moss at this point.

Survey techniques

The mosses treated in this section presented great problems for survey and analysis. They are mostly covered by recent tipping or urban growth, without any record having been made of any discoveries whilst this was happening. Of the two mosses treated in this survey, surface examination of the peat was possible only on Parr Moss (Fig 86). The fact that the surviving peat lies adjacent to large-scale industrial tipping means that little of its condition could be recorded. Prescot Moss is even less accessible to survey and identification as a result of the urban development and leisure facilities which now cover it. As landuse is so restrictive of a full appreciation of these mosses, no detailed maps have been provided for Prescot Moss, while those for Parr Moss have been amalgamated.

The mosses

Parr Moss

Here the original extent of the moss lies between the urban areas of Sutton on the west, Parr on the north, Bold colliery, the adjacent power station, and Burtonwood brewery to the south of the railway line, and a small amount of farmland to the east towards Collins Green in Burtonwood, Cheshire, although there are mines and small areas of coal tips covering parts of this area also (Fig 85). To a large extent this development, except for that south of the railway line, has avoided the historic core of the mossland, which amounted to *c*65ha. This has been severely damaged, however, by a large waste coal tip, reaching some 40m high, which has grown up in the centre of the mossland in this century, effectively destroying most of the potential the moss might ever have had. This has left only small surface areas of peat around the edges of the tip, totalling *c*12ha, being most extensive in Burtonwood Moss.

The nature of this landuse made survey of the surviving mossland difficult, as there was no way of recording the peat. The small area of arable land on the east side was also unavailable for fieldwalking at the time of the survey. Aerial photographic coverage has been poor in the past, and this source provides no extra information. In 1979, however, when there were plans to develop the rest of the site, a series of boreholes and pollen cores was taken as part of the Liverpool Museum programme of palaeoenvironmental survey, and these give some insight into the nature and condition of the peat in the area. These showed that the area of Burtonwood Moss was 3-4m deep around this south-eastern edge, and moderately well preserved. On the west side, however, in Sutton Moss, the peat was shallower (*c*1.5m) and its surface showed signs of having been cut and disturbed. The pollen preservation in the latter part of the moss was also poor, in comparison to that in the Burtonwood part of the moss. There is no indication of the depth or sub-surface contours of the centre of the moss complex, which lies beneath the coal tip.

The tip and adjoining mossland amount to *c*65ha, and are managed by The Groundwork Trust who are actively seeking to manage it as a recreational and educational resource, by creating a conservation area with a variety of vegetation and wildlife. Accordingly surveys of the vegetation and wildlife have been undertaken, and although the main focus is on restoring the tip, the surrounding mosses will also be included to provide a diversity of natural habitats. At the moment the moss vegetation mainly consists of *Molinia* and *Eriophorum* with dense willow scrub. Cheshire County Council are also investigating the possibility of managing the adjacent Burtonwood Moss (owned by British Coal) for nature conservation. Their survey has identified fen and bog communities on this portion of the moss.

Prescot Moss

The location of this moss is only known today from historic maps which show it lying to the north of the medieval town, where it covered an area of 40ha. The boundaries of the moss can be approximately traced from the tithe map as lying between St Helens Road to the east, the High Street to the south, the back of the tenements fronting Hope street on the west, and a line running to the north of, and parallel to, Eaton Street on the north. About three-quarters of this area now consists of late nineteenth century housing with gardens, bowling greens, and a football ground on the west and north-western perimeter, and an area of waste ground, used as a car park, on either side of Moss Street. The ability to locate the archaeology associated with the moss is therefore very limited. The only survey undertaken in this area consisted of a field visit to check the current nature of the landuse and identify any imminent threats to it.

Some documentary work was undertaken to attempt to identify the former bounds of the moss, as this seemed advisable in relation to future construction developments in the area, since the palaeoenvironmental work on the site indicated some potential of relevance to the county as a whole for the prehistoric period. This has produced the only information as to the condition of the peat on the moss, which at the sampling site was only 0.6m deep (Table 14), with a thick layer of modern building material dumped on the peat surface.

Archaeology and palaeoecology

Mesolithic *c*8000 BC – 4000 Cal BC

As no archaeological fieldwork has been possible due to the developed nature of the landscape in the area, the position of the mosses in their contemporary human landscape has to be viewed in general terms. Prescot Moss developed through hydroseral stages of open water, reedswamp, and alder carr during the Mesolithic and earliest Neolithic periods. During the Mesolithic, the woodland canopy around the site showed no changes other than a decrease of pine and birch and the subsequent increase of alder and oak woodland, with some lime, elm, and hazel (Table 14). There are no indications from the vegetation record that hunting communities might have been active in these forests, although in circumstances such as, for example, small groups being in transit between hunting stands, such indications need not be apparent.

The glacial hollow in which Parr Moss developed may have been dry during the Mesolithic period. Thus, in an area of extensive boulder clay soils, the site may not have been particularly attractive, although sandstone ridges, which may have provided less obstacles to movement, do penetrate the clay plain around the site. A previous programme of fieldwalking over a wider area, including the boulder clay plain *c*2km to the south of the mossland, produced no prehistoric material whatsoever (Cowell 1991a). About 5km to the east, however, several small Mesolithic sites were recorded along the upper part of the eastern sandstone ridge overlooking the Sankey Brook, and it may be that the major valleys, such as those of the Sankey and the Ditton (Cowell 1991a), were the major foci of activity in these southern areas.

The loss of the natural topography to the west of Parr Moss, with urban growth stretching virtually from St Helens to the outskirts of Liverpool, probably produces an artificial pattern to the distribution of such sites in the south. It might be expected that light exploitation of some of the river valleys in these sandstone areas was taking place, as was the case in Tarbock, *c*3km to the south (Cowell and Philpott 1994). The results of the NWWS, when combined with the results of previous survey more generally across the county (Cowell 1991a), suggests that the major emphasis for settlement in the county was placed on exploitation of the coastal areas and, to a lesser extent, the western central mosses.

Neolithic/early Bronze Age *c*4000–1650 Cal BC

The first indications for the possible human exploitation of these southern mosses is reasonably consistent at both the southern mossland locations. At Prescot Moss, human activity is suggested in the vicinity, with a phase of farming evident between 3630-3106 Cal BC (4650±80 BP; GU-2725) and 3630-2890 Cal BC (4520±140 BP; GU-2726), recognised by a limited fall in tree cover and the presence of *Triticum* cereal grains and some weed pollen (Table 14, phase 3). The presence of clay lenses in the peat dating from around this time suggests that soil erosion was taking place around the moss, which might have been associated with this farming activity. A subsequent phase of arable farming, woodland reduction, and disturbed ground is recorded. This may be late Neolithic/early Bronze Age in date, but as the moss is truncated above this level (Table 14, phase 4), probably because of medieval and post-medieval cutting, it should be regarded tentatively as contamination from pollen associated with later farming activity.

At Parr Moss, at about the same time, at 3898-3100 Cal BC (4740±140 BP; Birm-1224), there is evidence that very small, temporary openings to the forest cover took place, causing slight reductions in lime and elm and the appearance of weeds associated with clearance, although the woodland cover remained little affected overall (Table 15, phase 2). A second, similar phase is present, spanning the late Neolithic / early Bronze Age, terminating *c* 2279-1750 Cal BC (3620±90 BP; Birm-1226) (Table 15, phase 4). This is, however, a phase of more pronounced woodland disturbance, perhaps lasting several centuries, with all tree species, but particularly lime and elm, affected and many types of disturbed ground weeds present (Table 15, phase 4). After both these phases the surrounding oak woodland regenerated to become the dominant feature of the landscape.

The evidence from Prescot Moss forms the earliest

Table 14 Southern mosslands: Prescot Moss, palaeoecological summary

cms	Date (Cal BC)	Flandrian Zone	Local Pollen Zones	Ecology of Moss	Regeneration/ Disturbance Phases
8 9	3630-2890	FL III	**PR-e** Characterised by alder (dominant pollen type), birch, and oak with high hazel. Shrubs recorded. Ruderal and other herb types increase.		PR-4 More intensive clearance. Cereal-type pollen consistent. Range of herbs indicative of arable farming recorded. Tree pollen reduced.
10 20 22 23	3630-3160	FL III	**PR-d** Birch, oak, alder, and hazel characteristic types. Elm less, lime recorded. Ruderal and wetland herbs sporadic.	High proportion of clay in peat in zones PR-e and PR-d. Possibly due to erosion as a result of clearance in zone PR-b.	PR-3 Clear evidence of disturbance. AP (mainly alder) falls. Cereal-type pollen and *Plantago lanceolata* recorded. Increase in pollen of plants of open ground.
37 38 42 43	7051-6709 (1 δ)		**PR-c** Tree pollen as in PR-b. More heather and willow. Rise in bracken spores, grass pollen, and range of herbaceous types. Tree stump - no pollen.	Hiatus ?due to inwash	PR-2 Trees dominant with herb taxa and clay particles suggesting limited disturbance.
50 52 53	5957-5560	Fl II	**PR-b** Alder and oak dominant with hazel. Pine and birch low. Elm present in low values.	Floristically-rich fen carr dominated by alder with some birch.	PR-1 Mixed woodland with little evidence of clearance.
60		Fl II	**PR-a** Alder, pine, and birch characteristic with high hazel. Significant grass and sedge pollen. Some wetland and aquatic herbs recorded.	Reedswamp Open water	

indications of cereal farming in the interior of the county, in the second half of the fourth millennium Cal BC. This contrasts with the possible record of the first use of cereals on the coast as early as *c*4900-4530 Cal BC, and the potentially more secure interpretation of cereal use, on the North Wirral coast, at a similar date to the southern mosslands of *c*3690-3360 Cal BC (4740±70 BP; SRR-2924) (*see Chapter 7*). The chronology of, and processes leading to, the adoption of farming in the county are difficult to chart, but it seems likely that by the time of the clearings and possible cereal cultivation around Prescot Moss, Neolithic cultural and economic traits were fairly widespread across this part of the county.

It has been proposed for the coastal areas of the county that landuse and settlement changed little from the Mesolithic into the Neolithic period. Thus hunting, and probably gathering, were still an important part of the subsistence base and cultivation was small-scale, shifting, and not necessarily always a component of landuse in localised areas. Many of the herbs and weeds associated with the woodland reduction in the vicinity of Parr Moss suggest that a pastoral type of landuse may have been in existence (Behre 1981).

The archaeological evidence for this period is even sparser than that for the Mesolithic in these southern areas. A stone axe of this period was found in the nineteenth century at the south end of Newton-le-Willows, *c*4km to the east (Garstang 1906) and provides some evidence for continuity of the pattern of exploitation on the eastern ridge of the Sankey begun in the Mesolithic. An isolated stone axe has also been reported from Ashton, north-east of the Sankey Valley, slightly further afield (Cowell in press).

Although the evidence is very tenuous, it might be possible to see a continuation of the Mesolithic pattern in these southern areas. In the varied topography around Prescot, which is based on a series of relatively low sandstone ridges, clearings were made and cultivation took place in the vicinity of the moss during the middle Neolithic. It is also this kind of landscape, across a slightly wider area, that was being used by hunter-gatherers in the Mesolithic.

In contrast, around Parr Moss, which is adjacent to the boulder clay plain, where previous fieldwalking has produced no prehistoric material, there is only evidence for small-scale, potentially pastoral disturbances to the woodland in the Neolithic, although cereal-type pollen can be under-represented at pollen sites where thick woodland is dominant. This basic distinction between the two contrasting locations in the archaeology and pollen record might hint at a possible interpretation of landuse at this time. This would see the lighter soils on the sandstone as being more suitable for cultivation, in areas that had a tradition of exploitation by preceding hunter-gatherer communities. The use of the same areas, containing topographical diversity adjacent to rich fen carr resources, could then have been favoured for a whole series of reasons resulting from social (*eg* kin links to traditional hunting grounds), practical (*eg* unregenerated or partially regenerated clearings), or economic (*eg* the need for easily cultivated land) factors. In the absence of prehistoric archaeological evidence from the heavier, flatter claylands it might be implied that some of these factors may not have been present, so that use of this area might have been less intensive. This could have taken the form of a regime of extensive grazing, possibly complementary to the economy of the sandstone areas, and perhaps on a seasonal basis, with consequently less impact on the landscape.

Some parts of the national archaeological record are difficult to distinguish between the late Neolithic and the early Bronze Age. This is even more the case locally, as archaeological data from secure contexts is very scarce for these periods in the county. The area around the mosslands is devoid of any evidence relating to the Bronze Age, but *c*7km to the east, slightly to the north of the confluence of the Sankey and the Mersey (just outside the county boundary, at Winwick, Cheshire), is a small group of barrows (Chitty 1981). This group provides one of the greatest concentration of such sites in the lowlands and implies that this area of river valleys was of some significance. Two barrows have been excavated and shown to have been in use for cremation burials from *c*2200-1750 Cal BC (Freke and Holgate 1990).

Later prehistoric period *c*1650 Cal BC – AD 43

Towards the end of the early Bronze Age woodland regenerated with quite dense oak and ash cover in the vicinity of Parr Moss, and this remained the case for several centuries. This was succeeded by a relatively long phase where oak was reduced and weeds associated with pastoral ground appeared, although arable weeds were also present (Table 15, phase 6) suggesting a mixed form of landuse. It seems that rather than one long episode of clearance and farming in the vicinity the pattern was more of smaller peaks and troughs of clearance, although much more detailed palynological sampling is required to establish this. The surface of

Table 15 Southern mosslands: Parr Moss, Burtonwood, palaeoecological summary

cms	Date (Cal BC)	Local Pollen Zones		Ecology of Moss	Regeneration/Disturbance Phases	
					PMB-9	Major forest clearance (increasing). Cereal-type pollen recorded. Herb flora suggests possible mixed farming.
55			Oak and alder (80% AP) characterise zone.	*Sphagnum* moss dominates in later stages.		
					PMB-8	Clear evidence of human impact. AP falls, several ruderal types recorded, peak of bracken spores, and *Cannabis* recorded.
			Hazel unchanged.			
85						
			Birch values fall.		**PMB-7**	Woodland regeneration (oak and ash). Open areas may have persisted.
115	930-410	**PMB-b**	Ash more frequent.	Acid raised bog, with *Eriophorum*, heather, and *Sphagnum*, develops as birch declines.		
120			Heather, grasses, sedges, and bracken important.		**PMB-6**	Significant woodland disturbance (oak). Mainly pastoral weeds recorded but some arable ones. High heather.
155			Herbaceous pollen increases towards top of zone. Ruderals recorded.			
					PMB-5	Closed woodland returns in vicinity of moss. Oak with increased ash. Rise in heather pollen possibly result of incomplete regeneration.
195	2279-1750		*Sphagnum* values fluctuate.			
205					**PMB-4**	Some significant forest clearance. Elm and lime severely affected. Weeds of disturbed ground recorded.
230	2878-1986					
235						
240						
245						
280			AP >60% of TP. Birch pollen characterises zone with oak, alder, and hazel.	Rich reedswamp-fen carr community with birch, alder, willow, buckthorn, ivy, and honeysuckle.	**PMB-3**	Undisturbed oak/alder woodland. AP dominant. Earlier clearings colonised by variety of shrubs. Heather in woodland or raised mire.
305						
		PMB-a	Heather rises at upper boundary.		**PMB-2**	AP 90% TP. Possible small temporary clearings. Elm and lime decline. *Plantago lanceolata*, heather and thistle type pollen recorded.
327	3898-3100					
335			Herbs low throughout. High values of fern spores at base.		**PMB-1**	Dense deciduous forest, mainly oak and alder with some elm and lime
390			Aquatic pollen	Open water		

the mire had by this time developed into raised bog, with heather, grasses, and sedges prominent. This phase ended around 930-410 Cal BC (2590±100 BP; Birm-1227) with fairly complete woodland regeneration, thus placing this farming activity in the late Bronze Age or the earliest part of the Iron Age.

The archaeology of this period is rare across all of North West England and therefore it comes as no surprise that there is no contemporary evidence from around the moss to place against the economic activity, which suggests a degree of permanence in settlement may have taken place for the first time. One type of settlement suggested as being of this date, best identified from aerial photographs, took the form of single enclosures. These are generally of oval shape but are also found in a more rectilinear form, which group may be of slightly later, Romano-British date (Bewley 1994). A number of possible examples have been identified in the county, for example on the Wirral (Philpott 1991) and at Halewood (Cowell and Philpott 1994). The areas in which they would seem most likely to have been located in the area of these southern mosslands, *ie* the sandstone ridges, are now mostly lost to urban and industrial development.

Romano-British period AD 43–410

The settlement of the late prehistoric period in the area may be difficult to identify, but the few indications there are suggest that in form and location it may not have been too different from that recorded for the Romano-British period, which was based on single farmstead enclosures. There also seems little difference in the palaeoecological record between the two periods. After a regeneration phase at Parr Moss (Table 15, phase 7), which is dated to the later part of the Iron Age, there are two more phases of woodland clearance recorded from around the moss. Assuming that no hiatus has occurred in the peat profile, the position of the first of these in the profile (Table 15, phase 8) could be interpreted as belonging to the end of the Iron Age or Romano-British period. It is a very similar phase to the one dated to the late prehistoric period, with the presence of *Cannabis* pollen suggesting that arable farming may have been a feature of this late prehistoric/Romano-British activity.

There are no Romano-British finds from the St Helens urban area, and the scale of urban development on the lighter, better drained soils in the vicinity militates against discovery of sites through aerial photography. The sandstone areas to the south of Prescot have a very thin scatter of coins or metal artefacts, although the most significant site is a pottery scatter which lies on the south-facing slopes of the ridge south of Rainhill, 2.5km south of Prescot. This sparse pattern is in contrast to the distribution seen across the Liverpool urban area (Philpott 1991, fig 1). The general pattern shows a bias away from the mosses and towards the sandstone areas west and south-west of Prescot. The scale of urban and industrial development in the two conurbations of Liverpool and St Helens is reasonably similar, but more material has come from chance finds from the area towards the Mersey (Philpott 1991, 62).

Evidence is increasing, however, to balance this picture with a more developed landscape in the Roman period to the east of the southern mosses. It is based on individual farm enclosures, that have mainly been recognised through aerial photography. The eastern side of the Sankey Brook again provides the focus for the nearest evidence of this type. The Roman road from Wilderspool to Wigan runs along the valley for a short distance before heading northwards towards Ashton-in-Makerfield. This road must have attracted settlement along its course, and one such farmstead has been located *c*4km to the east of Parr Moss, south of Newton-le-Willows. Another site is located *c*4km to the north in the valley in Ashton which may be of this date also (Philpott 1991). Two more Romano-British farmsteads have been identified to the east of Winwick, *c*10km away (R Philpott, pers comm), and two Romano-British farmsteads have recently been excavated within 1km of each other in the sandstone areas 2km south of Prescot (Cowell and Philpott 1994), which suggests that this pattern may be fairly widespread across the southern part of the county.

It appears therefore that parts of the landscape were being cleared for these farmsteads in the general area of the southern mosses, and that this activity may be reflected in the pollen diagram at Parr Moss. Whether this farming activity was taking place directly around the moss or (perhaps more likely), over a slightly wider area in the Sankey Valley, is difficult to identify, but the trend in the present evidence, both here and more widely across the southern areas of the county, is towards the sandier areas away from the mosslands being more important.

Early medieval period AD 420–1066

The lack of suitable sources for the history of the development of settlement in the St Helens area during this period has already been discussed (*see Chapter 5*). The main points of relevance for the history of the two mosses in this section relate to the place name evidence, which suggests that much

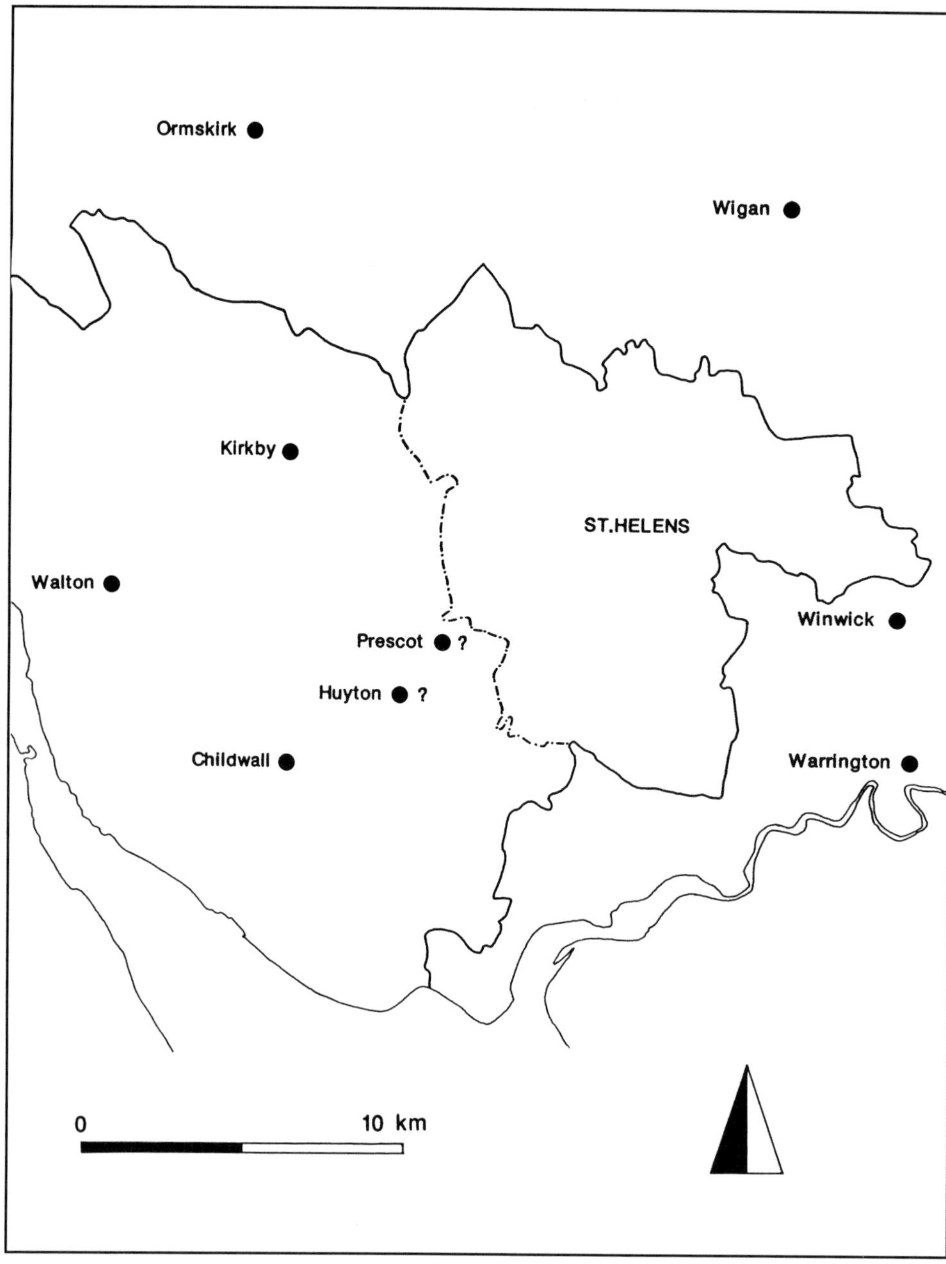

Fig 59 Southern and Central mosslands: Pre-Conquest churches and chapels (after Chitty 1981)

land remained uncleared before English influence after the sixth century AD. At the Norman Conquest, although woodland was not mentioned in the Domesday entry, it was probably recorded within the figures for the adjoining Hundred of Newton, where an area of nearly 150 square miles of woodland is recorded (Farrer 1899, 22). Only three berewicks are recorded in the Hundred, all lying well to the east of the southern mosslands, so that it is difficult to identify settlements in existence during the pre-Conquest period. A number of teamlands are mentioned, however, which through later documents can be traced back to Sutton and Eccleston in the survey area (Farrer 1899, 25), suggesting that arable land had been cleared here before the Conquest.

Over the wider area pre-Conquest foundations are likely to be represented by the four parish churches in that part of the county to the east of the Mersey (Cowell 1981, 1982), which are located in an arc to the south and west of the central mosslands (Fig 59). The size and number of the townships in these parishes (28 in Walton, Childwall, Huyton, and Prescot) is probably related to the low density of population in this part of the region from the early historic period into the late medieval period. The lack of mother churches in St Helens district suggests that the greater density of settlement within the area was to the west and south-west, as has been suggested for many of the preceding periods.

Prescot contains the mother church of the parish of that name, which roughly coincides with the Domesday Hundred of Warrington, and the place name, and form of the church enclosure, suggest a pre-Conquest foundation (Farrer 1899; Philpott 1988). At Prescot it seems likely that the sandstone knoll on which the town is built, with adjacent stream valley to the west, was the prime reason for the early location. Prescot appears to have been originally two manors, with the eastern half being the open field of Churchley, on the northern edge of which lay the moss, although the latter is actually located in the township of Eccleston.

In the late medieval period the typical pattern of settlement development in the eastern part of St Helens was based on small estates within the earlier vills. Some of these vills occur so early that it seems probable that they represent pre-Norman estates (Chitty 1981). It seems likely that woodland clearance played a limited part in the development of these estates, as at Domesday it has been estimated that less than 10% of the land lay under arable (Farrer 1899, 22-6). The pollen evidence from Parr Moss, probably covering the post-Roman period, may reflect one aspect of this. Here, phase 9 (Table 15) shows a dramatic increase in the rate of woodland clearance and associated introduction of grasses and other weeds, followed a little later by evidence for the cultivation of cereals which increases towards the end of the phase. If there is no later contamination of the upper peat levels, as is seen at the Flea Moss Wood site (*see Chapter 3*) then this would seem to represent the first major, increasingly arable, associated clearance in the area, as opposed to those earlier ones which are identified as being noticeable only in the context of the overriding wooded nature of the landscape. Although not dated absolutely, by reference to the depth of the peat above the dated late Bronze Age/ Iron Age phase (Table 15), it could tentatively be suggested that it might relate to the period around the Norman Conquest.

Late medieval period AD 1066–1500

The historical context for this period follows that already outlined in some detail for the central mosses (*see Chapter 5*), whereby rentals, surveys, leases, and place names from the thirteenth century onwards show the process of clearance of the widespread woodland that still survived at Domesday on a relatively extensive scale. This is to be seen against a background in this part of the St Helens district of a number of small estates developing within the Domesday vills. The most clearly identified of these early vills are found in Bold, where at least five such named 'lands' are known to have developed, each one associated with a moated site (Chitty 1981). The most northerly moated site, the Old Moat House, lay a little over 1km from the moss. A map of another estate at Burtonhead in Sutton, several kilometres to the west, shows the centre of the estate around the hall, and the moss is shown on the edge of the estate as a large area surrounded by a hedged boundary and called 'Merkelsse moss' (Fig 60). This represents the area of the Parr Moss complex.

In 1849, the township boundary pattern seen on the 1st edition 6″ OS map (Plate 6) shows the moss positioned on the edge of the four townships of Bold, Sutton, Parr, and Burtonwood, which all shared a portion of it. This pattern reflects a situation that had probably become established in the medieval period. About AD 1200 the boundaries of the Burtonhead estate with Parr are given as running '...thence in a line to Bale birch in Morkels moss...' (Farrer and Brownbill 1907, 358 n5). A boundary charter of *c*1254 of Bold also uses the same bounds on the moss between it and Sutton (Farrer and Brownbill 1907, 403, n13). This makes it clear that the moss was shared by the adjoining townships from an early date, probably as common land.

There is little information about the landuse of the moss from medieval documents. The palaeoecological evidence is limited; only at Parr Moss does it potentially survive for this period, although the later part of the core is not dated. The ecology of the mire, however, for the upper metre of its accumulation, after the dated late prehistoric clearance phase, shows that an acid raised bog, dominated by *Sphagnum* moss, was in existence. As with those townships around the central mosslands, the southern mosses in the medieval period may have been left as scrub woodland and mossy waste for grazing animals, providing wood, and turves for fuel. One reference of 1342 refers to a case before the manor court whereby ... 'Adam of Parr, his younger brother Richard and Hugh the fletcher killed a stag in Morkel's moss in the vill of Parr' (Farrer and Brownbill 1907).

In Prescot during the medieval period, the townspeople appear to have had no mossland within the bounds of the town. The township was originally made up of two parts: the glebe land around the church, and the vill of Churchley to the east. Subsequently Churchley was incorporated into Prescot, becoming the open arable field for the town that had started to develop after the granting of a mar-

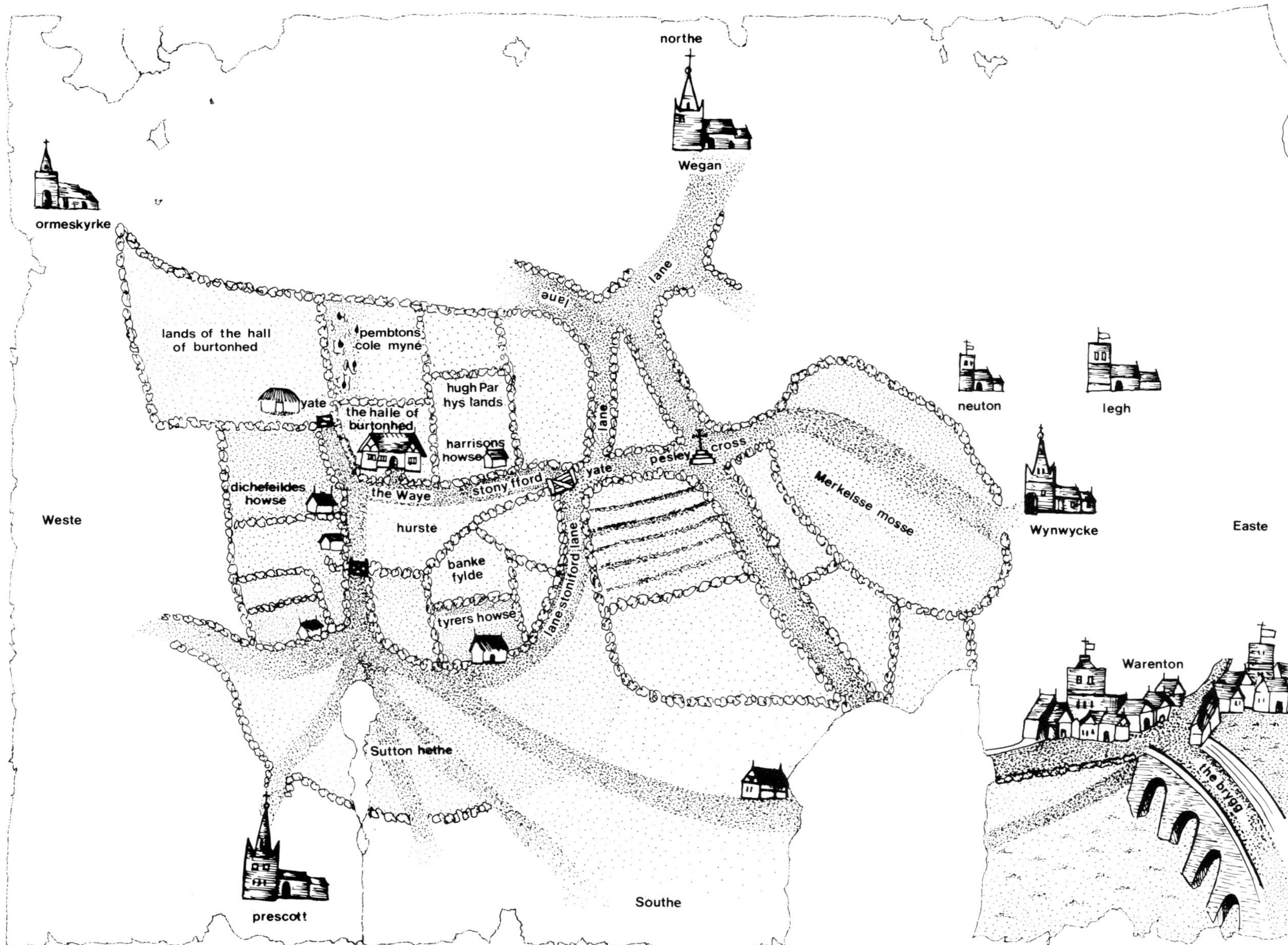

Fig 60 St Helens: Burtonhead Hall and Estate c1580 (after Chitty 1981)

ket and fair in 1323 (Bailey 1937). The township boundary explicitly avoided the area of mossland on the edge of the neighbouring township of Eccleston, although presumed dryland fields on either side belonged to Prescot. Whatever the reason for this, it seems likely that the townsfolk used the neighbouring mossland, for in 1508, when it was called 'Hackley Moss', it was granted to the people of Prescot and enclosed by a ditch. The moss was to be for grazing only, with many restrictions imposed so that the people 'should [not] delf upon said ground, no flay teorfes, no deep teorfes nay rigging teorfes, no cast no clay holes, break no ground, nay build upon it, nay close nay part of it, nay make no manner of waste upon it...' (Bailey 1937, 282).

Post-medieval period AD 1500 – 1850

In the late eighteenth century, most of the area of the southern mosses was at least as well populated as the rest of the county, with the exception of Liverpool, and presents a striking contrast to the thinly settled adjacent areas around the central mosslands of Knowsley and St Helens to the north and northwest (Plate 5).

From the eighteenth century, the pottery industry developed into large-scale production in this southern zone, in contrast to the earlier tradition of dispersed, farmstead associated, cottage industries around the mosslands to the north. These latter sites appear to have ceased production by the late eighteenth century, while large centres grew up in St Helens, Sutton (Marshalls Cross, Sutton Heath), and Prescot, drawing on the tradition of the earlier potters and their skills. By contrast the pipemaking trade remained dispersed and continued in the north, in Rainford and Eccleston, into the early twentieth century, in small centres of production within the family tradition of the area (Chitty 1981). In the nineteenth century, Prescot became internationally respected as a centre for watchmaking and associated tool industries. On the basis of these industries the town grew beyond its original medieval core along the sandstone ridge. By the mid nineteenth century the town mossland had been privately enclosed and a farm placed on the edge of the moss. At the northern end of the moss a gasworks and a pottery had also been built. The late nineteenth century saw the infilling of the farmland as industrial housing spread.

Seventeenth century records show that small-scale development was taking place around the Parr Moss complex, *eg* in 1634 Richard Bold built a house with two acres of ground at Morkels Moss, while elsewhere on the moss four acres of land were allotted to Richard Roughley (Duchy of Lancaster Pleadings and Depositions, St Helens Public Library). This kind of development, however, seems to have made little impression on the moss, which by the late eighteenth century was still open common land.

A series of long thin enclosures, shown on the tithe map of 1843, probably existed in the late eighteenth century, running from the triangular green of Parr Flatt to the north into the moss. Settlement lay around the green and along the road that marked the northern edge of the enclosures. It seems likely, from the disposition of the field boundaries, that this represents earlier enclosure of mossland. Between 1786 and the 1840s, the open areas of Bold Moss and Burtonwood Moss, in the southern half of the complex, were enclosed for agriculture, although almost half of the reclaimed Bold Moss enclosures were given over to woodland plantation. Sutton Moss had a few linear enclosures encroaching into the southern part of the moss by 1849, although the great majority of it still remained open. Two sinuous, roughly parallel boundaries along its western edge, in which the farm of Moss Nook lay, also probably represent late eighteenth or nineteenth century reclamation.

From before the middle of the nineteenth century, however, nearby developments were taking place that presaged the growth of the town of St Helens in the Industrial Revolution, and which slowly came heavily to affect the mosses. The Liverpool-Manchester railway line, the earliest passenger service in the country, was built through the southern half of the mossland in 1835. To the west, the early nineteenth century glass and copper works of Sutton, at Worsley Brow, had expanded and were now accompanied by the growth of adjacent industrial housing around St Helens Junction. Tileworks and rolling mills had been located right on the edge of Sutton Moss, while a little further away to the north the collieries of Parr Stocks and Ashtons Green had grown, and Bold colliery covered a large area immediately to the south of the railway line on the former moss. During the early to middle part of the twentieth century, the spoil from the colliery expanded onto the moss north of the railway line until, on abandonment of the mine, the tip covered over three-quarters of the former moss.

Palaeoecological survey

Parr Moss (PMB) (SJ 545940)

Parr Moss, south-east of St Helens, was chosen as

Plate 6 Parr Moss: 1st Edition 6″ OS Map 1849 (Photo R Cowell)

the pollen sampling site to represent the boulder clay plain of southern Merseyside, despite its having been very badly affected by human activity at various times in the past. The till plain of southern Merseyside forms a very important environmental zone, with a soil and vegetation history which is likely to offer a marked contrast with those of the sandy and peaty soils present in the other areas of the county studied so far in this report, *ie* Knowsley, Sefton, and North Wirral. Unfortunately it appears that very few deep peat deposits formed on the boulder clay plain, and the few which did exist, are either now totally destroyed, like Highfield Moss near Newton-le-Willows, or severely damaged, like Parr Moss itself. To the damage caused by drainage and cutting has been added more recent industrial depredation caused by the dumping of waste from the adjacent Bold Colliery until its closure in the last few years. Threat of continued dumping made this site a priority for rescue purposes at the time of its investigation in 1979. Although this threat of destruction has now receded, no better site was available in this area and the Parr Moss record seems to be the best that is available for the St Helens region. Two areas of peat survived which were far enough away from the waste heaps to avoid the problems of compression and deformation of sediments caused by the weight of the dumped material. At Sutton Moss to the west, the peat was shallow, perhaps being away from the original centre of the raised mire, and the surface showed clear signs of heavy cutting and disturbance. To the east, at Burtonwood Moss, the moss has been drained for grazing but otherwise seemed relatively undisturbed. This area was therefore chosen for sampling. The present day vegetation of the surviving mossland is dominated by coarse grasses and sedges (*Molinia caerulea* and *Eriophorum angustifolium*) and heather (*Calluna vulgaris*). The area is now regarded as ecologically important.

Peat stratigraphy

Nine borings were taken in the Burtonwood Moss section of the site to find the optimum point for sampling. Peat depths of 3-4m were recorded and the deepest profile (SJ 545940) was taken for analysis. The altitude of the core was 29.32m OD. Underlying the peat deposits at the sampling site (>3.91m) was a coarse white silty sand. The first organic deposits lay on top of this and were composed of a 0.04m layer of black gyttja with some wood and *Phragmites* fragments. Above this a monocotyledon peat dominated by *Phragmites*, but also with some wood fragments occurring between 3.87m and 3.20m. Brushwood peats followed this, dominated by the remains of *Betula* but including abundant *Phragmites* rhizomes. Above 3.09m, a change to less minerotrophic conditions is discernible as the peat changes to an *Eriophorum*-dominated type, still with frequent wood fragments. This continued up to 1.85m where a change to fresh *Sphagnum*-dominated peats occurred. The final 0.08m of the core were composed of highly humified amorphous peat.

Pollen analysis

Samples were taken for pollen analysis at 100mm intervals throughout the core and the resulting pollen diagrams are shown as Figures 61 and 62. Although changes in the tree pollen curves are not dramatic, the following two local pollen assemblage zones are recognised.

PMB-a 3.90-2.45m The zone is dominated by tree pollen, generally in excess of 60% of total pollen except at the base where high Coryloid values occur. *Betula* is the characteristic taxon. *Quercus*, *Alnus*, and Coryloid are present in moderate frequencies; *Calluna* rises near the end of the zone. Herb pollen values are uniformly low, although some aquatic types occur early in the zone. *Polypodium* is high in the basal levels before declining to very low values.

PMB-b 2.45-0.00m This zone is characterised by *Quercus* and *Alnus* pollen, which together make up over 80% of tree pollen, *Betula* falling from its previously high frequencies. Coryloid remains unchanged. *Fraxinus* becomes more frequent, while *Tilia* almost ceases to be recorded. *Calluna*, Gramineae, Cyperaceae, and *Pteridium* are all major contributors to the assemblage. Herb frequencies increase, especially towards the top of the zone, *Plantago lanceolata* and *Rumex* being most prominent. *Sphagnum* shows a number of peaks in frequency.

Ecological history

The stratigraphic record at the base of the succession reflects the infilling of a small hollow or the spreading of sediments from the central area of the bog which now unfortunately cannot be sampled. Gyttja accumulation took place under shallow open water conditions, and the presence of aquatic plants *Nymphaea* and *Nuphar* confirm the existence of ponds at the site at this time. The presence of the reed *Phragmites* and detrital alder wood suggests fen and reedswamp communities existed close by, and the organic infilling of the pond was underway. Above 3.87m the stratigraphic change to a *Phragmites* peat indicates the occupation of the sampling site itself by reedswamp conditions. Pollen of *Filipendula*, *Lythrum*, and *Typha angustifolia*, and seeds of *Menyanthes* reflect the establishment of a rich reedswamp/fen community. Both *Alnus* and

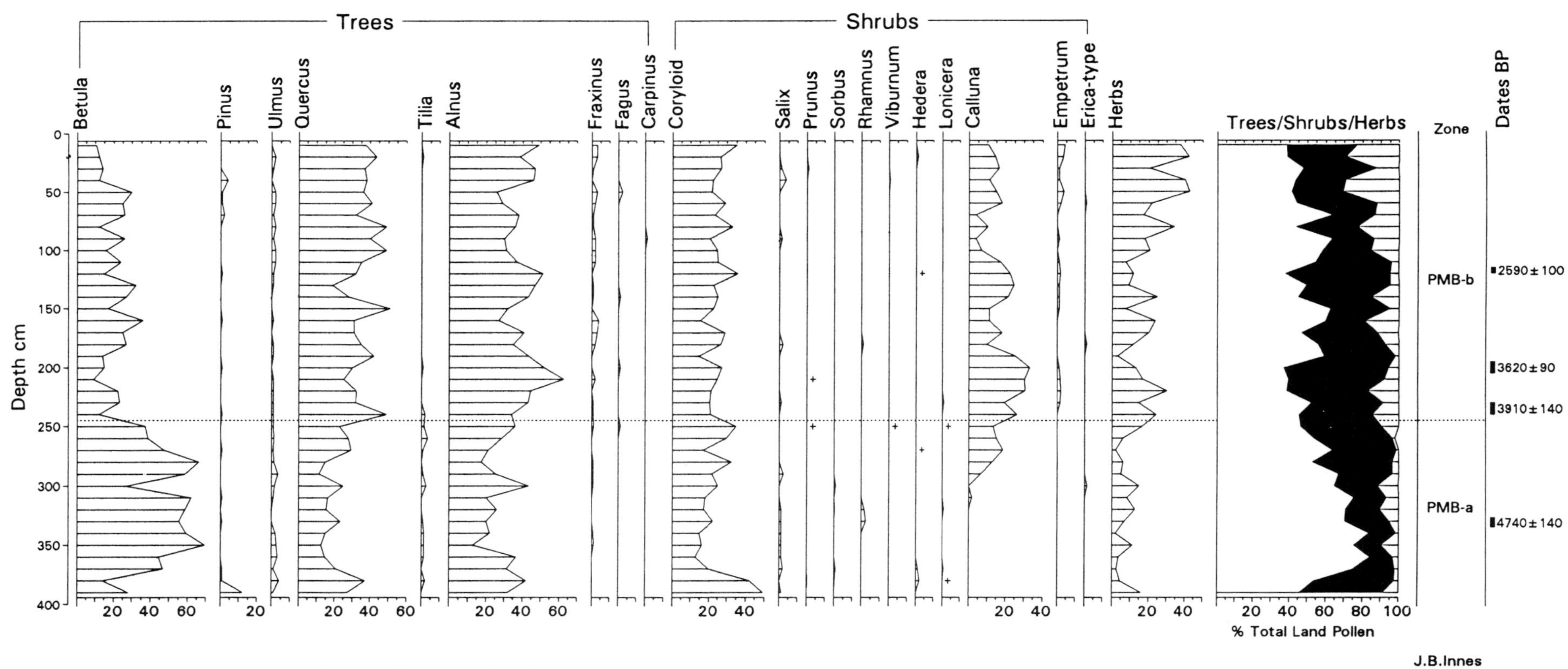

Fig 61 Southern mosslands: Parr Moss, Burtonwood, percentage arboreal pollen diagram (trees plus group)

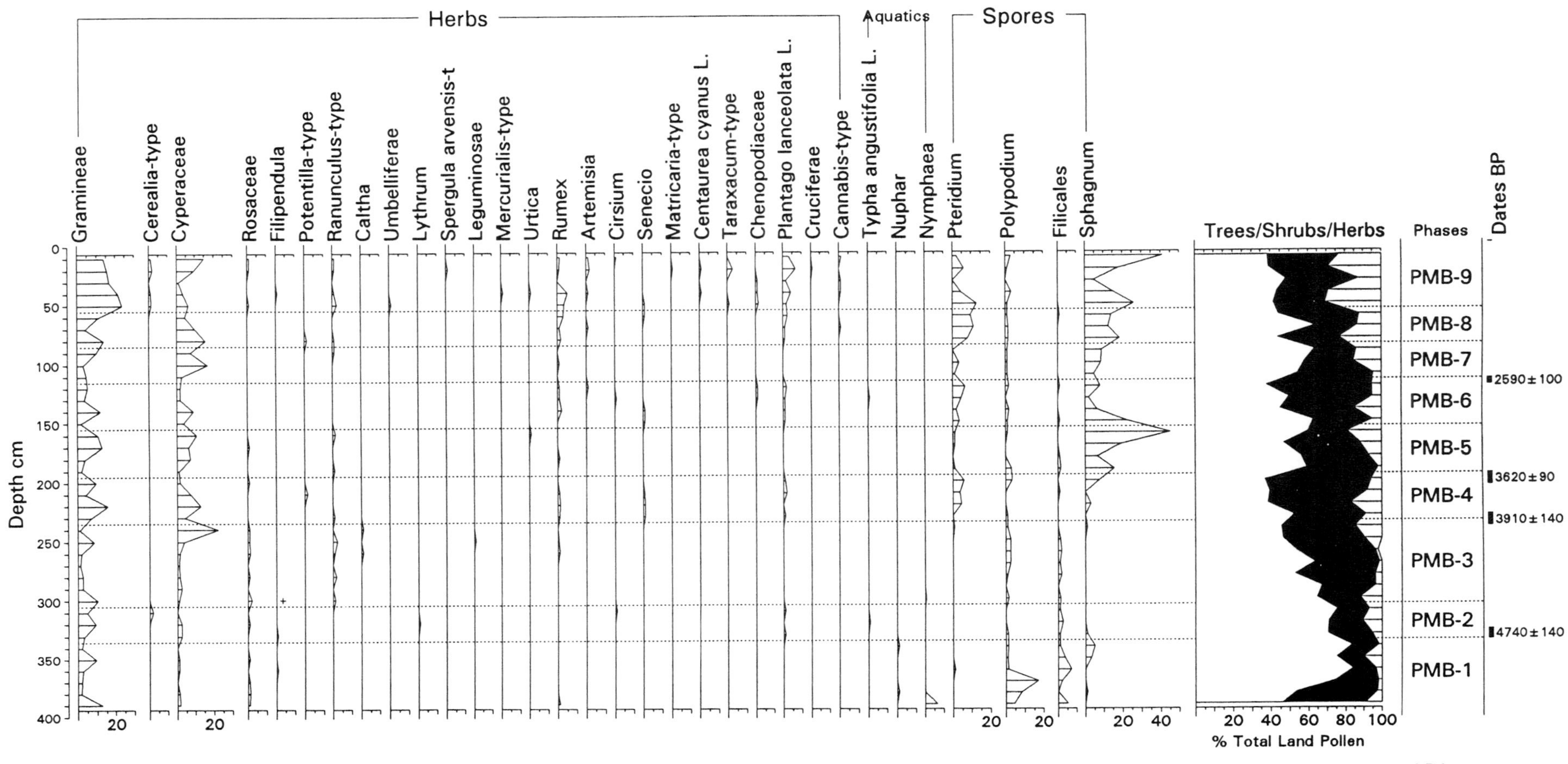

Fig 62 Southern mosslands: Parr Moss, Burtonwood percentage herb pollen diagram (trees plus group)

Betula wood are recorded, but both pollen and macrofossil evidence suggest that the latter was dominant as the mire became less wet and fenwood covered the site. Other taxa in the shrub community were *Salix*, *Rhamnus*, *Lonicera*, and *Hedera*.

Above 2.80m ombrotrophic conditions developed on the site, reflected in both the stratigraphy and pollen. *Eriophorum* became the main peat-forming material and the rise in Cyperaceae frequencies may be attributable to this taxon. Acid-tolerant plants like *Calluna vulgaris* and *Sphagnum* become common. The fall in *Betula* pollen frequency at the end of zone PMB-a coincides with the change to raised bog and the demise of the fenwood. Birch was probably not abundant in the woodland beyond the mire system, and its replacement by oak and alder shows these trees to have been of more regional significance. Later stages of mire development were dominated by *Sphagnum* and this persisted until the drainage of the bog in recent times.

Landscape history and forest disturbance

There are several parts of the pollen diagram which have indications of forest clearance due to human activity and the following nine phases are recognised.

PMB-1 3.90-3.35m There is no evidence of disturbance of the landscape in the lower part of the pollen record. The regional woodland appears to have been dense deciduous cover dominated by *Quercus* and *Alnus*, within which some *Ulmus* and *Tilia* occurred. While some high *Pinus*, Coryloid, and *Polypodium* values occur in the lowest levels, it is probable that these records are due to pollen grains derived from the basal silt or from mineral inwash. All of these types are resistant to corrosion and a high mineral fraction is included in the lower levels of this part of the core. There are no records of weeds indicative of clearance.

PMB-2 3.35-3.05m A minor phase of forest opening may be recognised, during which *Ulmus* and *Tilia* frequencies decline and some herbs indicative of forest clearance, especially *Plantago lanceolata*, are recorded. *Cirsium* occurs and *Calluna* rises at the end of the phase. These changes may be interpreted as due to the creation of small, temporary openings in the local woodland. The evidence is slight, however, and any clearings must have been very small. Tree pollen is reduced but still remains at 90% of total pollen. The start of this phase is radiocarbon dated to 3898-3100 Cal BC (4740±140 BP; Birm-1224).

PMB-3 3.05-2.35m A period of stability follows in which the *Quercus/Alnus* woodland is not disturbed. *Ulmus* reaches high values before subsiding, while tree pollen is totally dominant in the assemblage. *Tilia* and *Fraxinus* are present and *Fagus* occurs for the first time, indicating secondary recolonisation of the small clearings (Godwin 1975). Shrubs such as *Viburnum*, *Sorbus*, and *Prunus* were involved in the regeneration. The rapid rise of *Calluna* at this time may be mainly due to the more acid nature of the mire system, but heather may also have occurred in the regeneration community.

PMB-4 2.35-1.95m Some significant removal of woodland appears to have taken place in this phase, as tree pollen frequencies fall and several weed types indicative of disturbed habitats are recorded. All the major forest trees are affected, *Ulmus* and *Tilia* most severely. *Pteridium* spores are markedly increased and *Plantago lanceolata*, *Rumex*, and *Senecio*-type are all present. A more sustained type of clearance activity, perhaps of a pastoral nature, may be surmised. The rise of *Calluna* at this time may be due to a less than completely successful regeneration of tree cover. The start of this phase is radiocarbon dated to 2878-1986 Cal BC (3910±140 BP; Birm-1225) and the end to 2279-1750 Cal BC (3620±90 BP; Birm-1226).

PMB-5 1.95-1.55m Tree pollen values regain their former magnitude in this phase and although a few isolated herb grains occur, closed woodland conditions had returned to the vicinity of the site. The forest was dominated by *Quercus*, but *Fraxinus* had become more important after the previous disturbance. *Tilia* appears not to have recovered from its decline. There is no real evidence of continued open ground nearby, weed pollen and bracken spores being virtually absent, and regeneration seems to have been almost complete.

PMB-6 1.55-1.15m Significant woodland disturbance occurs in this phase, and *Quercus* frequencies are strongly reduced. *Plantago lanceolata*, *Rumex*, *Cirsium*, and *Senecio*-type pollen are consistently present, with a sharp peak in *Pteridium* values. These weeds are mainly pastoral indicators, but the appearance of *Artemisia* and Chenopodiaceae may suggest a more mixed form of land-use. *Calluna* again shows peak values. The end of the phase is radiocarbon dated to 930-410 Cal BC (2590±100 BP; Birm-1227).

PMB-7 1.15-0.85m This phase is one of tree regeneration and *Quercus* and *Fraxinus* are the main constituents of the woodland. Isolated *Pteridium* and *Rumex* records suggest that some open areas may have persisted. The landscape was probably mainly heavily wooded even on the most fertile soils, however, as *Ulmus* is consistently present and *Carpinus* occurs.

PMB-8 0.85-0.55m From this point until the top of the diagram, human impact on the landscape is clearly seen. A marked drop in tree pollen values occurs at the start of the phase and several ruderal weed types are recorded including *Plantago lanceolata*, *Rumex*, *Artemisia*, and *Senecio*-type. *Cannabis*-type pollen is present for the first time. A very great increase in *Pteridium* spores also occurs. A significant disturbance of the forest must have taken place.

PMB-9 0.55-0.00m In this final phase the intensity of the evidence for forest clearance increases dramatically. Tree pollen frequencies decline and at the start of the phase values for *Pteridium*, *Plantago lanceolata*, *Rumex*, and *Ranunculus*, with high grass frequencies, are very high. Later in the phase Cereal-type pollen is recorded for the first time, along with that of *Senecio*-type, *Taraxacum*-type, *Centaurea cyanus*, Chenopodiaceae, and *Cannabis*-type. These are types which may be associated with arable farming and a more mixed type of landuse. Near the end of the phase these possible cultivation elements of the assemblage are supported by Cruciferae and *Matricaria*-type pollen and by increased cereal pollen records. Major clearance of woodland for mixed farming seems to be represented in this final phase of the pollen diagram.

Prescot Moss (PM) (SJ 468929)

Peat deposits were discovered at St Helens Road, Prescot during the course of building development in 1979. The site is in the north of the town adjacent to the old Prescot Moss, formerly part of Hackley Moss (Bailey 1937), and the peat sediments discussed here almost certainly formed part of the southerly edge of this considerable area of peat, where it lay against the sandstone upland upon which Prescot stands. The site itself had been drained, cultivated, and subsequently built over in the nineteenth century (Bailey 1937; Tomlinson and Innes 1989). The deposits promised to provide information regarding the level of human activity on the sandstone outcrop, potentially an area attractive to early settlement.

Peat stratigraphy

The peat sediments lay beneath a thick layer of dumped material. They were sampled in a monolith tin and the top of the sampled profile was levelled to 76.52m OD. Underlying the organic deposits below 0.64m was a sticky grey clay with large pebbles. Above this clay, between 0.64-0.60m was a coarse red sand. The first organic deposits were composed of a fine black gyttja with some *Phragmites* which extended up to 0.54m. A *Phragmites* peat lay above this up to 0.50m and was succeeded by a *Betula* brushwood peat, terminating in a *Betula* tree stump between 0.36m and 0.44m. Overlying this up to the surface was a sticky grey-brown homogeneous peat with clay.

Pollen analysis

Samples were taken for analysis at 20mm intervals (Figs 63, 64) and the following four local pollen assemblage zones are recognised.

PR-a 0.60-0.53m The lowest zone is characterised by *Alnus*, *Pinus*, and *Betula*, but with considerable values for Coryloid also. *Quercus* frequencies are low. Some wetland and aquatic herbs are present and Gramineae and Cyperaceae are significant, but in general herbaceous values are low. Tree and shrub pollen dominate the assemblage, and *Ulmus* is absent.

PR-b 0.53-0.43m This zone is characterised by *Alnus* and *Quercus*, with considerable frequencies of Coryloid. *Pinus* and *Betula* are low. *Ulmus* is present but only in low frequencies. Gramineae and Cyperaceae frequencies are low and a very few isolated grains of ruderal and other herbs occur. The start of this zone (0.50-0.52m) is radiocarbon dated to 5957-5560 Cal BC (6850±80 BP; GU-2723). The pollen record is interrupted at the end of this zone by a *Betula* stump between 0.36m and 0.43m which is radiocarbon dated to 7051-6709 Cal BC (one sigma; 7980±60 BP; GU-2724). No pollen records are available for the levels occupied by the stump and, as an hiatus at this point seems likely, this pollen zone is terminated at 0.43m.

PR-c 0.37-0.23m The pollen record in this zone is similar to that of the previous one and is characterised by *Alnus* and *Quercus*. The percentages of the other tree types are also largely unchanged from zone PR-b, and Coryloid maintains its moderate values. *Salix* and *Calluna* are more in evidence, however, and *Pteridium* is considerably increased. Gramineae and Cyperaceae rise and the range of herb types recorded, including ruderals, is much increased.

PR-d 0.23-0.09m This zone is characterised by *Betula*, *Quercus*, and *Alnus*, with Coryloid also important. *Alnus* has declined from its dominant frequencies of the previous zone but remains high. *Ulmus* is reduced and *Tilia* becomes recorded for the first time. Ruderal and wetland herbs are still sporadically present. Consistent *Plantago*

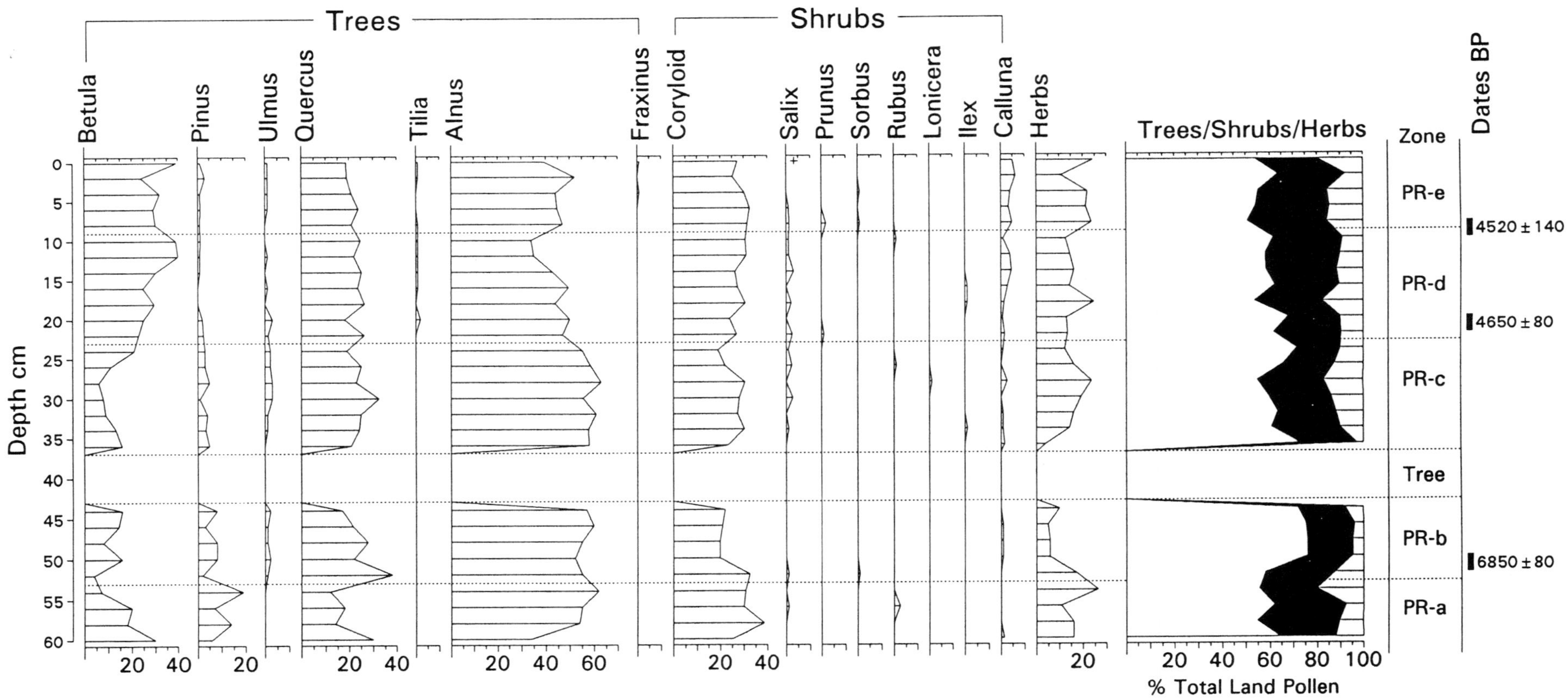

Fig 63 Southern mosslands: Prescot Moss percentage arboreal pollen diagram (trees plus group)

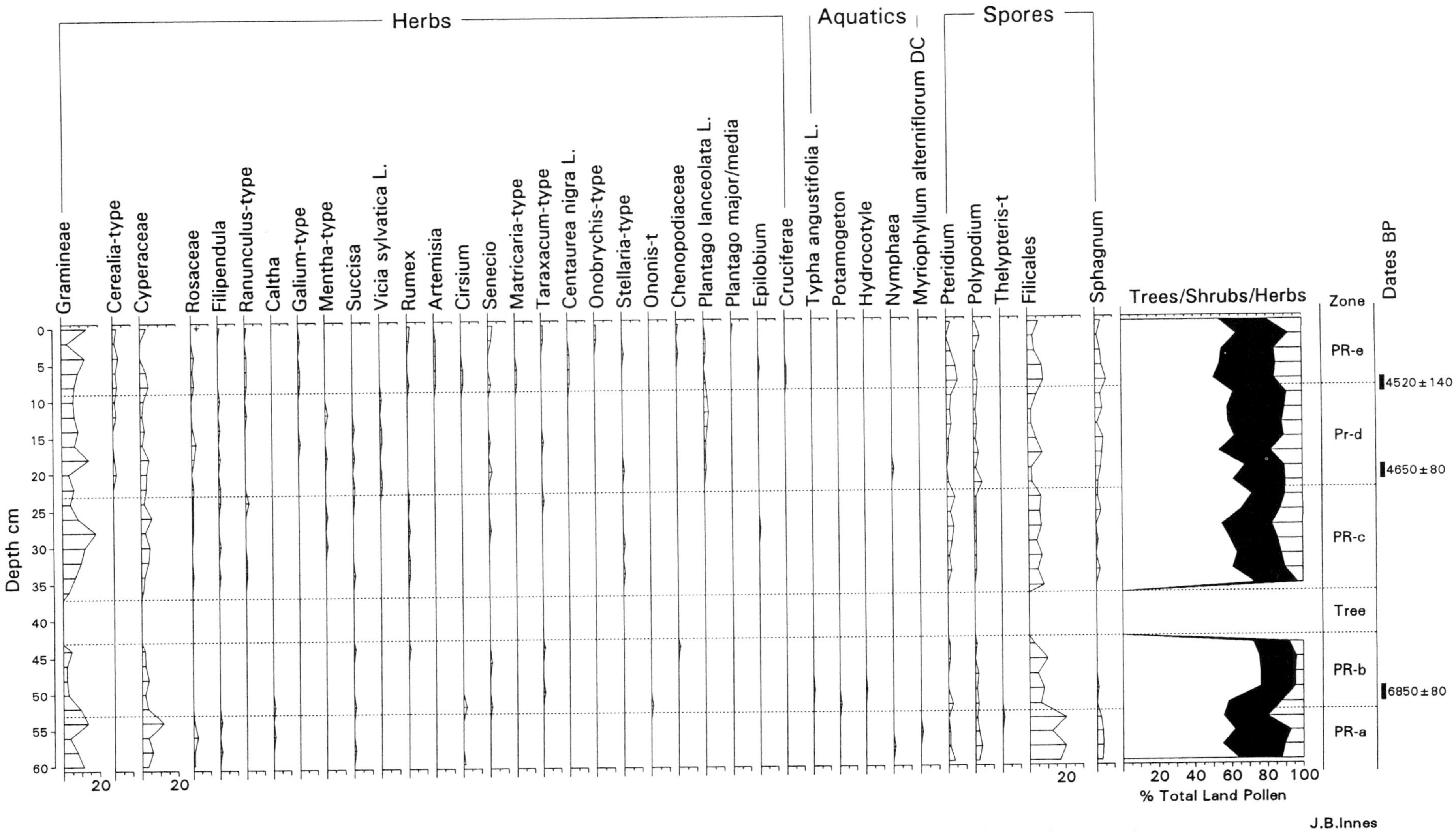

Fig 64 Southern Mosslands: Prescot Moss percentage herb pollen diagram (trees plus group)

lanceolata and *Calluna* curves occur and *Pteridium* is still recorded. The start of this zone (0.20-0.22m) is radiocarbon dated to 3630-3106 Cal BC (4650±80 BP; GU-2725).

PR-e 0.09-0.00m Characterised by *Alnus*, *Betula*, and *Quercus*, with Coryloid also in high frequencies. *Alnus* has replaced *Betula* as the dominant taxon. *Fraxinus* occurs and shrubs such as *Prunus* and *Ilex* are recorded. The range of ruderal and other herbs recorded increases greatly. *Pteridium* values increase again. The start of this zone (0.08-0.10m) is radiocarbon dated to 3630-2890 Cal BC (4520±140 BP; GU-2726).

Ecological history

The lowest organic deposit has been described as a gyttja, and its origin as a water-lain sediment is confirmed by the pollen of aquatic herbs being common within the lower part of the profile. *Nymphaea*, *Potamogeton*, *Myriophyllum alterniflorum*, and *Hydrocotyle* are all recorded. The site moved through the typical hydroseral succession of reedswamp to fen carr, shown by *Typha angustifolia* pollen and rhizomes of *Phragmites* in the peat itself. The fen carr was alder-dominated and its floral richness is shown by records of *Thelypteris*, *Filipendula*, and *Caltha*. Birch fragments are also present in this sediment and the *Betula* stump in the profile, which appeared in the field to be rooted and *in situ*, suggested some drying of the mire surface and colonising by trees in a fenwood phase prior to bog development. The radiocarbon date of 7051-6709 Cal BC (one sigma; 7980±60 BP), is however not in agreement with the other dates and the pollen data. The tree stump must therefore be regarded as intrusive, being older material incorporated in the profile. Perhaps it had survived within older peat deposits nearby and was carried into the Prescot sediments during a phase of erosion. The sediments above the tree stump all contain a very high proportion of clay within the peat, which supports the idea of soil erosion bringing material into the profile from this time onwards. The location of the site adjacent to the sandstone upland may indicate a source for this inwashed sediment. If true ombrotrophic peats ever developed at this site they must have been removed in some way. The radiocarbon dates suggest that there is almost certainly a hiatus at the level of the tree stump, and that much of the later Flandrian II sediments at the site are missing. The date of 5957-5560 Cal BC (6850±80 BP; GU-2723) at the start of zone PR-b shows the basal organic sediments to be of very early Flandrian II age, and the higher pine values of zone PR-a support this. The lower alder values in the basal counted level perhaps signify the very end of the *Alnus* rise itself. The absence of *Ulmus* in zone PR-a and the low values thereafter are therefore difficult to explain. Perhaps elm did not find suitable habitats in the mixture of sandy soils and peat which existed in this area at that time.

The radiocarbon dates of 3630-3106 Cal BC (4650±80 BP; GU-2725) and 3630-2890 Cal BC (4520±140 BP; GU-2726) for the clayey peat of the upper profile are likely to be correct, although the large standard deviation of the latter date reduces its usefulness. The *Ulmus* decline would in any case be difficult to define given the low elm frequencies recorded, but it is either at the start of zone PR-d or is within the hiatus between zones PR-b and PR-c, and is not recorded. The latter seems more likely as a date for the *Ulmus* decline of 3630-3106 Cal BC (4650±80 BP; GU-2725) would be very late for this area, although not unacceptably so. The profile above the tree stump is therefore regarded as of early Flandrian III date.

Landscape history and forest disturbance

There are very clear indications in this profile of the impact of human activity on the vegetation. Four phases are recognised, as described below.

PR-1 0.60-0.43m This phase includes the lower part of the diagram, which is of Flandrian II age and below the tree stump and hiatus. While there are sporadic records of ruderal herb taxa in this phase, such as *Rumex* and *Taraxacum*-type, they do not occur in discrete groups and the tree pollen values are not reduced at any particular level. There does not seem to have been any disturbance of the forest at this time.

PR-2 0.43-0.21m Tree pollen totals remain very high but there are some limited indications of more open conditions which may be attributable to woodland disturbance. Although still sporadic, records of taxa like *Rumex* are more common and both Gramineae and *Pteridium* increase markedly. The mixture of clay particles with the peat in this phase might suggest erosion from small-scale interference with the local vegetation. The interpretation of these pollen records as the result of human activity must, however, remain very tentative.

PR-3 0.21-0.09m This phase contains very clear indications of woodland disturbance caused by human activity. Tree pollen frequencies fall, cereal pollen is recorded, and a consistent *Plantago lanceolata* curve begins. Other likely open ground indicators *Senecio*-type, *Taraxacum*-type, *Calluna*, and *Pteridium* also occur. *Alnus* would appear to be the taxon cleared during this episode, but only a small impact upon the tree cover occurred. It is likely that this clearance and cultivation took place close to the sampling site, as cereal pollen is not transported far and the opening of the canopy was limited. The radiocarbon date of 3630-3106 Cal BC (4650±80 BP; GU-2725) shows that Neolithic farmers were responsible for this activity.

PR-4 0.09-0.00m A more intensive episode of disturbance occurs in this final phase and tree pollen totals are reduced. Cereal pollen is consistently present, and a range of herbs appears which may be associated with cultivation as well as disturbed ground. *Rumex, Epilobium, Artemisia,* Cruciferae, *Taraxacum*-type, *Centaurea cyanus, Matricaria*-type, *Senecio*-type, Chenopodiaceae, and *Cirsium* all occur. *Calluna* and *Pteridium* are also prominent, probably growing on areas where tree regeneration had not yet taken place after clearance. *Plantago lanceolata* is reduced, and so it is possible that this phase records landuse which was more strictly arable than previously, and produced little open bare ground or grassland. The date of *c*3630-2890 Cal BC shows this activity also to be of Neolithic age.

Conclusions

The main points of archaeological and palaeoecological significance from the survey in the southern mosslands that have particular relevance to research questions in the region or beyond are illustrated here, while the wider and more detailed discussion of these issues is left to Chapter 7, where they are seen in the context of patterns within the county. The main locations that may have the potential to develop this research are described here, as well as the threats that might destroy the information. This evidence can be used to set the criteria for preservation of wetland areas through management strategies, or for developing its research potential, through rescue archaeology if necessary.

The problems associated with understanding the relationship of these mosses to their contemporary human settlement and landuse are similar to those encountered in the survey of the Wirral coastal mosslands (*see Chapter 3*). In both areas, large extents of urban and leisure facilities make further discoveries difficult, yet the southern mosses do not have the advantage of a body of information from past finds to allow a proper judgement to be made on the potential of the surviving areas of peat.

How far the mosses were directly exploited in early prehistoric times is unclear, as so little prehistoric material has been either found or reported from this area. Surrounding areas of farmland on the sandstone ridges do suggest that Mesolithic hunters and gatherers were active in the woodland valleys here, and an area such as Prescot provides similar topographic attractions.

It is mainly the palaeoecological evidence that provides the best indications of the attractiveness of the area around these mosses as suitable locations for settlement. Prescot Moss provides the earliest example of possible cereal farming east of the Mersey, in the mid to late fourth millennium Cal BC, with clay inwash into the moss possibly being associated with this activity. Activity around Parr Moss is also suggested at this time from the palynological evidence, but possibly of a more pastoral nature. It has been suggested that these two elements may be part of a mobile system, similar in some ways to the medieval infield/outfield system, whereby the mosses such as at Parr on the flat clay lands were more suited to grazing, hunting, and collecting, while the infield, cultivation areas were based around the sandstone slopes. The proximity of fen carr environments would provide additional resources in the core areas. A similar pattern may have been present in the Bronze Age, although only at Parr is there contemporary palaeoecological information.

The late Bronze Age/Iron Age and possible Romano-British farming episodes seen at Parr Moss complement the palynological evidence from the central mosslands (*see Chapter 5*), providing the first evidence for concerted arable farming and associated clearance in the county. Over the wider area, evidence for a settlement pattern consisting of isolated farmsteads is starting to be identified from aerial photographic survey. Morphological parallels and some fieldwalking evidence suggest that a good proportion of these are likely to be of Romano-British date, although evidence from the Wirral and Cheshire suggests that some of these may have late prehistoric elements also. The lack of good dating evidence in the region for sites such as these, however, makes it difficult to identify whether the origins of some of these sites may correlate with the horizon of clearance and mixed arable farming seen in the eastern half of the county during the late prehistoric period.

Topographically, the incidence of the sandstone ridges which lie on the edge of the boulder clay lands to the south, and the sand and peat areas to the north, attracted some of the earliest historic settlement in the county, most notably in the survey area at Prescot, and a little to the north-east, at Eccleston (*see Chapter 5*). The mosses seem to have been peripheral to the initial settlement of the historic period, and they continued to be regarded as waste, in the medieval sense, up until the Industrial Revolution, when they started to be affected by enclosure and development.

Archaeological and palaeoecological potential

The potential for locating settlement associated with these mosslands is very limited, the more so since so much of their former area and margins has already been lost to development. They do act, though, as important palaeoenvironmental records, indicating the nature of early settlement in an area that is impossible to survey today, and has little palaeoenvironmental evidence in contrast to the northern part of the county.

The archaeological evidence that exists for settlement from the Romano-British period onwards across the wider landscape in this southern part of the county, suggests that closer to the mosslands favoured areas would have been the low, well-drained, sandstone ridges, rather than the immediate area of the mire itself. In the survey area, the potential for discovering such sites in this type of landscape is slight, as the sandstone ridges and hills to the west of Parr Moss and surrounding Prescot have largely been developed through the industrial growth of this area.

Prescot has perhaps been investigated more than any other built-up area of the county, yet no prehistoric archaeology has come to light, even in the form of chance finds. Most of this archaeological work has taken place around the historic core, on the southern slopes of the sandstone ridge, a little to the south-west of the moss. It is not clear how much of Prescot Moss has survived post-medieval development, but the site reported in this chapter suggests that even in built-up areas there may still be potential for peat survival. The palynological evidence suggests that the area may be archaeologically significant for the early prehistoric period, deriving from the potential cereal cultivation in the middle Neolithic. It has been suggested that settlement during this period was reasonably mobile, so that even if the cultivated areas lay off the mossland edges on the sandstone, the link with wetland resources may still have been strong, with the collection of wild plants and hunting playing an important role. This implies that wetland archaeology is more likely to be associated with this period than with the late prehistoric/Romano-British period, when the first evidence for settled farmsteads appeared, away from the mosses. In the late prehistoric period, however, some sites may have had a more intimate link with wetlands given the general trend across Britain for wet places being linked to features such as religion, economy, or security.

It is difficult to assess the archaeological importance of these mosslands during the historic period, but they would seem to have little potential for discovering associated archaeology, as they lay largely on the margins of the settled areas within the townships, and were used mainly for grazing, fuel, and wood. This continued quite late into the post-medieval period, when enclosure of the mosses started to take place. The medieval and post-medieval landuse of the mosses has also affected their palaeoecological potential, with cutting of the peat truncating historic development of the mires, and later farming leading to possible contamination of the upper peat.

Threats

The mosses covered in this chapter have both already suffered a high degree of development that has left little obvious potential. Prescot is a small town that is undergoing substantial redevelopment, although this has so far been mainly concerned with the historic core which lies a little to the south-west of the former moss area. Small-scale developments will undoubtedly take place in this area, but continuing development should be seen as a positive method of retrieving more information, particularly as the palynological evidence suggests significant early prehistoric activity may have been present in the vicinity.

Parr Moss is the subject of a management agreement sponsored by British Coal and administered through Operation Groundwork, which aims to secure the preservation of the surviving wetland on the east side of the site. It is the intention to return the moss to a wetland habitat through management of water-tables, so that the surviving peat should retain its palaeoecological potential.

7

THE MERSEYSIDE WETLANDS: SYNTHESIS AND DISCUSSION

The work of the NWWS in Merseyside has provided much new information that is invaluable in assessing the nature of prehistoric settlement in North West England, an area that has long seen little archaeological investigation. It has provided evidence for 112 new prehistoric sites and findspots, compared with the sparse total of 32 from previous fieldwork and chance reportings of finds in the survey area. However, for the Roman period, little new evidence has been forthcoming, with no finds of wetlands associated sites having been made, which means that no Roman site in the county has been located closer than *c*1km to mossland. A recent chance find from adjacent to Bickerstaffe Moss in West Lancashire suggests, however, that this may not be the case everywhere in the region. For later periods, the fieldwalking has produced little new evidence of sites, and documentary evidence remains the main source for understanding the exploitation of the mosslands in the historic period. In general, until Parliamentary enclosure the mossland areas were treated as common land that was exploited for a variety of resources and at some distance from the main settlements.

Two important trends can be identified from the results of the NWWS fieldwork. Firstly, it has provided significant new evidence for the early prehistoric period. The Sefton mosslands, an area that had not been surveyed previously, may now be regarded as one of the richest areas of lowland prehistoric settlement in the North West. This evidence is made all the more important in that 32 sites or findspots are located on the edge of or within peat deposits. This provides a suite of locations, relating particularly to the early prehistoric period, where there is good potential for the recovery of artefacts together with associated palaeoecological evidence which is normally absent from most 'dryland' sites. Little new evidence came from the coastal mosslands surveyed in North Wirral, due to unfavourable ground conditions for fieldwalking and the more intensively urbanised nature of the area. However, the implication of previously collected archaeological and palaeoenvironmental evidence from North Wirral makes it clear how important the coastal areas were as a focus of settlement throughout the prehistoric period.

The second trend from the survey confirms the pattern of prehistoric settlement that had been previously identified away from the coast (Cowell 1991a). This produced sparse settlement evidence in the dryland areas of the county, and the work of the NWWS has shown that this pattern also extends to the interior mosslands. Only on the Wirral is there a relatively dense pattern of prehistoric settlement on dryland areas away from the coastal wetlands. This may be due to the particular palaeogeography of the Wirral, whereby, after the early Mesolithic, no area in the heart of the peninsula was more than *c*8km from the major river valleys or estuaries of the Dee and the Mersey.

The following section presents an integrated view of the local archaeology incorporating NWWS evidence which is viewed, where possible, in the wider context of the region. Unless otherwise stated, the evidence for areas away from the wetlands derives from work largely undertaken through Liverpool Museum. This is summarised for the prehistoric period in Cowell (1991a), for the Romano-British period in Philpott (1991), and for the historic period in Chitty (1981), Cowell (1981, 1982), and Lewis (1982, 1991). The pre-Conquest background is covered by O'Hanlon (1991), and the environmental context provided by Innes and Tomlinson (1991).

Archaeological and palaeoecological synthesis

Early Mesolithic c8000 BC – 6800 BC

Prior to the NWWS, Mesolithic activity had been identified in a number of locations through fieldwalking, and contemporary patterns of natural vegetation successions had been established through

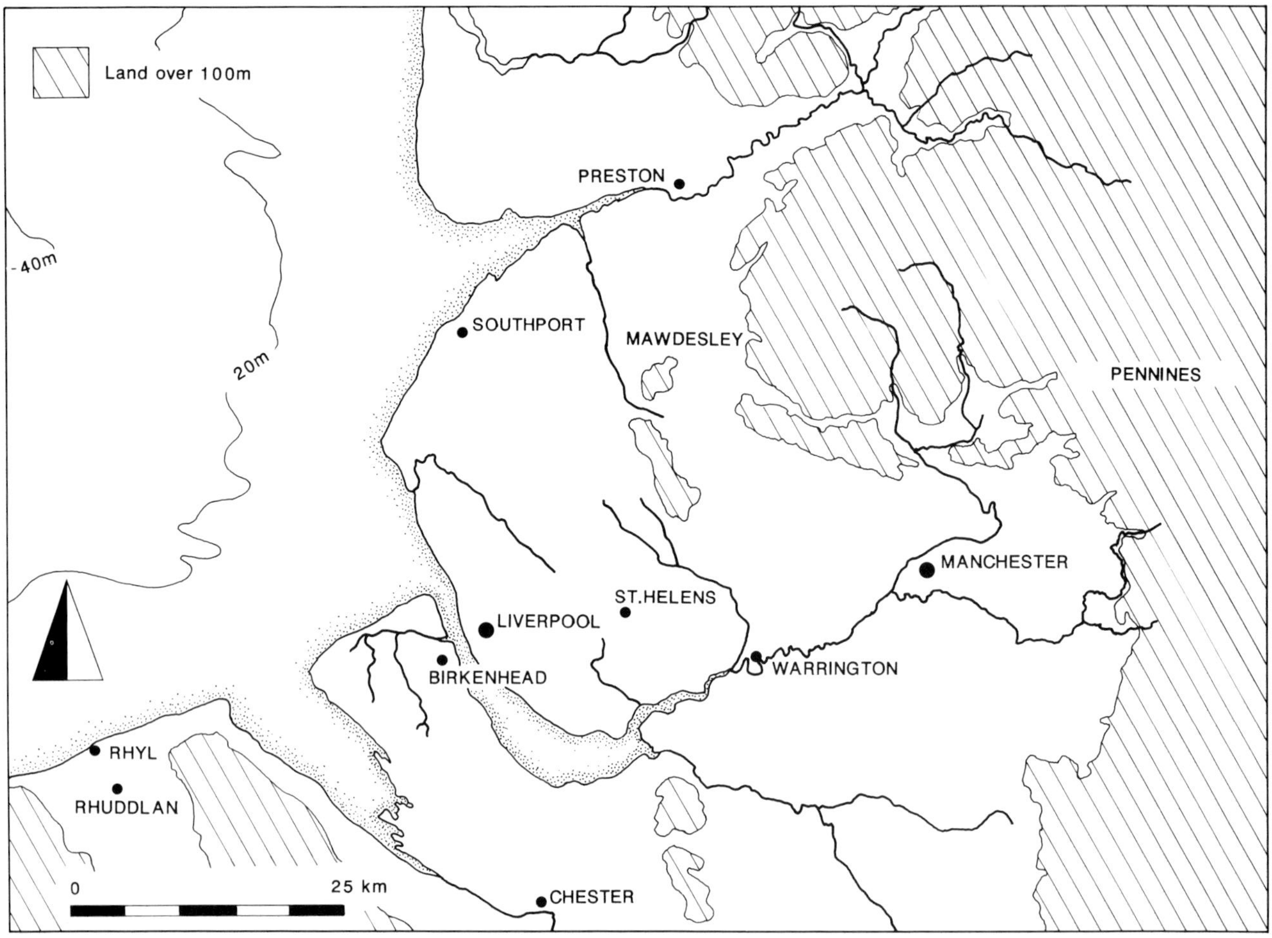

Fig 65 North West England: Topography

pollen analysis. The only wetland areas that had been fieldwalked as part of this earlier programme were the western fringes of the large central mossland complex, where a number of flint sites had been identified (Cowell 1991a). Elsewhere, the main locations for sites, mainly represented by small scatters of lithic material, were identified in the following topographical locations:

i) River valleys flowing through the sand areas to the west of the central mosses and across the boulder clay plain in the south, particularly near their confluences with the Mersey
ii) On the sandstone ridges of the North Wirral

The main exceptions come from the Wirral, where sites such as those at Greasby (the nearest to the North Wirral wetlands, lying *c*3.5km to the south) and Thurstaston are found, dated typologically to the early Mesolithic by the occurrence of obliquely blunted point chert assemblages (Cowell 1992c). If obliquely blunted point lithic assemblages are exclusively early Mesolithic in the region, then the sites are likely to be datable to *c*7000 BC (Jacobi 1987; pers comm), probably predating the onset of wet conditions affecting the Dee and the present coastal zone. The present evidence suggests that this did not occur until 6690-6410 Cal BC (*c*7360±60 BP; SRR-2926) at Bidston Moss but, on the evidence of Newton Carr (Table 3), perhaps a little earlier on the north-western coast. Raw material exploitation at Greasby and the similar, nearby site of Thurstaston, strongly suggests that the North Welsh coast was being used as a supply of chert for the manufacture of implements (Cowell 1992c). Regular movements across the Dee Valley, which may not have been tidal at this stage, would therefore have been a feature of the seasonal or annual cycle of local hunter-gatherers.

The sites from the sandstone areas on the Wirral are only one element of early Mesolithic activity across the North Wirral lowlands, which stretched beyond the present limits of the coastline (Fig 65). The early channels and more distant coasts are likely to have been the focus for much of this activity. In many places this will have been buried by the subsequent rise in sea-level and ground-water regimes.

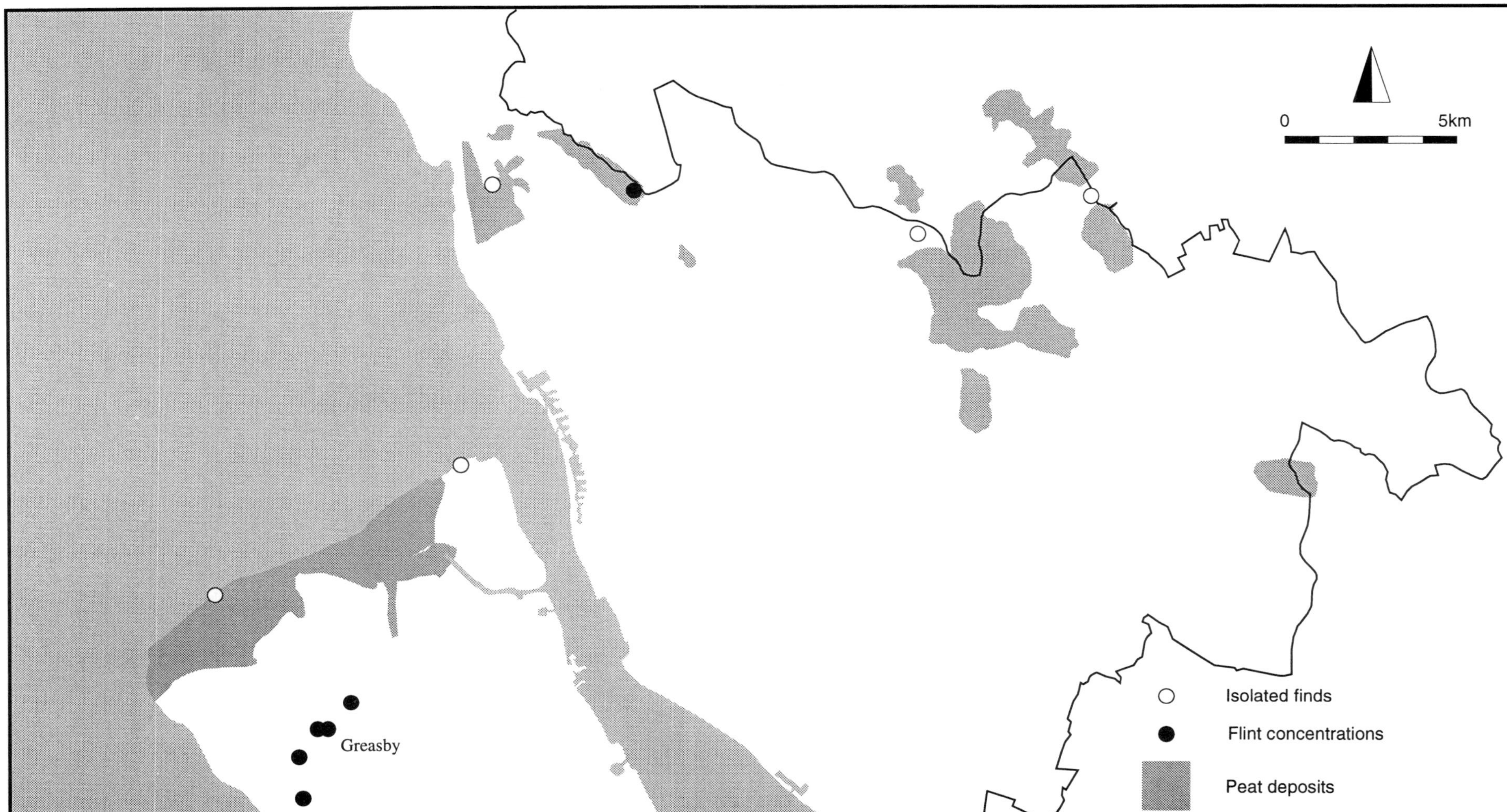

Fig 66 Merseyside: Early Mesolithic sites within the county

The early buried channel at the northern end of Newton Carr may have been one such focus, although the buried topography needs to be better understood before more accurate areas of potential can be isolated. The low sandstone cliffs along the north-east Wirral coast at New Brighton provided traces of early Mesolithic evidence in 1900 (A 6). This site, overlooking the now buried plain of Liverpool Bay to the north, may have lain to the east of a major channel flowing northwards into the Bay (Kenna 1986). Both these channels are deeply buried by marine and alluvial deposits and are not clearly defined.

New evidence from the NWWS implies that the Sefton area, across the Mersey, may also have lain within the annual territories of foraging hunters from the Wirral and North Wales. Two sites (Me 094, Me 038; Fig 76) produced material derived from the use of a similar (possibly Welsh) chert, and a similar on-site flint technology to that found on the Wirral (Fig 20). The Little Crosby/Ince Blundell area would have provided light, sandy soils, not available in North Wirral, which would have been accessible within a few hours' walk across what would then have been an essentially dry Liverpool Bay. The main site (Me 094) lies in the valley of the Alt, which, apart from providing riparian resources for hunters and gatherers, is also likely to have functioned as a route into the interior.

The only evidence relating to the vegetation of the early Mesolithic period comes from Knowsley Park and the eastern part of Simonswood Moss, lying in the central plain, and at Holiday Moss near the foot of the eastern Carboniferous ridge. They provide a picture of the regional vegetation which was probably quite typical of the sandy areas in the north-central part of the county. During this period the mires were dominated by reedswamp and carr, while the sandy soils of the wider landscape across the plain were clad with open pine, birch, and hazel woodland up until *c*6500 BC. Although there is no palynological evidence from the Alt Valley as early as this, it is probable that the sandy soils in the area carried similar vegetation.

Although the archaeological evidence in the Alt Valley suggests that it may have been a significant route into the woodlands of the interior, there is little from the latter area to confirm this, despite having fieldwalked over 2500ha of land, distributed reasonably equitably across all four non-coastal zones of the county east of the Mersey. There is sufficient similarity between isolated, possibly early Mesolithic, implements from Little Crosby (Me 038) and Rainford (Me 106; Fig 32.7), on the north-eastern edge of the central mosslands (Fig 66), to suggest some links between the two areas. Another possibly related site occurs in an intermediate location in Simonswood (A 27). However, because so little material is involved, it is difficult to know how far this pattern represents the extent of early Mesolithic activity around the sandy soils in the northern part of the county. Another problem is the lack of a good technological framework by which material of this date might be confidently recognised. This means that some material currently attributed to the later Mesolithic in this area (*see below*) could perhaps date to the earlier period.

However, if it is assumed that the lack of finds reflects a true picture, it suggests that early Mesolithic exploitation of these areas was quite light. On present evidence, there is a substantial difference between North Wirral and east of the Mersey in the early Mesolithic period. In the former area, surface sites provide the densest concentration of finds in the county, with between 100-500 surface pieces at a density of approximately 1-4 per m^2. Excavation has confirmed the nature of the sites as relatively large residential camps, where family groups came together for relatively long periods, within the generally mobile annual cycle, which have produced either subsurface features and/or a mixed chert assemblage, incorporating flint reduction material and a range of tool forms (Cowell 1992c). In the absence of radiocarbon dates, typological evidence, in the form of the presence of obliquely blunted microliths and the absence of triangular types, suggests the early Mesolithic date postulated. This appears to have been the standard early Mesolithic type of site as found elsewhere, possibly providing the base for intercept hunting strategies (Myers 1987).

The Sefton sites have not produced the same surface flint density as those in the Wirral, and the evidence from the sites occupying the sandy plain to the north of the central mosses is even less like that from the Wirral, consisting as it does only of single finds of implements. The latter are particularly likely to represent specialist foraging sites (Binford 1983) at some distance from residential areas, which may be represented by the larger sites in the major valleys to the west.

Later Mesolithic *c*6800 BC – 4000 Cal BC

Identification of sites that may represent exploitation of areas of wetland during the later Mesolithic is made less straightforward by the nature of much of the evidence that has been located. In contrast to the uplands and parts of coastal Cumbria (Bonsall *et al* 1986), microlith-dominated sites (the main typological method of dating surface scatters to the later Mesolithic) are absent from the county. Dating of the local sites therefore has to rely on aspects of technology associated with the waste material from flint knapping, which on most local sites is restricted in quantity (*see Chapter 1*). In the interior of the county very local, generally orangey-brown, boulder clay flint was used, while beach pebbles tend to dominate as a raw material in the present coastal areas. As no sites of the Neolithic period have been excavated locally, it is difficult to determine for how long such a late Mesolithic technology continued in this area. As mentioned above, some of these sites, particularly where boulder clay flint is not used (*eg* A 23 in Kirkby; Fig 32) could perhaps represent sites of the earlier Mesolithic.

Acknowledging these difficulties, the results of the NWWS have confirmed the pattern retrieved from previous survey in the county, whereby Mesolithic sites are the most common discovery resulting from extensive field survey. This is as true for mossland fringes as it is for the dryland areas of the county. Additionally, the NWWS results have revealed that the most intense settlement is found around the coastal mosslands south of the Alt estuary. Sites are mainly located along the valley of the Alt, and along the sandy ridge overlooking the mosslands of the coastal plain to the west and north, and the small channel running south-eastwards from the mosses. Despite the masking effects of modern development, there are faint indications from the wetland plain on the Wirral that suggest a similar pattern may have existed there also, although here there is less of an archaeological link with the mossland embayments than with the channels in the present coastal zone.

In the eastern part of the county, the dispersed nature of Mesolithic activity recorded during previous survey work has been confirmed by the more extensive work of the NWWS. Concentrations of struck flint are rare, but when taken in conjunction with scattered isolated finds, the emphasis is on Mesolithic activity around the north-western and western sides of the central mosslands. This contrasts with the almost complete absence of findspots around the east-central and to a lesser extent the eastern mosslands (Fig 67). This pattern may have been influenced to some extent by the loss of much of the sand subsoil in these latter areas through quarrying. A minimum of *c*350ha has been affected by quarrying, of which *c*20% was directly fieldwalked. A limited number of finds has been made in these quarried areas, both chance finds and from fieldwalking, but it is difficult to know how far they represent the former pattern.

On the previously surveyed dryland areas in the central and eastern part of the county, sites are found mainly in the major river valleys of the Mersey and Ditton, the smaller tributaries of the Ditton, and the upper reaches of the Alt. As with the wetland evidence, the further east one travels the sparser the evidence becomes, although small sites have been located in valleys overlooking the Mersey in St Helens, and in adjoining areas of Cheshire (Cowell 1991a). The results from the restricted amount of work carried out in Cheshire on the southern side of the Mersey do not contradict the pattern of small, river valley locations (Cowell 1992b).

Although the occurrence of blades, blade cores, and the occasional microlith, is indicative of Mesolithic activity in the county, a more precise dating within the period is less easy. Pollen evidence from a number of the mosses might suggest that some of the flintwork dates to the period of the sixth and early fifth millennia Cal BC, when the first evidence of possible anthropogenic woodland disturbances occurs at several sites across the county. However, this does not presuppose that earlier activity, unrecognised in the pollen diagrams, did not take place in these areas.

Site type and function

The nature of the flint evidence suggests that there may have been several landscape zones in the region which were exploited in different ways. This may equate with the pattern of hunter-gatherer exploitation of the landscape, established from ethnographic studies (Binford 1983). In this model different parts of the landscape are dedicated to different kinds of sites associated with a range of mobile hunting and gathering activities. There are two main categories: residential areas, places where a family, or group of families, might stay for a period of time, and various smaller, specialist sites associated with hunting and gathering expeditions (Smith 1992).

Around the coastal areas of Sefton, sites where reduction of flint pebbles and the manufacture of implements took place are reasonably common. These areas presumably provided a good supply of easily accessible raw material and so afforded good opportunities for sites associated with the prospection for flint and the 'industrial' working of the material. The evidence from the few Wirral coastal sites (Fig 67) suggests, from the imprecise records implying that they most probably contained high proportions of waste knapping material, that a similar pattern may have obtained here also. The Sefton sites do not include many tools or blades, which might imply that these elements were taken elsewhere to be used, although the very small size of the surface scatters makes it difficult to be sure how representative they are of the make-up of the total amount of material on site.

If the coastal sites are interpreted as specialised flint production sites, it implies a major component of the Mesolithic settlement pattern is missing, *ie* the residential sites. Although it is difficult to interpret with confidence from small surface scatters, this seems unlikely for a number of reasons: the raw material used for these assemblages is mainly a blue-grey flint which is probably derived from beach pebbles and has not been found on sites away from the coast, which might be expected if flint was being reduced on the coast and taken to hunting sites and base camps elsewhere. Rather, the flint evidence suggests that a variety of tasks might have been undertaken in a fairly restricted area of the county, so that the 'industrial' element at these Sefton sites may have been just one of a range of domestic activities. This may especially be the case at many of the larger sites, such as those on the Woodham Knoll ridge in Little Crosby (*eg* Me 022, 039), and in the middle part of the Alt Valley in Lunt (*eg* Me 093, Me 094). Residential sites that were also flint production sites might well occur in a coastal or estuarine location, which could provide a range of natural environments with resources to sustain larger groups for longer periods of time, as is suggested for Mesolithic sites in the Cumbrian coastal area (Bonsall 1981).

Inland, the composition of the surface assemblages differs. The material is concentrated at a limited number of locations which are more widely spaced across the landscape, compared to those on the coast (Fig 67). Sometimes these sites include a preponderance of blades, or more commonly, they are made up of general, small waste, flint scatters with the occasional blade fragment. Flint reduction sites, characterised by cores and primary waste debris, have not been found during the NWWS fieldwork around the central mosslands, nor previously in the dryland areas of the interior.

Previous excavation of a blade-dominated site, at Ditton, close to the Mersey, recorded a high proportion of tools in the small assemblage and a lack of on-site flint reduction (Cowell in prep b). This suggests that the basic working down of flint pebbles was undertaken elsewhere. It appears to represent a specialist hunting location where light dressing of tools and perhaps flint blanks took place, and ethnographic comparisons suggest it is likely that this was undertaken by a task based group (*eg* butchery site, hunting stand etc) rather than a larger family group. However, the flint at the Ditton site is mainly an orangey coloured boulder clay derived material, suggesting that the groups using this site were not bringing material with them from the coast. Croxteth Park, another excavated site from a river valley closer to the central mosslands, is the one inland site where the people may have been using beach flint, although the detailed analysis of the flint assemblage is still under way (Cowell in prep b).

It seems, therefore, that the pattern of specialist sites in the interior of the county may only represent one part of the larger pattern across south-west Lancashire, and that different types of site still remain to be found in unsurveyed areas. For example, unlike the Pennines (Barnes 1982) and coastal Cumbria (Bonsall *et al* 1986), there are no microlith-dominated sites known from the county, although this may reflect, to some extent, the difficulties of retrieving such minute pieces from fieldwalking. However, of the four blade-characterised sites which have been excavated in the interior of the county to date, the total number of microliths is less than ten. This might suggest that microliths were neither used nor discarded on sites such as these, but instead on residential sites or perhaps on different types of hunting sites, where whatever the nature of the activity on the site, it had little need of microliths. The local sites may therefore be specialist exploitation camps that differed in function from others found in the uplands and some other parts of the lowland areas.

Landscape use

The human exploitation of the landscape during the later Mesolithic would have taken place in an environment that had changed much from the earlier period. After the mid seventh millennium Cal BC the vegetation cover in the northern third of the county had developed, and large areas of swamp, and subsequently fen, started to form in hollows within the woodlands of the interior and the present coastal zones, and alder fenwood had spread through the coastal zone away from the Alt in Sefton and on North Wirral. The surrounding pine and birch woodland had also begun to be replaced by a more varied deciduous woodland, with trees such as oak, alder, and elm dominating the landscape by the early to mid sixth millennium Cal BC, and hindering easy access across large tracts of land.

Another factor affecting the late Mesolithic settlement in this part of the region would have been the position of the coast. It was generally a time of high sea-levels, both on the Sefton coast and on the Wirral (Table 6) (Tooley 1978a), and for much of the period the sea would have fluctuated close to its

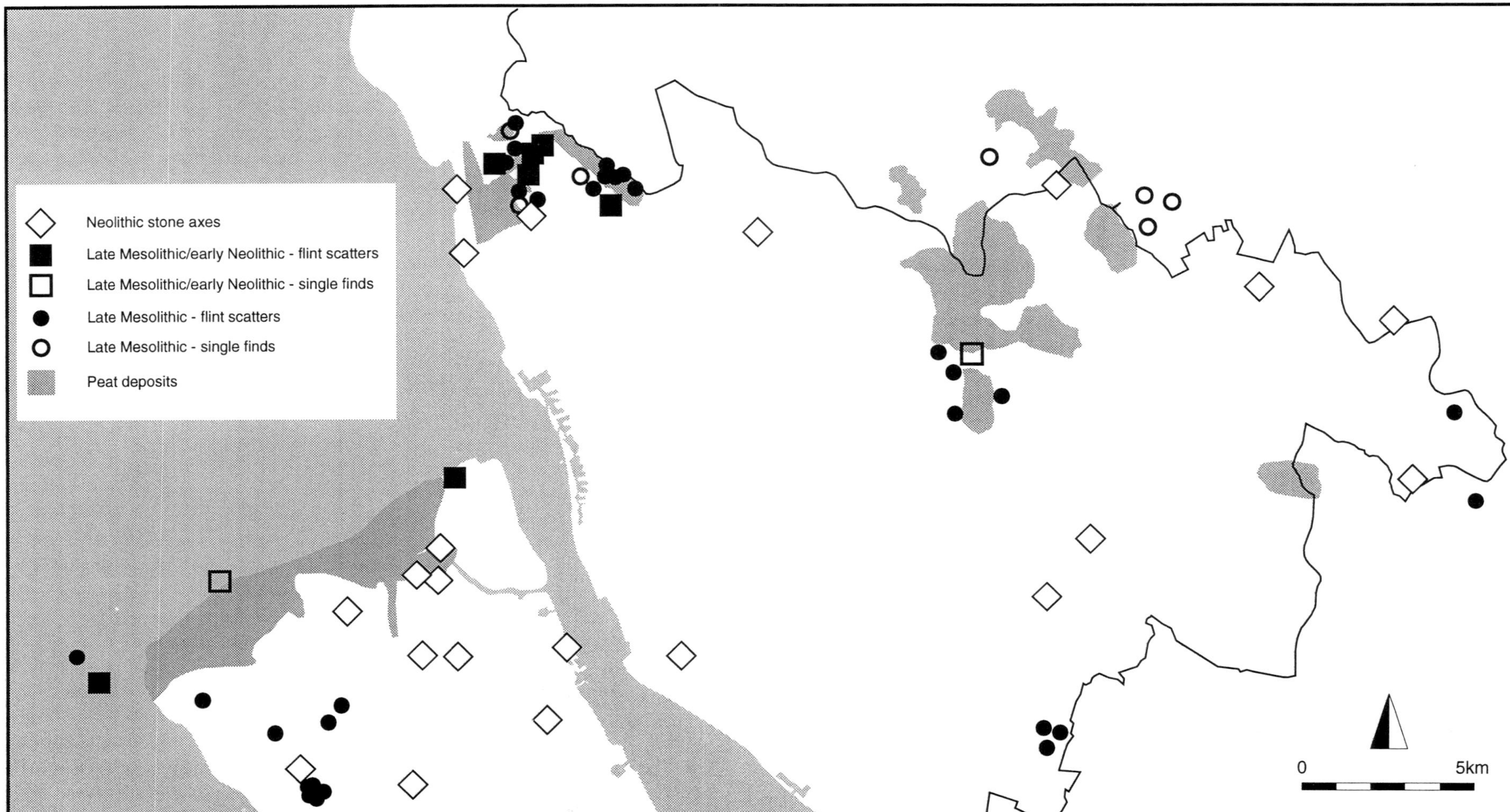

Fig 67 Merseyside: Later Mesolithic and Neolithic sites and finds within the county

present position, at times lying further to the east. To the north of the Alt Valley, the effects of a marine inundation for a few centuries from the early sixth millennium Cal BC saw mudflats and backing marshland extending over 2km inland of the present coast, while the modern day coastal fringe would have been covered by saltmarsh during the last few centuries of the late sixth millennium Cal BC. A similar situation obtained again for a few centuries around the beginning of the fifth millennium Cal BC. The present Altmouth probably marks the position of the prehistoric estuary and the route by which marine influence extended into west Lancashire during this period (Huddart 1992). Coastal landscapes also extended to the south of the Altmouth. It is not clear exactly when, but it seems very likely that it was during the late Mesolithic. During Tooley' s Downholland Moss (DM II) incursion, in the later sixth millennium Cal BC, estuarine conditions spread down the Alt as far as Lunt (Table 10). This suggests that the areas close to Flea Moss Wood and Orrell Hill would have been occupied by saltmarsh environments at this time, if not in the preceding DM I phase at the beginning of the millennium. On the Wirral, the present northwest coastal area was also affected by advancing marine conditions during the late Mesolithic. Estuarine flooding took place in Newton Carr (Table 16) and along the area around Meols after *c*5490-5240 Cal BC (Kenna 1986). The Newton Carr evidence suggests another incursion took place in the bay when the Sefton coast was inundated in DM III in the mid fifth millennium Cal BC.

The proximity of the coast to the fringing mires would have provided a rich variety of natural resources that would have been available almost all year round, within a small geographic area in Sefton and North Wirral. Estuarine muds would never have been too far away for much of the late Mesolithic, and these areas are likely to have been important for wildfowl. The present Sefton coastal area is one of the most important in Britain for wintering birds, and if it was as important in earlier times, it would have facilitated predictable hunting opportunities during a time of the year when other resources were limited (Bonsall 1981). The coasts would also have been attractive for sea mammals; dolphin bones, possibly burnt, have been found in marine silts on the Alt near Hightown, dated to the sixth millennium Cal BC (Lewis 1982, 18). Fish would also have been an important resource in shallow sea and estuarine areas, with spring shoaling in estuaries and sea trout and salmon providing predictable resources for most of the period from spring through to autumn, as they spawned upriver. These resources could have been enhanced by shellfish and low nutritional plants, which could have been collected from saltmarsh environments during the summer months (Bonsall 1981).

Backing onto these environments were freshwater reedswamps and later carr which would also have provided fish and other aquatic animals such as beavers and otters, while wildfowl and plants would also have been plentiful in these environments. The woodland growing on the dry soils around the wetlands would also have provided habitats for ungulates such as auroch, wild pig, and deer, and would have provided edible plants, berries, and nuts such as hazelnuts, that could be collected (Smith *et al* 1981). Other wetland plants could have been used for medicinal purposes, dyes, containers, and rope (Coles 1991, 8). The coastal zone and fens would also have provided a break to the dense woodland that had developed by the late Mesolithic. On the sandy soils this may have been less dense, and it is possible that such areas reached their climax of mixed forest vegetation several centuries later than adjacent claylands (Innes and Tomlinson 1991). The sandy areas could therefore have provided easier passage through the woodland than clay areas and would have been easier to clear if necessary.

In the 12 pollen diagrams presented in this study, woodland disturbance before the early fifth millennium Cal BC is only recorded from coastal areas, and mainly at Bidston on the north-east corner of the Wirral. If it is assumed that some of the archaeological evidence from the area is earlier still, it would appear that hunter-gatherers were active in the natural clearings around the edges of swamps or fens without affecting the woodland or disturbing the ground to any great degree.

One of the most interesting features of the evidence from the current survey relates to the economic nature of the Neolithic and its relationship to that of the Mesolithic. It is only during the fifth millennium Cal BC that disturbed ground and associated small gaps in the woodland cover become more common across the county. They occur both around the coastal areas and around the central mosslands (Table 17). At the coastal sites of Bidston Moss and Flea Moss Wood these phenomena are accompanied by cereal-type pollen at *c*4900-4500 Cal BC. Such pollen is recorded at a number of other sites in North West England and Northern Ireland around this time (Edwards and Hirons 1984; Wiltshire and Edwards 1993). The English evidence has been interpreted as the early adoption of farming by Mesolithic hunters (Williams 1985). If this is the case, cereal cultivation was taking place prior to the main Neolithic period characterised by burial

Table 16 Merseyside: Sea-level changes summary

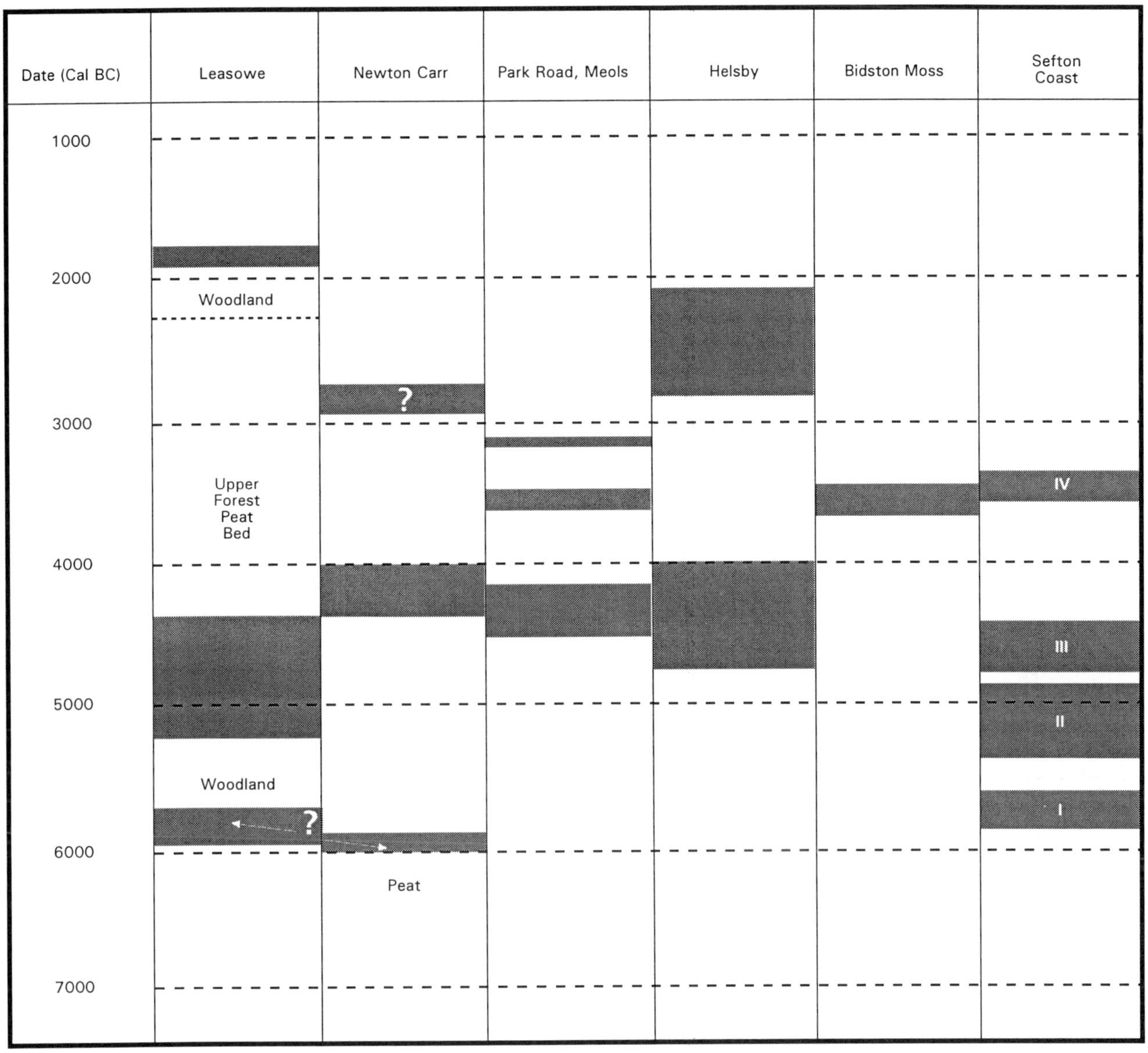

and ceremonial sites and associated cultural artefacts, dated to *c*4000 Cal BC in southern England (Zvelebil and Rowley-Conwy 1986).

If cereal-type, rather than positively identified cereal pollen is taken as representing the introduction of domesticated plants into the region at such an early date, the implications are that Mesolithic communities were adopting elements of a Neolithic economy (Simmons and Innes 1987). It is interesting that many sites where this phenomenon is found lie along the western seaboard of Britain (Edwards and Hirons 1984), suggesting that contacts might have been made with people along the western coasts who had access to domesticated plants. Subsequently, cereals might have become part of the Mesolithic repertoire, whereby small areas might have been planted as a slightly more predict able adjunct to the normal round of seasonal hunting and gathering. It is notable that the two local occurrences are both found in the coastal zone, where the stronger evidence for more sedentary residential Mesolithic sites is found. These are places where an investment of time in small paddocks or plots of sown plants may have been easier to accomplish than in a landscape such as the interior, where the range of resources may have been less predictable. It must be stressed that the evidence for this process of the adoption of cultivation is still very slight everywhere in Britain, and there are difficulties in the interpretation of the little evidence that exists (Zvelebil and Rowley-Conwy 1986; Kinnes 1988, 3).

In the central mosses the first good evidence for

Table 17 Merseyside: Potential woodland clearances summary

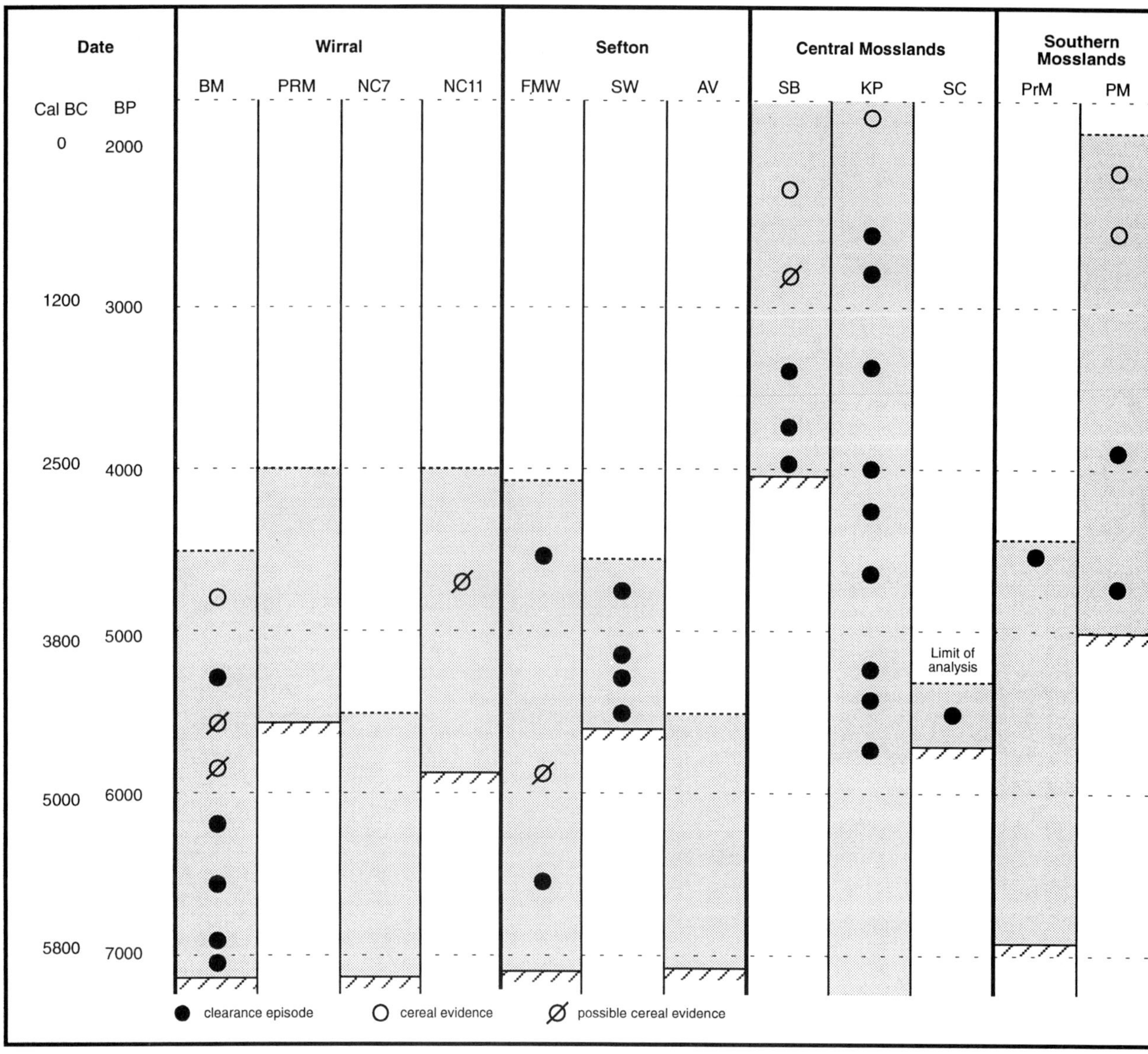

woodland disturbance is seen around Knowsley Park Moss, with three clearance episodes between *c*4998-4330 and 4340-3970 Cal BC. Four kilometres to the north, at Simonswood Moss, a similar phase is dated to 4669-3970 Cal BC (5440±160 BP; Birm-1222), associated with a charcoal band in the peat, and provides a rare regional example of evidence in the lowlands for clearance associated with burning in close proximity to archaeological evidence.

Woodland clearance involving the use of fire at the forest margins is quite extensively recorded for upland areas of northern England (Mellars 1976). This is thought to have provided open areas encouraging the growth of new shoots attractive to game, which would have made prediction of animal movements and their interception easier. One possible interpretation of the Simonswood evidence is that a similar activity was being undertaken in the lowland mire systems as well. Carbonised plant remains are found near the base of a number of mosslands, particularly in the east of the county, but dating of these is generally poor. Perhaps related to these observations is the record of a charcoal band in a similar stratigraphic position at Hoscar Moss in West Lancashire, 10km to the north (Cundill 1984). However, the question of whether the burning has a natural or anthropogenic cause remains unresolved, although the balance of circumstantial evidence seems to point to the latter cause.

If the adoption of cultivation did take place in some locations in the North West at this early date, it did not lead to an increasing exploitation of the woodland for farming. Although the pattern of woodland disturbances noted during the fifth millennium Cal BC becomes more widespread throughout the county and at some sites is a little more frequent than previously (Table 17), the nature of each disturbance episode is still limited in extent. Each episode is also succeeded generally by long periods of forest recovery. Even in areas where cereals may have been part of the seasonal round, hunting and gathering must still have played an important and dominant role throughout this millennium and the following one, by which time Neolithic cultural artefacts had been introduced into the county.

The earlier Neolithic period *c*4000–3200 Cal BC

In England, Neolithic cultural material dates from *c*4000 Cal BC. Locally, the lack of chronology makes it difficult to identify how far the economic activity of the second half of the fifth millennium Cal BC is associated with the introduction of Neolithic cultural material. However, because pollen evidence suggests that the exploitation of the Merseyside landscape in the early Neolithic differed little from that in the preceding phase, similar types of site in similar locations might be expected. Despite this, the archaeological evidence for the fourth millennium Cal BC is as difficult to identify from around the Merseyside wetlands as it is from the dryland areas of the county to the south.

Prior to the NWWS, knowledge of the settlement pattern for the period was much sparser than for the Mesolithic, but included the identification of three possible early Neolithic burial sites found along the upland fringes, spread between Cheshire (Longley 1987), Lancashire (Bu'Lock 1959), and southern Cumbria (Clare 1979). Settlement sites are extremely difficult to identify and Neolithic artefacts are rare, apart from stone axeheads. Nationally, although there are quantifiable differences between the flint technology of the early Neolithic and that of the Mesolithic (Pitts and Jacobi 1979), it is still one based on blades, and sites when identified are still relatively small, discrete entities (Healy 1987).

Confidence in identifying these types of site in the North West is weakened by the lack of a regional technological context. In Merseyside, typical Neolithic implements are rare and much emphasis has had to be placed on waste material. The possibility exists, therefore, that some of the sites identified through the NWWS as Mesolithic may, in fact, belong to a slightly later period. Notwithstanding these difficulties, the NWWS results have, for the first time, identified a number of potential settlement sites of the earlier Neolithic, which lie in the coastal areas of Sefton. One site in Ince Blundell (Me 065), a small homogeneous group of local reddy-brown boulder clay flint, included a leaf arrowhead in the same material. Without the latter piece the material showed a number of similarities to other material interpreted as Mesolithic. The adjoining discrete site (Me 066) is made up of the same material, and therefore may be of the same date, but includes a microlith. These are the only two assemblages in Sefton which record the use of reddy-brown boulder clay flint, although a few isolated pieces have been located (*eg* Me 071). In contrast, the raw material found on most sites in Sefton is normally either beach flint or grey boulder clay pebbles, some of which might also have been collected from the beach.

A number of sites found along the Alt Valley, and set a little way off the edge of the Orrell Hill wetland in Ince Blundell, may also belong to this period. They are characterised by a number of flake-like blades which contrast with the typical small, parallel-sided blades found commonly at other sites in the area. They also exploit a distinctive, grey-blue beach pebble flint. This material is found in the same areas as the Mesolithic sites and occurs in similar small clusters. Given the difficulties of distinguishing between the material of the two periods, the implication from the archaeological evidence, which complements the pollen evidence, is that little changed in the nature of the exploitation of the coastal landscape through the early fourth millennium Cal BC.

The archaeological evidence is marginally better on the Wirral compared to the inland areas to the east of the Mersey, and is based almost wholly on previously known finds. Slightly to the south of the coastal wetlands, the inland sandstone ridge overlooking the Dee valley around Irby, which was important in the Mesolithic, continued to act as an important focus for early Neolithic sites (Philpott and Cowell 1992). Even allowing for the limited scope of field survey on the Wirral as part of the NWWS, and the poor to moderate retrieval conditions of those areas that were surveyed, it does seem probable that the human activity around the northwestern wetland areas of the Wirral was less intensive than around Bidston Moss. Archaeological sites from the ridge bounding Newton Carr to the south are almost absent, apart from one extremely small site which may be of this period (Me 002). The same areas that were occupied in the Mesolithic form the focus for settlement during this period. These are concentrated on the small islands to the west of the large channel near West Kirby, where arrowheads are found (A 8), and to the north-east of the channel on the low sandstone areas near Hoylake (A 10), while the only piece of Neolithic pottery from the county was found in the Dove Point area (A 5).

The trend to relatively substantial clearance and cultivation in the north-east of the peninsula appears to have taken place against a background of generally wet, marine-dominated conditions within North Wirral (Table 16). Elsewhere, only at Newton Carr 11 is palaeoecological evidence established that may be of this approximate age. Here, a small clearing is recorded with cereal-type pollen, dated roughly by the pollen assemblage to the middle Neolithic.

Little new flint material which might be ascribed to this period has been found from around the central and central-eastern mosslands. Only Me 098 on the southern edge of Kirkby Moss, and Me 107 and Me 108 from the lower slopes of the high Billinge ridge, overlooking Kings Moss, possibly relate to this period. Previously recorded stone axeheads may be used in addition to the meagre flint record which, when complemented by the palaeoecological evidence for the fourth millennium Cal BC, might allow a more complete pattern of landscape exploitation to be constructed for the Neolithic.

The distribution of axes in Merseyside, found mainly in the coastal areas of the North Wirral and Sefton, tends to reinforce the pattern of flint finds from the NWWS. In Wirral they are found mainly in the north-eastern wetland area, around the former courses of the Rivers Birket and Fender and particularly on the lower slopes of the sandstone ridges overlooking Bidston Moss (Fig 67), where pollen evidence shows the greatest impact was made on the vegetation at this time. A number are also found further along this ridge to the south-east in Birkenhead, overlooking the lower Mersey valley to the east. There is a slight extension southwards onto the boulder clay areas of central Wirral around Barnston, possibly connected with the valley of the former River Fender (Cowell 1991a).

To the east of the Mersey the archaeological pattern is less concentrated, in keeping with a dispersed pattern suggested by the palaeoecological evidence around the interior mosslands (*see below*). It implies a fairly even exploitation, concentrated on the major river valleys in the area. There is a thin scatter of artefactual material along the Alt/Ditton Valley, while east of the Mersey, the important locational factor appears to be the Sankey Valley on the eastern edge of the county in St Helens. Here the distribution is particularly strong at the southern end, around its confluence with the Mersey. This distribution is continued to some extent south of the Mersey, where axes are found along the upper slopes of the River Weaver in North Cheshire, near its confluence with the Mersey (Cowell 1991a).

It is around the north-eastern corner of Wirral, where most of the archaeological evidence is found, that the palaeoecological evidence points to a substantial effect on the landscape, which contrasts with the rest of the county. After the elm decline, dated 4034-3790 Cal BC (5120±50 BP; SRR-2929) on the North Wirral coast at Park Road, Meols, the pattern of vegetation changes seen during the early and middle Neolithic continued much as before. Only at Bidston Moss are the vegetation changes seen to be different in scale. The difference between the Wirral and elsewhere is, however, one of degree. Woodland regeneration is found after each disturbance, as in other parts of the county, but each successive clearance phase is slightly more intense and the regeneration less complete than in each previous phase. By the middle of the fourth millennium Cal BC the general woodland cover around Bidston Moss would have been more open than anywhere else in the county.

The archaeological evidence in Sefton is perhaps better than from the Wirral, due to the discovery of potential flint sites during the NWWS fieldwork. In addition, a number of stone axes have been previously recovered around the estuary of the Alt, concentrated around the coastal area of peat (Fig 67). One comes from the Little Crosby Shirdley Hill Sand ridge (A 18), and two examples from the coastal fringe to the west (A 19, A 21), although the position of their original location is vague. Palynology gives us fewer clues to human activity here, however, than in North Wirral. Neither of the two main Sefton sites reveal evidence of clearance in the woodlands around the coasts. The late Neolithic stratigraphy in the peat core from the Flea Moss Wood site (which covers a more extended period than that at Sniggery Wood) is probably contaminated by modern pollen, although no evidence similar to that on the Wirral is noted before this. The peat from Sniggery Wood is terminated by blown sand after 3370-3040 Cal BC (Table 9), as marine conditions approached during the middle Neolithic and sand dunes became established near to their present position or slightly inland. The evidence is slightly stronger here, although the general scale of woodland clearance through the period is not as extensive as on the Wirral, and there is no evidence for cereals being part of the strategy, although this could be a result of the small pollen counts at these sites.

Although the evidence is rather unsatisfactory there

are, perhaps, hints that the vegetation changes on this side of Liverpool Bay were not as comprehensive as they were on the north-east Wirral. The coastal areas, however, record slightly more woodland clearance activity than the inland mosses (Table 17). Even so, to the east of the Mersey the nature of woodland disturbances during the fourth millennium Cal BC differed little from those found during the Mesolithic period. This is in fact the case for the succeeding two millennia. It is not until the late Bronze Age that the pattern of small, fairly widespread, but infrequent woodland disturbances alters. Occasionally cereal-type pollen is identified associated with the woodland disturbances in the Neolithic, as at Prescot Moss in the south central mosslands, where the end of this phase is dated to 3630-2890 Cal BC (4520±140 BP; GU-2726). It is possible that cultivation may have been more widespread than this, however, as cereal pollen is restricted in its dispersal, and it may therefore be under-represented at some sites. However widespread cultivation may have been during the Neolithic, it seems still to have been practised within a pattern of small woodland disturbances which soon regenerated. These are found around the southern mosses, the central mosses and the mosses in both the main coastal areas during the fourth millennium Cal BC. At each site, however, there is often a minimum of several centuries between each disturbance phase, during which time woodland regeneration was fairly complete. This implies a system of mobile communities operating in these woodlands and around the mosslands, for whom gathering and hunting still played an important role.

The late Neolithic and early Bronze Age *c*3200–1650 Cal BC

Nationally, archaeological evidence shows a number of developments by which material and sites of each period can be distinguished. These are most obvious in the development of pottery styles, the introduction of metalwork, and new burial styles (Burgess 1974). Flint technology, however, which shows strong contrasts with earlier Neolithic styles, changes little from the late Neolithic period into the Bronze Age, and a technological continuum has been noted for this period (Healy 1984). It is possible to differentiate the flintwork of these two periods in certain geographic areas where a good, dated technological framework exists from excavated sites (Pryor 1978, 1980; Healy 1984).

In the North West context these differences are less obvious, mainly because of a lack of material. The NWWS fieldwork has not produced pottery of either period, although one late Neolithic sherd is recorded from a wetland context at Dove Point, Wirral (A 5). Only two occurrences of Bronze Age pottery are recorded within the county, both close to the wetland area of the north-west coast of Wirral (A 9 and A 11). No metalwork has been located from the NWWS and only one example of early Bronze Age metalwork is known from the county, from outside the wetland zone on the Wirral (Cowell in press).

There is one burial site in the county, the Calderstones, probably of late Neolithic date (Cowell and Warhurst 1984), on the southern side of the Triassic sandstone plain south of Liverpool, but there are no convincing sites of this type elsewhere in the lowlands. The distinctive early Bronze Age burial tradition, based on earthen mounds, is only a little more widespread, although none are in a wetland context. Three sites are recorded from the county, with another three being located immediately over the south-eastern county border in Cheshire at Winwick (Freke and Holgate 1990), within several kilometres of Highfield Moss in Greater Manchester. Another three are suspected within the county from documentary and aerial photographic evidence (Merseyside SMR).

Flintwork, the means by which these two periods might most easily be recognised locally, is difficult to differentiate because of the lack of a technological framework for Merseyside or the North West in general. This problem is of particular relevance to material collected during the NWWS, as the fieldwork produced one of the largest groups of flintwork retrieved from the region that is datable to the later Neolithic or early Bronze Age. The material is mainly found on the Woodham Knoll ridge in Little Crosby (Me 007, 018, 019, 036, 037), overlooking the coastal peat on the west and north. It differs markedly in its technological characteristics from all the material so far discussed (Fig 22), and is the first substantial assemblage of this type to be found in the region. As this is at present the only material of its type known from the area there is some difficulty in dating it more precisely than to a general late Neolithic/early Bronze Age period. Only a couple of isolated finds have been located in the Sefton coastal area away from the Woodham Knoll ridge (Me 063, Me 068). The only other site of this type in the area is one previously located on the sandy ridge on the north-eastern side of the Alt Valley, at Lydiate (Lewis 1982) about 4km away. No material of this type came from around the central mosslands (Fig 68).

The NWWS fieldwork in Sefton also discovered flint material which is formally undatable. It consists mainly of natural pieces or knapping waste retouched into various cutting, scraping, or piercing tools. These pieces are often found singly or in small clusters of two or three pieces. At some locations they may be associated with burnt pebble fragments. Comparison of form and raw material with other assemblages from the survey suggests that it may be later than the above material, which would imply a possible middle or even late prehistoric date. It is found on the sand overlying the coastal plain peat, and along the edge of the peat channel to the east of the Woodham Knoll ridge in Little Crosby. A few locations with burnt flint nodules are also recorded in the Alt Valley floodplain (Me 072, Me 079). This type of material is very occasionally found around the west-central mosslands, as at A 28, to the south-west of Knowsley Park Moss in the small valley of the Mill Brook.

Elsewhere in the county the only other flintwork evidence for this period comes from chance finds of barbed and tanged arrowheads. These are found in the same areas as the major burials of this period, which lie on the Triassic sandstone east of Liverpool (Smith 1868) and on the sandstone ridge to the south-west of Newton Carr, Wirral (Mayer 1849, 153).

The palaeoecological evidence largely ceases in the coastal area by this time, although the lower sea-level of the late Neolithic (Tooley 1978a) may have continued to affect the north-eastern corner of Wirral for a while, as the Upper Peat/Forest Bed on the North Wirral coast shows that woodland extended beyond the present coastal area (Table 16). After this date wetter conditions returned (Kenna 1986). There is even less information for the Sefton coast, although in general the present coastline may have marked the approximate prehistoric position for much of this period (Innes and Tooley 1993).

In the central mosslands, as seen at Simonswood Moss and Knowsley Park, the surrounding forest remained relatively dense during the early Bronze Age. The pattern of vegetation disturbance is similar to that of the Neolithic, with small infrequent woodland disturbances (Table 17). In the south similar phases are recorded at Parr Moss, Burtonwood. These may reflect relatively weak attacks on the forest, as they are mainly characterised by the appearance of weeds, plantain, bracken, and grasses, with only very limited reductions in woodland cover, although the occasional phase such as at Knowsley Park in the mid third millennium Cal BC (Table 13, phase 12), or Parr Moss at the end of the third millennium Cal BC (Table 15, phase 4), witnesses more substantial woodland reduction.

An economic pattern based on mobility and cattle grazing is sometimes invoked for the early Bronze Age (Fleming 1971), although it is likely that hunting still played an important role. Pastoralism is very difficult to prove archaeologically, but the pollen evidence in the county could be interpreted in this way. It follows essentially the same pattern as in the early prehistoric period, when mobility for hunting and gathering is seen as the basis for the Mesolithic and early Neolithic economy.

The flint material, which is almost exclusively found in coastal Sefton, is difficult to date, but it seems very likely to belong to the period after the landward migration of sand dunes over the coastal fens in Little Crosby (Innes and Tooley 1993). It is not clear if these reached all the way to the foot of the Shirdley Hill Sand ridge of Woodham Knoll. None of the possible late Neolithic material (Fig 68) which clusters on the knoll is found on the sand, unlike the *ad hoc* flint material, which is more likely to date to later in the period. There may therefore have been fenland in the vicinity of the knoll during the late Neolithic period, possibly backing onto the dune system. The only other material of this type is found on the inorganic Shirdley Hill Sand close to the Orrell Hill fen.

Although the evidence is slight, one explanation may be that the fens were being used for summer pasture for cattle, and the associated settlement lay on these sandy soils set a little off the wetland, as might be found in several similar areas (Pryor 1976; Coles 1978). During the winter months the fens and nearby estuary could have been used for wildfowling and fishing, while the cattle were moved to higher land. The distribution of the flint material suggests only a limited area would have been used for settlement activities, with quite wide ranging grazing taking place. However, pollen analysis from the coastal site of Mount Pleasant, Waterloo, *c*3km to the south of Little Crosby, indicates a late Neolithic or early Bronze Age phase of clearance and cereals (Innes and Tooley 1993). This suggests that the dune fringe was also attractive for cultivation.

In the interior there is less archaeological material attributable to the later Neolithic or early Bronze Age, and all of it is represented by single finds (Fig 68). If the same pattern is evoked for this area, the light dispersal of finds and the palynological evidence for possible clearings might suggest wide-ranging, low intensity pastoralism being practised around the edges of the mosses. No convincing evidence has yet been found, in all the fieldwalking undertaken, that might suggest home range areas where sites with tool types of this period might be found, other than around Little Crosby. One possibility is that these areas were located further to the south on the sandstone areas, many of which have suffered modern urban and industrial spread. One area recently excavated *c*4km to the south of Prescot provided some evidence of Bronze Age activity in this type of landscape (Cowell and Philpott 1994), and the sandstone ridges to the south of the wetland zone on North Wirral have provided surface evidence that suggests widespread Bronze Age activity (Philpott and Cowell 1992).

The fens in the Alt Valley seem not to have been used for settlement at this time. The *ad hoc* flint tools and burnt flint pebbles do, however, occur more frequently in the valley, the Little Crosby peat channel and across the blown sand areas to the west. It has been suggested above that this material is later and may be of middle Bronze Age date or later. The dispersed distribution of the material may indicate temporary exploitation of the fens, carrs, and dunes, perhaps associated with grazing rather than representing settlements, although this must remain conjectural until better evidence is forthcoming.

Late prehistoric period *c*1650 Cal BC – AD 43

Inland, in the central mosses around Simonswood Moss, there is some evidence for drier conditions during the second and early first millennia Cal BC, with cotton-grass and heather-dominated mires fringed and partly occupied by birch scrub. Traditionally the Bronze Age ends with the introduction of iron, at *c*700 Cal BC. However, the paucity of evidence from the North West makes it more convenient to treat this later prehistoric period as one unit.

In the absence of known examples of prehistoric ironworking, the main artefactual evidence comes from metalwork of late Bronze Age date. A common feature of the national distribution of such metalwork is its bias towards wetland areas and rivers, particularly in the form of hoards (Bradley 1984a). This has been interpreted as occurring at or shortly before a time of worsening climate, which in turn has been suggested as being a major factor determining the changes seen at the end of the early Bronze Age, involving the disuse of barrow burials, stone circles, and alignments, and most pottery styles (Burgess 1974).

In the North West hoards are rare, the more usual pattern being single finds (Davey 1976). Locally there are too few finds to allow a discussion of metalwork production and use, and the finds are of limited use for understanding the contemporary pattern of settlement. Although finds tend to be located in the middle to upper reaches of the main rivers in the region (Davey 1976), within the county there is no noticeable connection of late Bronze Age metalwork with wet places (Fig 68).

Other evidence for settlement is limited. The only pottery known in the region comes from a hill settlement at Beeston in Cheshire, which includes late Bronze Age and Iron Age vessels (Keen and Hough 1993). However, a number of recently excavated sites provide evidence of settlements which existed towards the end of the prehistoric period. One of these is located on the edge of a wetland just over the county border in Cheshire (Nevell 1989), although the pottery from this site is very amorphous. Recent aerial photography has identified a number of additional lowland sites in the county that on morphological grounds may belong to the end of the prehistoric period. These are in the form of oval or subrectangular enclosures of putative farmsteads, although they are all located away from the edges of the mosslands, being mainly confined to valley sites in the Triassic sandstone areas of the county (R Philpott pers comm). Two of these sites have been excavated, and show possible Iron Age material succeeded by Romano-British evidence on the same site at Irby on the Wirral, and at Halewood in the south of the county (Philpott 1993; Cowell and Philpott 1994).

Palaeoecological evidence, particularly relating to the late Bronze Age, suggests that the areas around the central mosslands were the focus of relatively concerted farming activity for the first time during the early first millennium Cal BC. Episodes are recorded at Simonswood Moss and at Knowsley Park (Table 17), representing the first relatively substantial clearance of woodland in these areas and the best indications of the adoption of mixed agriculture, with the presence of cereal-type pollen and weeds associated with arable cultivation.

It has been suggested that the beginning of the first millennium Cal BC may have heralded a decline in the climate of the North West, which could have led to increased rainfall and a fall in temperature of *c*2°C, so that the summer average was *c*0.5°C lower than that of today (Lamb 1981). To the east of the county, in the Mersey Basin, a reduction of clearance activity is noted at this time (Howard-Davis *et al* 1988). The shift of the pattern of metalwork finds in the region from the lower Mersey to the upper reaches of the rivers might therefore be interpreted as a response to the worsening of conditions on the lower ground (Davey 1976). However, the local pollen evidence seems to contradict this to some extent.

If the apparent extra investment in farming can be used to imply that settlements were for the first time becoming more fixed in the landscape, it might be expected that sites of the late Bronze Age and the Iron Age would be more identifiable. One interpretation might be that sites are located within the wetlands, just as they are found elsewhere in the country (Coles and Coles 1986; Pryor 1991). Settlements were located within lake-associated environments, mirroring to some extent the importance of wet areas for votive offerings in the late prehistoric period (Bradley 1984a). The stratigraphic evidence from the central mosslands, however, suggests that such flooded environments were not widespread at this time. Simonswood Moss had developed into an ombrotrophic mire by the beginning of the first millennium Cal BC (Table 11) and Knowsley Park Moss by the early Neolithic (Table 13). Nor is there evidence that these locations did contain settlement sites, for despite very detailed survey of the Simonswood Moss upper peat, for example, no structures were located. The cotton-grass and heather-dominated plant communities which characterised the mires at this time would have made them wet underfoot, although relatively accessible, while any grazing land they afforded would have been of poor quality.

Sites may therefore have been placed away from the mires. Aerial photographic coverage of these areas around the central mosslands has not, though, produced any potential sites in the light sandy soils surrounding them. The relatively large areas surrounding the central mosslands that have been subject to sand quarrying may partially account for this. Another explanation may be that the farmsteads themselves were still relatively slight structures, particularly in the late Bronze Age, as the earliest evidence for substantial farmsteads in the region belongs to the late Iron Age, although the dating evidence is still very slight for these types of site.

Romano-British period AD 43–410

The main evidence for the Romano-British period is found on the Wirral. The large coastal settlement of Meols appears to have acted as a focus for the growth of a series of isolated farmsteads around the sandstone ridges in the hinterland, identified as oval or rectilinear enclosures through aerial photography and, as at Irby, through excavation (Philpott 1993). A number lie on the sandstone ridge a little to the south of Newton Carr, although it is unlikely that these settlements exploited the moss to any extent on account of their distance (*c*4km) from it. These sites have produced Black Burnished Ware, a little Samian, and at one site a large number of bronze *fibulae*, suggesting a developed Roman landscape of individual farmsteads existed. Dove Point, Meols, which was potentially one of the most important sites in the North from the Roman into the medieval period, may have acted as an entrepot for this agricultural hinterland. It probably also played a part in the development of the town of Chester, linked with trade along the Irish seaboard during this period (Philpott 1991).

The Romano-British settlement of the county to the east of the Mersey has traditionally been more difficult to identify, but the pattern of an area almost empty of occupation between the military centres of Chester, Wigan, and Walton-le-Dale is now being redressed. Pollen evidence implies that much of the area must have been left as woodland and waste, but scattered cropmark enclosures are now recognised on the heavier soils in the south-east of the county. These have come from previous programmes of aerial photography and, to a lesser extent, from fieldwalking by Liverpool Museum staff (Philpott 1991). Recent excavations have shown that this distribution may greatly underestimate the extent of Romano-British rural settlement in the county. Excavations carried out in the summer of 1993 on a presumed Mesolithic site in an area of Triassic sandstone ridges in Tarbock, Knowsley, produced evidence of a rural settlement of the second to fourth centuries AD. Further traces of Romano-British settlement were found *c*1.5km to the south in excavations on a medieval farmstead, and on a late prehistoric farmstead in Halewood, in the boulder clay plain *c*3km to the south, close to the Mersey (Cowell and Philpott 1994). The pattern of medieval estates and fields in the township of Tarbock would now seem to be related to elements of the Roman landuse pattern.

In the coastal and sandy areas to the north and north-west of the county, such sites are so far proving more difficult to locate. This may be because the light soils do not provide the right conditions for aerial photography, but fieldwalking and chance finds are also very scarce in the area. The NWWS has provided no new evidence for activity from around the mosslands, so that the main mossland evidence comes from place names. Bury Hill in Knowsley Park, a little south of the moss (Fig 84), might indicate a Roman farmstead, although no Roman pottery has come from fieldwalking in the vicinity. A second possible site may be located near the moated medieval site of Mossborough Hall, which lies on a spur of boulder clay between Simonswood and Reeds Mosses. The 'bury' element in both names signifies a 'fortified' or significantly enclosed site (Ekwall 1922), an element which is often associated with Roman farmsteads. The

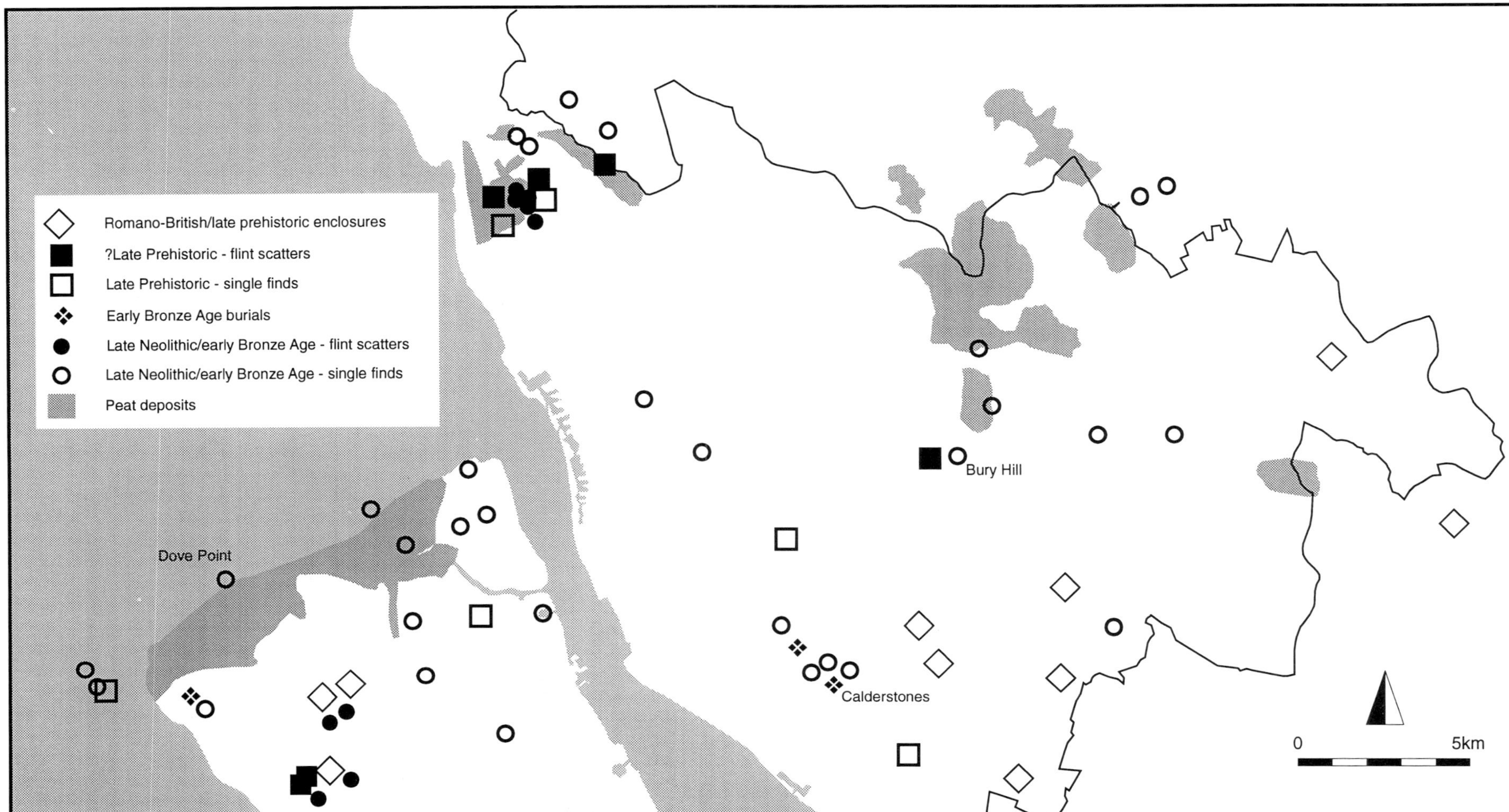

Fig 68 Merseyside: Later prehistoric and Romano British sites within the county

pollen diagram from Knowsley Park Moss, c1.5km to the south, suggests mixed farming was being carried out in the vicinity between AD 240-440 (Table 13). This implies that the mossland exploitation might have been restricted to grazing, and wood and turf for various purposes might have been available. The reason for settlement location, however, would seem to be the conjunction of a number of complementary resources rather than the monopoly of a wetland resource. There may be a tendency for late prehistoric sites to lie in the same locations as the Roman ones, suggesting that this pattern may have earlier origins (Fig 68).

Early medieval period AD 410–1066

For the medieval period, the relationship of wetlands to the community that utilised them can be more objectively defined than in earlier periods, through the identification of territories and boundaries from ground survey and documentary sources. Prior to the Norman Conquest the area of the central mosses, particularly on the eastern side, lay in a heavily wooded region with a sparse population (Chitty 1981; Ekwall 1922). In the whole of St Helens district there was not a single pre-Conquest church (Fig 59). After English influence reached the area, probably after the seventh century AD, the survival of Celtic place names suggests that a large area was avoided. This is partly explained by the amount of woodland cover on the relatively large expanse of boulder clay soils, although the mossland areas dominated by wet *Sphagnum* bogs must also have been a limiting factor in the northern half of the district. Throughout this period the region seems to have been in a marcher area between the kingdoms of Northumbria and Mercia. When it passed from the former to the latter kingdom in the ninth century it still remained separate from Mercia, being royal demesne. It has been suggested that this status and remoteness may have saved it from the punitive ravaging of the North in William's time, as no areas are recorded as being waste.

Later medieval period AD 1066–1500

This wooded landscape formed the basis for the Royal Forest of Lancaster east of the Mersey in the late medieval period (Cunliffe Shaw 1956). The western-central mosslands formed part of the hunting forest in Simonswood, and the now lost mosses to the west of Knowsley lay within the royal demesne of West Derby, where much of the land was given over to hunting ground. Although adjacent areas lay in the purlieu of the Forest, *ie* subject to forest law but not included within the hunting forest, this status did not prevent the growth of farming communities, separated into townships. Post-medieval maps show that the mosslands were located on the fringes of the townships and held as common land. This pattern probably derives from the medieval period, when numerous court rolls show that although the mosslands were peripheral in location they were not unimportant, and their resources were carefully managed.

In the coastal fens seasonal flooding provided rich grazing lands for cattle and sheep, leading to monastic holdings of land in these areas. Township boundaries in Sefton run at right angles to the course of the Alt, so that each can share in this valuable resource. The wetter parts of the fens would have been important for eeling, fishing, and fowling, while a swannery is mentioned on Newton Carr in the thirteenth century, leading to monastic holdings of land in these areas. Rushes and sedges could be collected, and the peat itself could be cut for fuel (turbary). This is the most common reference to the use of the wetlands in the late medieval period.

The only example of settlement in a wetland context at this time comes from the Wirral, where the Dove Point site came to full prominence in the post-Roman period. Unfortunately little is known of its immediate environment other than that it was close to the sea. Many examples of Saxon and Viking artefacts are recorded from the area of the present beach, and settlement drift during the late medieval period is visible along the beach (Hume 1863). Wooden remains of buildings recorded on the beach during the nineteenth century may relate to this.

Post-medieval period AD 1500–c1850

With the increased political and economic stability of the post-medieval period agriculture generally continued to expand, and until the nineteenth century the landscape remained predominantly rural. Even after the late eighteenth century expansion of population in response to the full development of industry in the St Helens and Liverpool areas, the central mosslands remained largely undeveloped. The contrast is strong between here and the areas to the south and west of the mosslands on the heavier soils around the pre-Conquest centres of Huyton and Prescot and around St Helens. It was at this time that the southern mosslands began to be developed.

The scattered nature of settlement is a feature of the remaining undeveloped rural areas today, much of it only originating in the post-medieval period, and in some places there has been a noticeable shrinkage in settlement as the populations were drawn to the industrial towns during the nineteenth century. This has allowed large areas of the central mosslands to avoid heavy exploitation. Medieval nucleated settlements never grew up in the area except at Rainford, which lay far from the moss areas. The economic role of the mosslands as common land on the margins of the townships, first recognised in the medieval period, continued into the nineteenth century. In many cases this mossland often remained unenclosed into the late eighteenth and nineteenth centuries.

The coastal mosslands lay in areas that had seen more intensive medieval settlement, although nucleated settlements lay away from the mossland areas. Along the coastal fringes, the deposits of blown sand over the peat areas dictated that they be used for grazing and were mainly regarded as common land. In areas in the Alt Valley it was not until large-scale drainage in the late eighteenth century that the wetlands came under an intensive arable regime. Townships in the Wirral which included mosslands away from the sandy coastal fringe saw their common mosses remain unenclosed into the late post-medieval period. It was not until the late nineteenth century industrial development that the first intensive exploitation of these areas took place.

Archaeological and palaeoecological potential

A major contribution of the survey to archaeological knowledge is the retrieval of the extensive pattern of prehistoric settlement from around and within the mosses of the Sefton area. This evidence is of great potential in furthering the research into the nature of prehistoric sites and the interpretation of settlement patterns, as discussed in Chapter 1. The Sefton evidence (*see Chapter 4*) provides the greatest concentration and variety of prehistoric sites in the county, matched only perhaps by the sandstone ridges of North Wirral (*see Chapter 3*), deriving from work unconnected with the NWWS (Cowell 1991a). The importance of the Sefton evidence is potentially greater, as a proportion of these sites are in wetland contexts. Many of these sites may also have been afforded a greater degree of protection from plough damage due to the depth of post-depositional layers potentially overlying them. These factors could be of great value in investigating the nature of prehistoric sites largely unaffected by subsequent distortion due to ploughing, which would be of great benefit in interpreting surface scatters in the region, while potential organic survival will also enhance the interpretation of such sites.

Even in the areas where surface flint material is found on the mineral soils, the range of surface evidence found potentially allows the investigation of the changing pattern of early prehistoric settlement systems across the landscape, which can be integrated into their palaeoecological context at both site and sub-regional level. Even in the central mossland areas, where the amount of archaeological evidence is much reduced, the palaeoecological evidence from the NWWS hints at distinctive patterns of landuse that may be represented by very little surface material. This pattern is interesting in its own right, as it provides complementary evidence for the interpretation of palynological evidence in areas where settlement type or density is light. This raises questions as to whether important environmental changes, such as the widespread burning found in the central mosslands, result from minimal human involvement, as witnessed by the artefactual record, or from other, perhaps natural factors such as climate. If early prehistoric settlement patterns involve different characteristics and scales of activity in different environments, then such evidence as is found in the central and east-central mosslands (*see Chapter 5*) must be viewed as integral to an understanding of the coastal areas, where a greater range of site types is found.

The NWWS in the wetlands of Sefton has now extended the pattern of sites, whilst identifying over a dozen sites that lie beneath the peat and are therefore potentially waterlogged to some extent. In a number of cases the peat is protected from plough damage and surface desiccation by a layer of blown sand in the coastal plain, and alluvial or marine clay, or both, in the Alt Valley. This obviously provides the potential for recovering a greater range of archaeological and palaeoenvironmental information for sites of the early prehistoric period.

Sefton

The archaeological pattern in the Wirral coastal area mirrors that of Sefton in many respects. The area of the Sefton mosslands has several advantages over the North Wirral. These derive from the fact that much of the landuse here is still agricultural, within the Green Belt, and therefore not affected by the same housing and industrial pressures to which the wetland areas of North Wirral continue to be subject. Agricultural erosion of the Sefton archaeology may be constant, but its eventual destruction is occurring more gradually than on the Wirral, allowing phased recording and management strategies to be devised in advance of its demise. The landuse pattern has also facilitated the discovery of more sites than it has in the Wirral coastal areas, thus providing a better predictive model in Sefton for the location of other sites in areas of deeper stratigraphy. Also, unlike the North Wirral wetland areas, a number of significant archaeological sites have been located through the survey within and around the edges of the mosses.

The coastal group

The Sefton coastal peats are reasonably well protected by blown sand and marine clays in the farmland areas. Relatively high water-tables also often keep the lower peats moist. There is no archaeological evidence from within the coastal plain peats in Little Crosby, as the pre-Flandrian sand subsurface, wherever it could be recorded in section, is generally *c*1m below the plough surface, and the upper peat surfaces are generally *c*0.6m beneath the modern ploughsoil surface. Potential for waterlogged evidence in these areas is high, but their archaeological potential is unknown because the gently shelving subsurface is buried too deep for the easy discovery of sites through surface survey. The incidence of Mesolithic and Neolithic flintwork on the adjacent Woodham Knoll ridge makes it extremely likely that evidence for activity of these periods lies either under the peat or on now buried former margins of the wetland.

Later Mesolithic settlement may well have concentrated around the margins of the coversand ridge, as fluctuating coastlines and wet fen landscapes may have precluded settlement activity from the areas to seaward. The best indication of this is the extensive pattern of small flint sites on the low sand islands to the north of Flea Moss Wood, where agriculture has eroded much of the peat that formerly covered them (Me 050, 052, 054) probably preserving some organic evidence on the sites.

Where the peats lie at the surface in the Little Crosby channel, they are less protected from ploughing and drainage than is some of the peat elsewhere in the coastal group. This is illustrated by the greater length of the channel in which the peat is now an organic soil (Fig 75). Peat depth seems to be on average *c*0.2-0.3m, and where it survives is generally quite humified. The Little Crosby channel is particularly important for a range of archaeological sites on both the mid to upper slopes of the channel and along the peat, or organic soil boundary with the Shirdley Hill Sand slope. It seems possible that a number of potential sites may have some organic survival (Me 023, Me 024, Me 029, Me 032).

There is scattered archaeological evidence of probable Bronze Age date from the Flandrian blown sand above the peat across the plain, postdating the contraction of wetland conditions on the plain. The sand is generally *c*0.6-0.8m deep over the peat, and its lower layers could preserve settlement horizons within the former dunes, while the sand protects the underlying peat.

Alt Valley

In the farmland along the Alt Valley, drainage is having a greater effect on the upper peats, which are drying and causing mineral sand ridges to emerge from beneath them. This is a particular problem southwards of Carr Houses (Fig 75), and is accentuated the further south one goes. It is in these areas that the archaeological evidence is strongest, and there are grounds for believing that most of the sites in the valley are still in a partially waterlogged state. Nine sites have been identified as belonging to this category (Me 076-079, 081-085).

The northern part of the Alt Valley, as far south as Searchlight Wood, is in some ways similar to the Little Crosby coastal peats. This area has been less productive of sites, although the overlying alluvial and marine clays found here reach a greater depth than the sand covering in Little Crosby. Finds in this area consist of single flints, mainly located towards the edges of the peat near its junction with the pre-Flandrian sandy river terrace. It is not clear from the available evidence if the single finds represent the initial erosion of larger sites, or are isolated pieces associated with 'off-site' activity (Foley 1981). The peat in the areas where the finds come from is generally quite humified as it has lost its covering layer of alluvial clay. The potential for the discovery of more archaeological evidence adjacent to these findspots further into the valley floodplain is reasonable, while the survival of associated palaeoecological evidence is probably more predictable given the depth of covering layers to the peat.

The relative density of archaeological evidence from the surrounding dryland (particularly in the southern half of the valley) suggests that the clay and peat in the valley are likely to hide further prehistoric archaeological evidence in this zone. In the valley floor this will probably depend on how suddenly the sub-Flandrian surface slopes, and whether small eminences occur across it. However, it was not possible to define this from the survey.

In the southern half of the valley, where the alluvial clay thins in extent and depth, the main concentrations of flint lie on the valley floor or on the valley slopes. It is in these areas where the largest sites are located, due to the more advanced state of erosion of covering sediments. The main part of

Me 093, on the terrace overlooking the river, is mainly found on the Shirdley Hill Sand, and although once covered by peat, little organic material is likely to have survived on the ridge, even though archaeological features in the sand may have done so. A part of this site is, however, found at the foot of the slight ridge which marks the present edge of the organic material, and this may have a little more potential. The peat appears fairly humified here and is unlikely to be much deeper than *c*0.3m. The occasional flint piece is found a little further out into the peat, where physical and organic preservation may be a little better. At the other large site in Lunt, Me 094, it seems very likely that although *c*1ha of the site is partially subject to plough erosion, other elements of the site are still largely protected by peat and clay (Fig 21.B).

The terrace slopes around Lunt would appear to have some potential for the location of more sites, as the 4m contour appears to have been crucial for sites in the Mesolithic. These slopes were under-represented in the fieldwalking, due to large amounts of land here being left waste close to the landfill tip, while unsuitable field conditions prevailed in a number of other fields. The potential for the discovery of more sites in this zone is therefore high. The crucial contour is, however, generally a minimum of *c*20m from the edge of the peat, which reduces their potential for the survival of organic material quite strongly.

Nine smaller sites lie to the north of Me 94 in a similar topographic situation (Me 076-079, 081-085), as well as a number of dispersed single findspots around them. These may be functionally different from the two larger sites, where less debris has been discarded, or they may be similar sites to the larger ones but are better protected from ploughing. In most cases the smaller groups of material are located right at the edge of the peat/clay and sand interface, rather than on the top of the sand islands. This implies that they are still mostly unaffected by ploughing and are covered by layers of alluvial clay and peat, consequently with good archaeological potential.

In this southern part of the valley floor it is difficult to assess the depth of the peat associated with these sites, as they generally lie 50m or more from the ditches where the sections are available for recording the peat. In the ditches peat is found in two bands *c*0.4-0.6m deep, with a similar thickness of marine clay, whose inception began between *c*5200-4800 Cal BC, separating them. The peat is in reasonable condition in the central part of the valley where the best sections were available, and pollen preservation was good in the sample taken close to Me 089. The closeness to the River Alt also makes it seem likely that the lower peat, which mainly lies beneath the field drains, is still essentially waterlogged. The potential for the survival of organic material *in situ* at these valley floor sites, which are mainly likely to be Mesolithic in date, is consequently quite good.

Orrell Hill

This peat area exhibits characteristics of both the preceding areas. A small area of peat lies at the surface, within which occasional pieces of struck flint, probably of Mesolithic date, found in arable areas (Me 060) imply that larger sites which exist within organic contexts are being actively eroded. Adjacent areas are under pasture and display similar topographical features, implying that the archaeological potential is reasonably high in this area. The Orrell Hill peat is also important in that marine clays lie within 200-300m of the Shirdley Hill Sand ridges where sites are also located, suggesting that the area has potential for investigating the link between marine episodes and human exploitation of the area.

North Wirral

On the Wirral the problem is different. Although preserved peats are present beneath the later marine deposits, a far greater area is more inaccessible. This is through a combination of very large landfill tips and construction over the peats, making them both impossible to survey and perhaps affecting the actual characteristics of the peats in unknown ways. The dryland margins in the potentially most sensitive areas, around the Wallasey ridge and the Bidston ridge, are largely developed. The problem therefore is one of predicting which of the wetland areas could contain archaeological evidence. Any sites that may have lain around the sandstone slopes to the north-east and south-west of the moss will have suffered badly from housing developments, with the growth of Wallasey and Bidston. The archaeological link between the moss and the activity seen in the pollen diagrams must therefore rest on the survival of sites in lower, more deeply buried, environments around the moss and its immediate coastal hinterland, but this is very difficult to predict. The evidence at the moment suggests that the buried channels could be as, or more, important for settlement than the areas of open wetland. On the Wirral the lines of some of the channels have been partially reconstructed from the work of Kenna (1978, 1979, 1986), but discovery of sites will still rely on chance.

There is a formidable problem in archaeological research within the coastal areas of Wirral and Sefton, and it is one well known to many wetland and also conventional archaeologists. It is only when development takes place, deep sections are cut and sediments exposed, that the discovery of archaeology or associated evidence becomes feasible, but a blueprint for predicting its occurrence has yet to be devised. The only possibility of retrieving this evidence lies with monitoring developments in these sensitive areas. The problems are immense, however, with many developments only affecting the upper metre or so of the ground surface. Where service trenches cut the pre-Flandrian levels the trenches are so deep that safety considerations and construction programmes make it difficult to identify anything of value.

Central mosslands

The evidence from the inland peat of the central mosses suggests that archaeological potential is more restricted. The NWWS has reaffirmed the results of previous survey in showing that the archaeology of the central mossland areas is sparse, particularly in the eastern part. This shows that mosslands *per se* need not always be archaeologically rich. The early archaeology in the central mosslands is not absent, but in comparison with the coastal areas it is on a much reduced scale. The fact that these mosslands were probably exploited in a limited way presupposes that occasional well-preserved finds await discovery, but their location is, however, likely to be random.

The potentially more sensitive west-central mosslands, centred on Simonswood Moss, are threatened with eventual destruction through commercial milling, and here a number of islands and ridges are noticeable in parts of the moss not too far beneath the surface of the cut peat. There is reasonable potential therefore for the discovery of further material in this area, particularly as much of the surrounding Shirdley Hill Sand has been removed by quarrying. To the east, the lack of finds from NWWS fieldwork seems to suggest that there is less potential for the discovery of such sites.

As these central and east-central areas have avoided development, and even intensive farming until modern times, the depths of peat are often quite impressive, with some mosses reaching depths of 7m (Knowsley Park), or 5m (Reeds Moss, Bickerstaffe Moss). Water-tables are generally relatively high, and sizeable areas of peat have been left in a semi-natural state, showing that recent farming has accounted for up to *c*1.5m loss of surface peat in the farmed areas. In some cases it is less than this.

The peat areas therefore remain important regional reservoirs of palaeoecological information. The little pollen work that has been done, mainly in the west-central peats, shows a coherent pattern of potential human interference to the woodland during the fifth and fourth millennia Cal BC. This provides a context for the early prehistoric finds and sites from the area, and is of value when interpreting this material and also when comparing and contrasting the situation in different areas of the county, and indeed the region. The widespread evidence for burning in the central and eastern mosses during the late Mesolithic is one particularly important feature of this area.

Southern mosslands

The peat in these areas has been much affected by recent development which has also greatly affected the surrounding mineral soil areas. The two pollen sites presented in this volume have produced palaeoecological evidence of value to an understanding of the landuse of the areas in the south, which differ topographically from the main wetland areas in the north of the county. Their value may increase as more excavation and survey takes place in these areas (*eg* Cowell and Philpott 1994), and they may be particularly helpful for understanding the developing pattern of late prehistoric and Romano-British settlement.

Future work

The next stage for the NWWS work in Merseyside will be to build on this solid framework of archaeological and palaeoecological information. This will primarily comprise assessment of the survival and importance of many sites that could be crucial for advancing archaeological knowledge, and will mainly relate to the potential for waterlogged early prehistoric sites. These may be of particular relevance to understanding the nature of early settlement and landuse and to what extent, if any, it changed with the introduction of agriculture and other elements of Neolithic culture. The partially waterlogged sites in the Sefton area, lying on mineral sand ridges that are just emerging from eroded peats, display obvious potential of regional, if not national, importance in this regard.

The Wirral coastal mosslands were probably as well settled in the prehistoric period as those of the Sefton coast. The potential for their discovery is much reduced because of present landuse, but monitoring of development, particularly around the buried margins of Bidston Moss and Newton Carr, should be regularly undertaken to locate sites that could be directly associated with the palaeoecological evidence.

In the central mosslands the large area of Simonswood Moss will need to be regularly monitored, as the few concentrations of early prehistoric flint material in the central mosslands are found in this area. The monitoring of freshly uncovered sand surfaces, undamaged by sand quarrying, would also provide some comparative data by which more accurately to assess the level of site destruction through the mining of the Shirdley Hill Sand elsewhere in the central mosslands.

The wealth of palaeoecological detail presented in this volume is of great regional and even national importance for landscape interpretation studies. The particular nature of the settlement and landuse in some areas of the county, and the difficulties of identifying it archaeologically, have made the palaeoecological evidence from the survey of even greater importance. In several areas of the county it provides the main evidence for an interpretation of the human settlement of the area.

If the settlement and landscape patterns that have emerged from the work of the NWWS are to be developed, then this aspect of the survey must be enhanced. Much needs to be done in stratigraphic ally and chronologically relating the development of the mires in Sefton, the richest area archaeologically, to the profound changes that the landscape underwent in the early prehistoric period from changes in sea-level. The area around Orrell Hill is particularly important in this respect. The peat cores from Newton Carr on the Wirral also urgently need dating, so that the wealth of important detail for sea-level changes in North West England can be understood better.

In the central mosslands, the palaeoecological evidence so far collected for the west-central mosslands will be enhanced if the stratigraphic work undertaken for the east-central mosslands can be dated better. The central mosslands are the deepest and best preserved and contain a wealth of palaeoecological information. It is perhaps in these areas that the eventual aim of the project in Merseyside, the management of the best preserved peat deposits, will be realised.

This survey has shown that the archaeology of North West England has characteristics that distinguish it from many other areas of the country in terms of the visibility of the earlier part of its past landuse. It has also hopefully laid the basis for illustrating that its particular character need not preclude it from taking part in the national assessment of the development of human settlement and landuse. It is suggested that the wetland evidence presented here, for the Mesolithic and early Neolithic periods particularly, shows that the region has the potential to provide evidence of research value on a par with anywhere in the country. Although North West England has long been neglected for studies of this kind, the work of the NWWS can be seen as the first step in allowing its true potential to be realised.

Glossary

Basin mire	Mire developed in enclosed, waterlogged depressions which have become colonised by peat-forming vegetation. Most often found in areas of glacial deposition where local irregularities of relief such as kames and kettle-holes provide the necessary geomorphological environment of hollows with enclosed drainage. Usually characterised by relatively small surface area (<50ha) compared to depth (often >5m).
Biostratigraphy	Stratigraphy based upon changes in biological assemblages in the sediment (*eg* pollen zones).
Bi-polar	A technique of detaching a piece of flint from its parent nodule (or core) by placing on a stone anvil and striking with a hard stone.
Blade	A small narrow flint with approximately parallel sides, carefully struck from a parent piece or core. Generally is at least twice as long as it is broad. Usually indicative of the Mesolithic or early Neolithic periods.
Bog	General term for ombrotrophic mires.
BP	Before present. In terms of radiocarbon years present is defined as AD1950.
Cereal-type	Large grass grain of between *c*40 and *c*60 microns in size.
Chronozone	A chronostratigraphic unit, defined as the time-span of a selected stratigraphic unit of another kind, *eg* a pollen zone, the boundaries of which have been radiocarbon dated (*eg* Flandrian II).
Core	The parent flint nodule or pebble from which pieces are struck to produce flakes. Pieces are usually struck from the core repeatedly from around the same flat plane or platform. One core can have several platforms.
Dendro-chronology	Tree-ring counting. Can give an absolute dating.
Devensian	Last glacial stage recognized in Britain ending *c*10,000 BP.
Diatom	Unicellular alga with silica frustule ('shell') which is readily preserved in lake and marine sediments. Analysis of fossil frustules can yield information about environmental factors such as water quality at the time of deposition.
Ecotonal	Pertaining to areas at the boundaries between differing types of environments.
Elm decline	A fall in elm pollen frequencies as percentages of tree pollen, dated throughout north-western Europe to around 5000 BP. Defines the boundary between the Flandrian II and III chronozones and Godwin's pollen zones VIIa and VIIIb. Exact cause undetermined but may be related to climate change, disease or human activity, or a combination of these factors.
Fen	General term for minerotrophic mire.
Flandrian	The present interglacial stage in Britain, from 10,000 BP to the present.
Floodplain mire	Minerotrophic mire frequently developed in lowland alluvial plains traversed by sluggish rivers. Characterised by fen communities.
Foraminifera	Group of mainly marine rhizopod Protozoa which form shells, mainly of lime. Analysis of their fossil remains in marine sediments can yield information on past sea-levels.
Grenzhorizont	Literally, 'boundary horizon'. A prominent recurrence surface found in many European ombrotrophic mires. First identified by the German stratigrapher C A Weber and frequently found to date to around 1350-1750 BP.
Gyttja	Fine detrital mud.
Holocene	The most recent epoch of the Quaternary period, being the ten thousand years since the end of the Pleistocene.

Humification	Degree of decomposition (of peat).
Hydrosere	A plant succession commencing in waterlogged sites.
Landnam	Temporary episodes of woodland clearance of variable duration followed by regeneration. Traditionally associated with slash-and-burn agriculture in the Neolithic period. Recent work has doubted the applicability of this model of subsistence to temperate Europe.
Late Neolithic/ Early Bronze Age	From 3200-1650 Cal BC spanning the gradual introduction of the first metalworking amongst existing Neolithic communities.
Macrofossil	Usually taken to mean *plant macrofossil.* A term which covers everything that cannot be considered a microfossil. It effectively includes all plant remains which can be recognized by the naked eye or with the aid of a low power microscope.
Mesolithic	From *c*8000 BC – 4000 Cal BC. The period characterised by mobile societies based on hunting and gathering wild plants.
Microlith	A small flint arrowhead or projectile point, typical of the Mesolithic period.
Minerotrophic	Used to describe a mire whose surface receives water from outside that mire's own limits.
Mire	Peat-producing ecosystem which develops in sites of abundant water supply.
Mire macrotope	Mire complex which has been formed by the fusion of isolated mire mesotopes which originated from separate centres of mire formation.
Mire mesotope	Mire systems developed from one original centre of peat formation.
Neolithic	From *c*4000-2500 Cal BC. The period of the first farmers. Distinctive items of this period include polished stone axes and leaf-shaped arrowheads.
Ombrotrophic	Literally, 'fed by rain'. Used to describe a mire which receives water only from the atmosphere.
Ontogeny	Biological or ecological development through time.
Phase-shift	Change in macrofossil assemblage indicating a change in surface hydrological conditions (*ie* from a wetter to a drier situation or vice versa). See also **Recurrence surface.**
Pollen zone	A body of sediment with a consistent and homogeneous fossil pollen and spore content which differs in type and frequency from that of adjacent bodies of sediment.
Post-glacial	Used to denote the temperate period since the end of the Devensian at 10,000 BP.
Raised mire	Ombrotrophic mire characterised by a very low amplitude convex profile and usually occupying topographical situations such as level floodplains of river systems and alluvial deposits of estuaries. The ontogeny of raised mires usually indicates that ombrotrophic plant communities have developed over minerotrophic or aquatic ones as upward growth of peat has insulated the surface from groundwater influence.
Recurrence surface	Stratigraphical feature found in many ombrotrophic mires consisting of a boundary between two peat layers of markedly differing humification, the upper layer being less humified. Thought to be related to climatic changes (particularly increased rainfall) affecting the wetness of the mire surface leading to an increased rate of mire growth and peat accumulation. Recurrence surfaces have been classified into a sequence in Europe by the Scandinavian palaeoecologist Granlund and are often given the abbreviation prefix 'RY' (from the Swedish *'Recurrenstyor'),* eg RYIII being the 'Grenzhorizont'.
Sere	Plant succession.
Swamp	Mire with a water-table permanently above the peat surface.
Valley mire	Occurring in small, shallow valleys or channels which are not enclosed so that movement of water along the long axis is possible even with only a slight gradient. Typical of wet, elongated depressions in acidic, lowland heathland and moorland areas at moderate altitudes.

Appendix 1: Gazetteer

Previously recorded sites

Wirral

A 1	**Wallasey**
SMR No	2692.01
NGR	SJ 236908 to 266922
Site Type	Submerged Forest
Period	Mesolithic
Source	Field Visit

Remains of Lower Forest Peat Bed have been recorded lying on the boulder clay exposed on the beach at Leasowe. The peat is now very poorly preserved on the beach, although two stumps of in situ ?oak were visible at the time of a visit in 1981. In the late 19th and early 20th centuries photos show a large expanse of large tree stumps belonging to the same bed, on the beach at Meols several kilometres to the west. A visit in 1985 showed that only small pieces of wood and tree stumps survive and the peat is also very poorly preserved. Sections of the deposit survive behind the dunes along the coast.

A 2	**Bidston**
SMR No	2991.01
NGR	SJ296915
Site Type	Findspot
Period	Neolithic
Source	Chance find
Location	Wallasey Library

A polished stone axe with straight, facetted sides, found in a garden in 1964. Petrological analysis suggests it is Group XXI (Mynydd Rhiw, North Wales).

A 3	**Bidston**
SMR No	2890.01
NGR	SJ282903
Site Type	Findspot
Period	Neolithic
Source	Chance find
Location	Liverpool Museum Acc No 1969.219

A polished stone axe, made of basalt, was ploughed up in 1916 from the southern slopes of the sandstone ridge backing onto Bidston Moss.

Dimensions: L. 129mm; B. 44mm; Th. 27mm

A 4	**Bidston**
SMR	No 2890.20
NGR	SJ 283902
Site Type	Findspot
Period	Neolithic
Source	Chance find
Location	Whereabouts unknown

A polished stone axehead is recorded as coming from near Bidston Hill. It has been identified from records at the Botanic Gardens Museum, Southport.

Dimensions: L. 125mm; B. 60mm; Th. 32mm

A5	**Great Meols**
SMR	No 2390.01
NGR	SJ231906
Site Type	Findspots
Period	Mesolithic; Neolithic; Bronze Age; Iron Age; Roman; Medieval
Source	Chance finds
Location	Liverpool Museum Acc No 8.11.74 (Lost); Acc. No 42.17.287; 21.10.1901: others Grosvenor Museum

A number of finds have been made since the mid 19th century in the vicinity of Dove Point, Meols. Some of these have been associated with the submerged forest (A 1) or the associated peat layers: eg 21 'flint arrowheads and other small instruments' in black, white and red 'flint' are noted as coming from the 'upper woody deposit or the thin superincumbent bed of clay' (Smith 1866a).

The terminology for much of this material is vague but there is enough information to suggest that it spans all periods of prehistory. An illustration in Shone (1911) includes two apparently Mesolithic types, one an obliquely blunted point, the other a notched blade. Varley (1964) mentions a number of Mesolithic (Irish) Bann flakes from this collection in the Grosvenor Museum (not located), and a rough scraper from this vicinity identified as Mesolithic (Chitty and Warhurst 1977), is in Liverpool Museum.

There are also a large number of recorded items from the 'Cheshire Shore' of which the majority probably come from the Dove Point area. Most of the terminology, however, is vague and correct attributions cannot be verified, though one or two items have perhaps a little more interest. These include two that could have a prehistoric context, a 'bone pin', and a 'bone arrowhead' (Smith 1867).

Other types recorded from this vicinity include 'arrowheads and scrapers...'. A plate in Shone (1911) shows three certain leaf arrowheads. There is also a reference (Jackson 1936, 74) to 'fragments of polished stone axes with flattened edges associated with numerous flint flakes and implements' coming from 'Hoylake'. This may refer to the Dove Point vicinity, although perhaps a more likely location is that at Red Rocks Point *c*4km to the south-west, on the edge of Hoylake (see A 11) from where struck flints have been reportedly excavated (Glenn 1914), although Varley (1964) shows axe findspots at both these locations, without giving any authority for their attribution. A fragment of coarse gritty pottery with lines of whipped cord impressions has also been found on the shore at Meols (Varley 1964: Grosvenor Museum), probably of late Neolithic Peterborough type.

Among the 'arrowheads' recorded from the site are included 'two finely barbed... in flint' (Smith 1871). A plate in Shone (1911) also shows two and probably four barbed-type arrowheads as having come from this locality. A small piece of bronze measuring 32x29x15mm has been found on the shore (Grosvenor Museum).

In the historic period the site at Meols became one of the most important in the region as a trading centre, up to its desertion in the late medieval period. The start of this rise to eminence may be represented by two Celtic coins from the site (Liv Mus Acc Nos M4069-70). They are 1st century BC billon (base silver) staters of the Coriosolites, a tribe from Brittany and the Channel Isles.

Over 80 Romano-British coins were also found along the beach during the 19th century, in addition to over 70 brooches, suggesting occupation of the site from the AD 1st-4th centuries.

Two circular buildings with stone footings in which the remnants of timber stakes were found in the 1890s, *c*2 feet below the level of high tide on the shore at Dove Point and 'about a foot below the medieval floor level' (the level at which medieval artefacts occurred along the beach) (Cox 1895). The best indication of the relative stratigraphic position of these features is perhaps Smith's (1866) a section of the strata at Dove Point which would suggest that the more likely date of these features may be historic.

A 6	**Liscard**
SMR No	2994.01
NGR	SJ299940
Site Type	Flint findspots
Period	Mesolithic, Neolithic, Bronze Age
Source	Chance finds
Location	Liverpool Museum Acc No 21.10.1901 (Finds lost): others Manchester Museum

Approximately 600-700 pieces of worked and unworked flint were found in the late 1890s on the west side of 'Red Noses'. They comprised 'spoiled and chipped cores, splinters, flakes, knives, scrapers, spear and arrowheads and burnt flakes...'. Illustrations (Roeder 1900) show a number of scrapers, which could be of any prehistoric date. Blade pieces are also shown, however, and two illustrated obliquely blunted microliths (Varley and Jackson 1940) are reported as coming from the site. They also include an illustration of a barbed and tanged arrowhead.

The flint floor, extending *c*140m (460 ft) westwards from the Red Nose outcrop, was found when the overlying sand-dunes were being eroded away. The surface on which the flints were found was strewn with 'small pebbles, stones and middle-sized erratic blocks', beneath which was located a black sandy layer one inch deep at its thickest lying over boulder clay to the west and red sandstone to the east.

A 7	**Little Meols**
SMR No	1986.01
NGR	SJ1999867
Site Type	Flint findspots
Period	Mesolithic
Source	Chance finds
Location	Liverpool Museum Acc No 1983.203. 01-22: others private ownership

For over a century flints have been recorded eroding out of a narrow dark band of soil on the Little Eye island near West Kirby. Most of these descriptions have been too imprecise to identify, eg 'knives, scrapers, arrow points ... protruding from the bank on the landward side' (Smith 1871). Other descriptions suggest a reasonable sized site lay here. Modern collectors have also produced a small quantity of mainly small chips and spalls, while a rod microlith and a small blade core suggest that at least part of the occupation here was of Mesolithic date.

A 8	**Little Meols**
SMR No	1887.03
NGR	SJ184879
Site Type	Flint findspots
Period	?Mesolithic, Neolithic, Bronze Age, Roman.
Source	Chance finds
Location	Liverpool Museum Acc No 18.11.74 (Finds lost), others Grosvenor Museum

In the late 19th century many instances of flints were recorded as coming from the vicinity of Hilbre Island. The descriptions are for the most part vague, eg 'knives and scrapers, arrow points, 2 arrowheads, chisels, a thumb flint, dagger blade, 2 notched articles'. Other descriptions of finds being located on the 'Eye', (which is distinguished from the 'Little Hilbre Islands', and therefore must refer to Hilbre Island) may, however, be tied down to the Neolithic and more specifically the late Neolithic '... six arrowpoints in flint, ... one ... unbarbed but well worked to a triangular Irish type and possessing a shank or stem, three are single winged ...' (?oblique arrowheads). Others are more generally Neolithic, '... the others are of oval form' (leaf arrowheads?) (Smith 1872).

A barbed and tanged arrowhead, is recorded from the 'Eye, Hilbre Island' of which only an illustration survives now (SMR 1887.03: Liv. Mus. Acc. No 18.11.74.2, lost).

Other descriptions of finds being located on the 'Eye' are less easy to interpret but the 'knife or scraper of minute proportions, a thin and narrow instrument of light coloured flint beautifully edged by a single stroke' (Smith 1872) may conceivably refer to a microlith.

Excavations on the island in the 1920s produced 23 sherds of 3rd-4th century pottery from an area of burnt soil overlying a single posthole.

A 9	**Little Meols**
SMR No	1887.06/1887.11
NGR	SJ189875
Site Type	Findspots
Period	Bronze Age
Source	Chance finds
Location	Whereabouts unknown (axe), Grosvenor Museum (pottery)

A bronze oval-sectioned socketed axe with loop, was recorded as coming generally from Middle Eye island in 1926, while in 1976 a bucket urn was found eroding from the boulder clay cliff on the northern tip of the island. No trace of a cremation was found in the stony soil fill surrounding the pottery (Petch 1976).

A 10	**Little Meols**
SMR No	2088.01
NGR	SJ202885
Site Type	Flint findspots
Period	?Mesolithic, ?Neolithic, Bronze Age
Source	Chance finds
Location	Grosvenor Museum; others Lost

About 100 flints were found in the late 19th century during coastal erosion of the Red Rocks area of Hilbre Point, Hoylake. A small excavated area in 1914 produced a further *c*250 lying on the boulder clay which lies beneath the blown sand. This clay surface was fire-blackened in places. Associated with the flints were animal bones 'pig, small ox and sheep/goat' and shells, although the excavator cautions against them being necessarily prehistoric for some unstated reason. The flints are described as 'lanceheads, a barbed arrowhead (slightly fractured), round scrapers, knives and borers.' as well as 'chips, splinters, and pieces of waste flint, two flint cores and a well pitted hammerstone ...'. A plate in Shone (1911) illustrates a number of pieces which are difficult to identify, although 1 blade and 3 or 4 blade-like chips are included. There is a possibility that 'fragments of polished axes with flattened sides' (Jackson 1936, 74) may have come from this site, although they are only provenanced to 'Hoylake'. The arrowhead would suggest activity in the early Bronze Age although part of the site may also have been occupied in the 3rd and 4th millennia Cal BC.

A 11	**West Kirby**
SMR No	2286.06
NGR	SJ221868
Site Type	Burial Site
Period	Bronze Age
Source	Chance find
Location	Whereabouts unknown

'Several' inverted urns, one of which was a cordoned urn, were found on Grove Hill whilst enlarging a garden in 1840. Filled with burnt bones, charcoal, and ashes and one at least contained 'edged flint' (Mayer 1849).

A 12	**Wallasey**
SMR No	2891.01
NGR	SJ280917
Site Type	Perforated Adze Head
Period	Bronze Age
Source	Chance find
Location	Wallasey Central Library

A perforated adze head was found in 1936 in the blue silt of the old course of the river Birket to the north-west of Bidston Moss.

A 13	**Wallasey**
SMR No	2792.04
NGR	SJ27709291
Site Type	Dagger Findspot
Period	Bronze Age
Source	Chance find
Location	Private Ownership

A small bronze dirk or dagger was found on the beach at Leasowe in 1982 by a metal detectorist. It is probably of middle Bronze Age date.

A 14	**Wallasey**
SMR No	2692.04
NGR	SJ267922
Site Type	Brooch Findspot
Period	Romano-British
Source	Chance find
Location	Warrington Museum and Art Gallery

A bronze fibula of Claudian-Neronian date (AD 43-68) was found on the beach at Leasowe, close to the Roman findspot A15, two years after that find was made.

A 15	**Wallasey**
SMR No	2692.03
NGR	SJ26909220
Site Type	Coin Findspot
Period	Romano-British
Source	Chance find
Location	Liverpool Museum Acc No 1979-632

A silver denarius of Trajan, dated AD 115-117 was found on the foreshore at Leasowe in 1979.

A 16	**Great Meols**
SMR No	None
NGR	SJ 212886
Period	?Neolithic
Source	Chance find
Location	Liverpool Museum

A large scraper on a water worn, dark brown boulder clay flake was found in *c*1991 in the blown sand of the Royal Liverpool Golf Course.

A 17	**Wallasey**
SMR No	2891.01
NGR	SJ280917
Site Type	Bone Midden
Period	Bronze Age
Source	Chance find
Location	Unknown

Antlers and bones of deer and oxen, a rib of one of the latter producing a date of 3980±70 (Birm-1013) (2861-2310 Cal BC), were found in grey silt below the Upper Peat Forest Bed on the foreshore at Leasowe in 1967 (Kenna 1978).

Sefton

A 18	**Little Crosby**
SMR No	2902.05
NGR	SD 314023
Site Type	Stone axe head
Period	Neolithic
Source	Chance find

A rough-out of a stone axehead, probably made of a rock found in the Lake District, was donated to Liverpool Museum in 1975 (Acc No MCM 1975.340). The present farmer of the land provided the location for the findspot; on the brow of the Woodham Knoll sand ridge. The Grid Reference is only approximate as he was not the finder.

A 19	**Little Crosby**
SMR No	2902-01
NGR	SD 290020
Site Type	Stone axe head
Period	Neolithic
Source	Chance find
Location	Unknown

A stone axe is recorded as coming from Blundellsands, near Altmouth.

A 20	**Little Crosby**
SMR No	2902-01
NGR	SD 290020
Site Type	Flint Findspot
Period	Neolithic
Source	Chance find
Location	Lost

A leaf shaped arrowhead was found pre-1886 in the Altmouth area and donated to Liverpool Museum (Acc No MCM 4.12.16.4.) It was lost during the war when the Museum was bombed.

A 21 **Little Crosby**
SMR No 2902-05
NGR SD 294028
Site Type Stone axe head
Period Neolithic
Source Chance find
Location Liverpool Museum (Acc. No 1975.340)

A polished stone axehead was found on the beach at Hightown in 1975, on peat overlying glacial clay. A macroscopic inspection suggests it may be a siltstone.

Dimensions: L. 116mm; B. 62mm; Th. 32mm; Wt 341g

Central mosslands

A 22 **Kirkby**
NGR SJ 44449768
Site Type Flint findspot
Period Mesolithic
Source Fieldwalking
Conditions Moderate

A blade segment in a brown boulder clay flint came from the adjoining field to A23, *c*50m to the west of the main concentration. It may represent a separate site from that one or be part of the activity centered on A23. Another piece, a single boulder clay flint pebble, possibly tested along one edge was found *c*60m to the north.

A 23 **Kirkby**
NGR SJ 44529764
Site Type Flint concentration
Period Mesolithic
Source Fieldwalking
Conditions Good

Seventeen pieces of struck flint were located during fieldwalking in 1982. They covered a core area of *c*25x35m on a slight sandy knoll, *c*600m to the south of the moss. The knoll forms the highest point of the belt of Shirdley Hill Sand lying between Kirkby Moss and Knowsley Park Moss further to the south. The raw material is mainly a white or off-white opaque flint, except for two pieces of boulder clay flint and one blade in a dark grey chert.

A 24 **Kirkby**
NGR SJ 44549723
Site Type Flint findspot
Period Prehistoric: uncertain
Source Fieldwalking
Conditions Good

A very thick irregular flake from a light grey flint glacial pebble, with possibly some limited earlier flaking was found on the lower slopes adjacent to Knowsley Brook in a belt of Shirdley Hill Sand between Kirkby Moss and Knowsley Park Moss. The field has been subjected to commercial sand extraction in the 1970s which would suggest that no subsurface features survive.

A 25 **Simonswood**
NGR SD 42340108
Site Type Flint findspot
Period Prehistoric: uncertain
Source Fieldwalking
Conditions Good

A single narrow, primary flake in grey-brown boulder clay flint was found in 1982 on Shirdley Hill Sand next to a small brook.

A 26 **Simonswood**
NGR SD 42270109
Site Type Flint findspot
Period ?Neolithic
Source Fieldwalking
Conditions Good

A single, end and side scraper on a large snapped ?blade in reddy brown boulder clay flint was found during fieldwalking in 1982 on Shirdley Hill Sand close to a small brook to the north of Simonswood Moss. Another piece, a natural flake with artificially sharpened point, was found *c*20m away.

A 27 **Simonswood**
NGR SD 42780140
Site Type Flint findspots
Period Mesolithic
Source Fieldwalking
Conditions Moderate

Three pieces of flint, two of which are struck, and a large bashed pebble were found in 1982 in an area of *c*15x15m on a small, almost imperceptible rise in the wide, Shirdley Hill Sand ridge lying between Simonswood Brook to the north and Simonswood Moss to the south. The size of the blade suggests a possible early Mesolithic date for a small occupation site at this location.

A 28 **Knowsley**
NGR SJ 43999599
Site Type Flint findspots
Period ?Later Prehistoric
Source Fieldwalking
Conditions Good

Two natural flakes, one retouched into a scraper and the other with a retouched edge, were found close together during fieldwalking in 1982, near the foot of the fairly steep slope on the southern side of the Mill Plantation Brook.

A 29 **Rainford**
SMR No 4703.01
NGR SD 475032
Site Type Findspot
Period Neolithic
Source Chance find
Location Private ownership

A polished flint axehead was found in a field by the farmer in about 1978. It came from a sandy subsoil which had previously been subjected to commercial extraction. The method of excavation and re-instatement means that the location quoted is very likely to have been near to its original findspot. The findspot is located at a height of *c*65m OD at the foot of the Billinge Ridge, within *c*350m of the western edge of Holland Moss.

Dimensions: L. 140mm; B. 64mm; Th. 21mm

A 30 **Kirkby**
NGR SJ 44539790
Site Type Flint findspot
Period Mesolithic
Source Fieldwalking
Conditions Moderate

A single small flake in a light grey boulder clay flint, largely patinated white, came from an area of Shirdley Hill Sand *c*200m from the southern edge of Kirkby Moss. The field has been subjected to commercial sand extraction which will have destroyed all trace of subsoil features.

A 31	**Knowsley**
NGR	SJ 44389570
Site Type	Flint findspot
Period	Neolithic
Source	Fieldwalking
Conditions	Moderate

A single transverse arrowhead made of dark brown flint was found during 1982 on boulder clay soil at the foot of the western slope of Berry Hill which forms the south-western limit of Knowsley Park Moss.

A 32	**Knowsley**
NGR	SJ 44609670
Site Type	Flint findspot
Period	Prehistoric uncertain
Source	Fieldwalking
Conditions	Good

Two finds were found widely separated across a field on a belt of Shirdley Hill Sand lying between Kirky Moss and Knowsley Park Moss. The field has been subjected to commercial sand extraction in the 1970s although the topsoil is largely unaffected. These two pieces complement one and possibly two struck pieces that were recovered from the south-eastern part of the field from fieldwalking in 1982. The material is very undistinguished, including one piece of natural grey boulder clay flint with apparent working to a point on one edge, and a small boulder clay pebble which may have been struck twice. This may reflect some prehistoric activity in the vicinity that has survived the sand extraction in the ploughsoil, although no subsurface features are likely to have survived.

Sites discovered during NWWS fieldwork 1990/92

Wirral

Me 001	**Grange**
NGR	SJ239880
Site Type	Flint concentration
Period	?Mesolithic
Source	Fieldwalking
Conditions	Poor to Moderate

A blade core, 3 waste pieces, and a small retouched flake were found during fieldwalking in an area of *c*20x10m. They lay just below the break in the north-west facing slope of the boulder clay ridge overlooking Newton Carr, at a height of *c*16m OD. A narrow depression running downslope skirts the flint area to the north-east, possibly marking a former channel. A sixth piece, a blade, lay *c*10m away on the eastern side of the channel, in the adjoining field. The flint is all boulder clay derived.

Me 002	**Grange**
NGR	SJ238878
Site Type	Flint findspot
Period	Neolithic
Source	Fieldwalking
Conditions	Poor to Moderate

A single discoidal core was found during fieldwalking *c*100m to the south-west of Me 001. It was in a similar topographical position, except that the slope here also trended slightly to the west as well as to the north-west. The flint is boulder clay derived.

Me 003	**Grange**
NGR	SJ 23788785
Site Type	Flint findspots
Period	Uncertain
Source	Fieldwalking
Conditions	Poor

A small burnt chip and a small boulder clay pebble with three incomplete blade scars were found within *c*50m of each other just below the highest part of the boulder clay ridge overlooking Newton Carr.

Sefton

Me 004	**Little Crosby**
NGR	SD 31340130
Site Type	Flint findspot
Period	?Mesolithic
Source	Fieldwalking
Conditions	Good

A single, small blade in light grey flint was found on the blown sand surface of the coastal plain which here probably overlies a belt of peat.

Me 005	**Little Crosby**
NGR	SD 31440138
Site Type	Flint findspot
Period	Bronze Age
Source	Fieldwalking
Conditions	Good

A single, large, edge retouched flake, in a medium grey flint, found on the coastal plain to the west of the Woodham Knoll ridge. It comes from a blown sand surface *c*15m from the edge of an area where a patch of very strong white sand may imply a former, possible prehistoric, dune. A possible bipolar core in brown boulder clay flint was also found, *c*75m to the north-west and may also represent later prehistoric activity on the plain.

Me 006	**Little Crosby**
NGR	SD 31030136
Site Type	Flint findspot
Period	Prehistoric uncertain
Source	Fieldwalking
Conditions	Good

A single burnt fragment of a moderate sized flint nodule from the surface of the blown sand which overlies peat on the coastal plain west of the Woodham Knoll ridge.

Me 007	**Little Crosby**
NGR	SD 31580156
Site Type	Flint concentration
Period	LN/EBA
Source	Fieldwalking
Conditions	Good

Two pieces of struck flint, an end scraper, and a retouched flake came from the foot of the western slope of the Shirdley Hill Sand ridge of Woodham Knoll. The surface from which these finds came may have been covered in blown sand, dateable to after *c*2500 Cal BC, which may be concealing traces of a larger site. The technology and raw material is identical to that of the group of sites on the brow of the Knoll *c*500m to the north eg Me 018.

A single scraper, of the same technology, came from *c*40m to the north of Me 007 on the mid-western slope of the Woodham Knoll ridge. The surface here appears to be Shirdley Hill Sand rather than the later, blown coastal sand.

Me 008	**Little Crosby**
NGR	SD 31610166
Site Type	Flint findspots
Period	?Mesolithic
Source	Fieldwalking
Conditions	Good

A single platform core with a few small scars on a brown boulder clay pebble with washed cortex, and a similar nodule which may have been tested, found within *c*60m of each other. A chip, in grey boulder clay flint, lies *c*50m to the north of the core.

Me 009	**Little Crosby**
NGR	SD 31100153
Site Type	Flint findspot
Period	Prehistoric uncertain
Source	Fieldwalking
Conditions	Good

A single moderate sized fragment of a burnt flint nodule *c*300m to the north of the burnt piece at Me 034 on the blown sand overlying peat on the coastal plain west of the Woodham Knoll ridge.

Me 010	**Little Crosby**
NGR	SD 31210160
Site Type	Flint findspot
Period	Prehistoric uncertain
Source	Fieldwalking
Conditions	Good

A single flint flake in the same material as the retouched flake at Me 005, from the coastal plain west of the Woodham Knoll ridge where the surface blown sand layer overlies peat.

Me 011	**Little Crosby**
NGR	SD 30990157
Site Type	Flint findspot
Period	Prehistoric uncertain
Source	Fieldwalking
Conditions	Moderate

A single, moderate sized, elongated pebble fragment, heavily burnt was found on the western side of the coastal plain. The surface is wind blown sand, dateable after *c*2500 Cal BC, which overlies peat. Another burnt chunk of flint was found *c*110m to the west.

Me 012	**Little Crosby**
NGR	SD 30850156
Site Type	Flint findspot
Period	Prehistoric uncertain
Source	Fieldwalking
Conditions	Moderate

A single small flake, in light brown boulder clay flint found *c*125m to the west of Me 011, on the blown sand surface of the coastal plain. A chunk of burnt flint was found *c*80m to the south-east of this location.

Me 013	**Little Crosby**
NGR	SD 30800195
Site Type	Flint findspot
Period	Prehistoric uncertain
Source	Fieldwalking
Conditions	Good

A single flint chip in a honey boulder clay flint from the coastal plain to the south-west of Woodham Knoll. The findspot is on a slight rise in the field surface which is blown sand, and which post-dates *c*2500 Cal BC.

Me 014	**Little Crosby**
NGR	SD 30770193
Site Type	Stone Concentration
Period	Uncertain/?Medieval
Source	Fieldwalking
Conditions	Good

A concentration of sandstone blocks on the south-eastern edge of a small ridge on which Me 013 was found may represent the footings of building. Two sherds of late medieval pottery (Me 015), which is rare in the fields on the coastal plain, were found *c*130m to the north-east and may provide a context for this site.

Me 015	**Little Crosby**
NGR	SD 30900194
Site Type	Pottery Findspots
Period	Late medieval
Source	Fieldwalking
Conditions	Good

Two sherds of pottery were found within an area of *c*10x10m to the north-east of Me 014. Such pottery is quite rare in this part of the coastal plain. It may represent debris from an occupation in the vicinity, which may be represented by Me 014, rather than from manuring of fields belonging to the sites of existing farms.

Me 016	**Little Crosby**
NGR	SD 31270186
Site Type	Flint Findspots
Period	Mesolithic
Source	Fieldwalking
Conditions	Moderate

A single blade in orange-brown boulder clay flint was found on the western mid-slopes of the sand ridge of Woodham Knoll. A burnt, small flake of uncertain prehistoric date was also found, *c*45m to the south-east, near the foot of the slope.

Me 017	**Little Crosby**
NGR	SD 31400190
Site Type	Flint Findspots
Period	Prehistoric uncertain
Source	Fieldwalking
Conditions	Moderate

Three small burnt nodules of flint were found on the upper part of the western slope of the Woodham Knoll ridge *c*130m to the north-east of Me 018. A fourth burnt piece, a small flake, found *c*150m to the south is included as part of Me 016.

Me 018	**Little Crosby**
NGR	SD 31550198
Site Type	Flint concentration
Period	Late Neolithic/Bronze Age
Source	Fieldwalking
Conditions	Moderate

Twenty one pieces of struck flint were found on the west facing brow of the Little Woodham Knoll ridge. They covered an area of *c*40 x 60m. They are all, except three pieces, of the mid brown flint raw material and technology described in Cowell (1991b), ascribed to the LN/EBA.

1-11:flakes. 12-16:retouched implements. 17-18:chips. 19:primary flake. 20:retouched flake. 21:core fragment. The three latter pieces are of dark grey lustrous boulder clay flint.

Me 019	**Little Crosby**
NGR	SD 31400216
Site Type	Flint concentration
Period	Late Neolithic/Bronze Age
Source	Fieldwalking
Conditions	Moderate

Three pieces of struck flint and four medium sized beach pebbles were found across an area of *c*90 x 60m on the east facing slopes of the Woodham Knoll ridge. This possibly signifies another site of the LN/EBA period, as the struck flint belongs to the same raw material and technology as outlined in Cowell (1991b).

1-2:retouched implements. 3:flake.

Me 020	**Little Crosby**
NGR	SD 31770209
Site Type	Flint findspot
Period	?Bronze Age
Source	Fieldwalking
Conditions	Moderate

A single small irregular flake, with robust retouch along one edge. It may represent activity associated with the flint material found in the field to the west, although technologically it appears more rudimentary than that material which is largely assigned to the LN/EBA.

Me 021	**Little Crosby**
NGR	SD 31830214
Site Type	Flint findspot
Period	Mesolithic
Source	Fieldwalking
Conditions	Poor

A single struck flake in the same material as the bulk of the flint assemblage from Me 022, found on the mid-slope of the sandy Woodham Knoll ridge overlooking the peat-filled channel, although here the peat is very degraded. Conditions in this part of the field were worse than those where Me 022 was located and so it is likely that this find represents a similar site to the latter.

Me 022	**Little Crosby**
NGR	SD 31920221
Site Type	Flint concentration
Period	Mesolithic
Source	Fieldwalking
Conditions	Moderate

Sixteen pieces of struck flint were found over an area of *c*20x3m on the south-western edge of the peat-filled channel adjacent to the Woodham Knoll sand ridge. The peat in this section of the channel is very degraded This probably represents a reasonably prolific flint knapping site.

The raw material is mainly an opaque, pale blue-grey flint, except for a medium grey chert (4 and 7) and a brown lustrous boulder clay flint (8).

1-3:flakes. 4-6:blade chips 7-16 chips.

Me 023	**Little Crosby**
NGR	SD 31870224
Site Type	Flint concentration
Period	?Mesolithic
Source	Fieldwalking
Conditions	Moderate

Four pieces of struck flint from the south-western edge of the peat-filled channel adjacent to Woodham Knoll sand ridge. The peat is very degraded in this section. A fifth piece (No 5) was located *c*50m to the north-east on the lower slope of the channel. There is no homogeneity in the material which makes dating difficult but No 1 and possibly No 5 suggest Mesolithic may be the most likely date.

1:blade. 2:flake (burnt) 3-4:chunks (1 burnt). 5:small core with crude blade scars. The unburnt material is a brown boulder clay flint (1), and medium grey boulder clay flint (5).

Me 024	**Little Crosby**
NGR	SD 32090220
Site Type	Flint findspot
Period	Prehistoric uncertain
Source	Fieldwalking
Conditions	Good

A single, burnt, flint chip was found in the peaty sand surface near the edge of the channel lying to the north-east of Woodham Knoll. A second, moderate sized burnt chunk of flint was located *c*50m to the north-east in a similar location.

Me 025	**Little Crosby**
NGR	SD 32230231
Site Type	Flint findspot
Period	Prehistoric uncertain
Source	Fieldwalking
Conditions	Good

A single, two-platformed, small core, made from light brown boulder clay flint, with water-worn cortex, found on a slight sandy ridge adjacent to the peaty channel to the north-east of Woodham Knoll. The findspot is located *c*140m from the channel edge where the peat is now very sandy and degraded.

Me 026	**Little Crosby**
NGR	SD 31750225
Site Type	Flint findspot
Period	Prehistoric uncertain
Source	Fieldwalking
Conditions	Moderate

A single chunk of burnt flint from the mid slope of the Woodham Knoll sand ridge overlooking the peat channel *c*100m to the north-east. Single finds such as this are relatively common around the channel and may be prehistoric.

Me 027 **Little Crosby**
NGR SD 31960256
Site Type Flint findspot
Period ?Bronze Age
Source Fieldwalking
Conditions Moderate

A single natural flake with retouch along one edge was found midway along the very low gradient slope, *c*200m to the north-east of the peaty channel adjacent to Woodham Knoll. The material is a mid-brown boulder clay flint with water-worn cortex.

Me 028 **Little Crosby**
NGR SD 31820244
Site Type Flint Findspot
Period ?Mesolithic
Source Fieldwalking
Conditions Good

A small blade in a pale grey flint and a burnt flint chunk found *c*20m apart on a low sandy ridge marking the northern side of the peat-filled channel to the north-east of the Woodham Knoll ridge. The finds lie within *c*60m of the peat. About 40m to the south-east a primary flint flake, from a beach pebble, was located within *c*20m of the edge of the peat channel.

Me 029 **Little Crosby**
NGR SD 31700240
Site Type Flint concentration
Period Mesolithic
Source Fieldwalking
Conditions Moderate

Five pieces of struck flint were found in an area of *c*65x40m on the low ridge marking the south-western edge of the peat-filled channel and from the north-eastern side of the channel (Nos 4 and 5), although the peat in this section is now largely decomposed.

The raw material consists of three varying shades of brown boulder clay flint (1 to 3) and two light grey, possibly beach flints.

1-2:small scrapers. 3-4:chips. 5:chunk

Me 030 **Little Crosby**
NGR SD 31780254
Site Type Flint Findspot
Period ?Bronze Age
Source Fieldwalking
Conditions Moderate

Two pieces of struck flint lying next to each other came from the sandy ridge *c*70m from the north-eastern edge of the peat-filled channel, although the peat has largely decomposed in this section.

Me 031 **Little Crosby**
NGR SD 31580247
Site Type Flint findspot
Period ?Neolithic
Source Fieldwalking
Conditions Moderate

A single large, thin flake with retouch along the upper part of one side, was found on the lower slopes on the south-western side of the peat-filled channel, although the peat has not survived in this section. It is a dark brown, grey-patinated flint with slightly coarse cortex.

Me 032 **Little Crosby**
NGR SD 31530257
Site Type Flint concentration
Period Mesolithic
Source Fieldwalking
Conditions Moderate

Seven pieces of struck flint were found over an area of *c*50x 60m on the lower slopes of the peat channel adjacent to Woodham knoll. The peat is moderately humified here but the site may have some potential for waterlogged material.

1:micro blade. 2-3:blade fragments, one burnt. 4:blade. 5:retouched flake. 6:thinning flake. 7:burnt pebble fragment.

Me 033 **Little Crosby**
NGR SD 31400215
Site Type Flint concentration
Period Mesolithic
Source Fieldwalking
Conditions Moderate

Ten pieces of struck flint were found on the western brow of the Woodham Knoll ridge. They covered an area of c.150x70m, with another similar piece (1) found c.80m to the east. The raw material is mainly a brown boulder clay flint, except for a light grey boulder clay flint(7), two beach pebble flints (4 and 8) and two pieces of light grey chert (5 and 9).

1-3:small scrapers 4:primary flake 5-6:chips 7-10:small chunks

Me 034 **Little Crosby**
NGR SD 31420277
Site Type Flint findspot
Period Mesolithic
Source Fieldwalking
Conditions Good

A small flake in a light grey-brown boulder clay flint, was found in the organic sand of the peaty channel adjacent to Woodham Knoll, *c*110m to the south-west of the flint group of Me 042. The peat in the vicinity is an organic sand, the nearest surviving peat in the channel lies *c*35m to the south-west.

Me 035 **Little Crosby**
NGR SD 31230210
Site Type Flint findspot
Period Prehistoric uncertain
Source Fieldwalking
Conditions Good

A burnt flint chunk and a moderately water-worn pebble with some flaking were found within *c*30m of each other just below the western brow of the Woodham Knoll ridge.

Me 036 **Little Crosby**
NGR SD 31280232
Site Type Flint concentration
Period Late Neolithic/Bronze Age
Source Fieldwalking
Conditions Moderate

Eleven pieces of struck flint were found on the west facing brow of the Little Woodham Knoll ridge, covering an area of *c*80 x 50m. They are all of the mid brown flint raw material and technology described in Cowell (1991b), ascribed to the LN/EBA.

1-4:retouched implements. 5-10:flakes. 11:chip

Me 037 **Little Crosby**
NGR SD 31350240
Period Neolithic/Bronze Age
Site Type Flint concentration
Source Fieldwalking
Conditions Moderate

Sixty seven pieces of struck flint were found on the east facing brow of the Little Woodham Knoll ridge, overlooking the peat channel *c*200m to the east. They covered an area of *c*90 x 150m. They are all, except three pieces, of the mid-brown flint raw material and technology described in Cowell (1991b), ascribed to the LN/EBA.

1-20:retouched implements. 21-63:flakes. 64-67:chips

Me 038 **Little Crosby**
NGR SD 31300246
Period Mesolithic
Site Type Flint concentration
Source Fieldwalking
Conditions Moderate

A further six pieces were found within Me 037 which have been ascribed to a separate site as they are probably Mesolithic.

68:blade, light grey chert. 69:core, very pale ?chert. 70:flake, light grey-blue beach flint. 71:chip, brown boulder clay flint. 72-73:chunks (1 burnt).

Me 039 **Little Crosby**
NGR SD 31190244
Site Type Flint concentration
Period ?Upper Palaeolithic/Mesolithic/Neolithic
Source Fieldwalking
Conditions Moderate

Thirty pieces of struck flint were found over an area of *c*110x60m on the sloping brow of the Woodham Knoll ridge overlooking the Flea Moss Wood peat, which lies *c*30m to the north. The subsoil is Shirdley Hill Sand. The material is characterised as largely Mesolithic from the incidence of the blade cores. Despite the high proportion of flakes to blades this is judged to be a factor resulting from the small flint pebble raw material rather than a chronological feature. The raw material is mainly a combination of brown and grey boulder clay flint and pale grey beach flint. A single piece (30), a long thick blade in a non flint material with oblique truncation on one upper edge, may be of Upper Palaeolithic date (J Wymer pers comm). It came from near the mid-point of the slope down to the moss. The final piece (20) is made from the raw material that is typical of the LN/EBA material found across the field.

1:small scraper. 2-3:blades. 4-10:small flakes. 11-20:chips. 21-24:chunks. 25-29:blade cores. 30:large, obliquely truncated blade.

Me 040 **Little Crosby**
NGR SD 31060232
Site Type Flint concentration
Period Mesolithic, LN/EBA
Source Fieldwalking
Conditions Moderate

Three pieces of struck flint were found within *c*10m of each other on a slight rise on the edge of the coastal plain. It lies at the junction of the plain and the north-west slope of the Woodham Knoll Shirdley Hill Sand ridge. The surface is blown sand, dateable to after *c*2500 Cal BC, although there is a surface scatter of natural gravel pebbles on the west side of the rise and another scatter *c*140m to the west. A gravel deposit can be seen overlying the Shirdley Hill Sand and beneath the blown sand in the ditch section on the west side of the field. This may signify the course of an abandoned channel, further traces of which may be present in ditch sections to the south-east (at SD 31070189) running immediately to the west of the Woodham Knoll ridge. A further small flake from a water-worn pebble of the same material as the two Mesolithic pieces was not precisely located within the south-western part of the field. The Mesolithic material is earlier than the blown sand which may be masking a larger site. The LN/EBA piece (No 3) is, apart from the raw material, technologically similar to the group of sites on the brow of the Knoll (eg Me 018).

1:small single platform core in light grey boulder clay flint. 2:chip, similar raw material. 3:edge retouched irregular flake, light grey, poor quality boulder clay flint.

Me 041 **Little Crosby**
NGR SD 30930232
Site Type Flint findspot
Period Prehistoric uncertain
Source Fieldwalking
Conditions Moderate

A single bi-polar core in light brown boulder clay flint with water-worn cortex found on the coastal plain *c*125m to the north-west of Me 040, adjacent to a surface patch of gravel pebbles which may signify an abandoned channel.

Me 042 **Little Crosby**
NGR SD 31420277
Site Type Flint concentration
Period Mesolithic
Source Fieldwalking
Conditions Good

Five pieces of struck flint from an area of *c*50x40m on a slight sandy ridge on the north-eastern side of the peat-filled channel which lies adjacent to the Woodham Knoll ridge. The flint lies *c*50m from the current edge of the peat although here it is quite sandy and degraded. The nearest surviving peat in the channel lies *c*90m to the south.

1:micro-point. 2:broken blade, opaque light bluey-grey flint. 3-4:small flakes. 5:chip. With the exception of No 2, the raw material is a light grey-brown or medium brown boulder clay flint.

Me 043 **Ince Blundell**
NGR SD 32120275
Site Type Flint findspot
Period Prehistoric uncertain
Source Fieldwalking
Conditions Good

The findspot reference is centred midway between two isolated struck flint pieces, lying *c*45m apart, on a slight sandy ridge.

1 flake (possibly serrated). brown/grey flint with dark reddy brown mottling. ?Neolithic. 2 small chip, light grey/white flint.

Me 044	**Ince Blundell**
NGR	SD 32020299
Site Type	Flint Findspot
Period	?Mesolithic
Source	Fieldwalking
Conditions	Good

Two flint pieces found in adjacent fields *c*50m apart. One is a small, single platform flint core, with three blade removals, in a yellow-brown boulder clay flint, with a water-worn, white cortex. The other is a small scraper in a light grey-brown local flint. There is no obvious topographic reason for the location of the site at this point and the sand is quite clayey here.

Me 045	**Ince Blundell**
NGR	SD 31740302
Site Type	Flint findspot
Period	Prehistoric uncertain
Source	Fieldwalking
Conditions	Good

A single, light grey flint flake found in sandy soil on the west-facing slope of a very slight linear depression *c*70m wide.

Me 046	**Ince Blundell**
NGR	SD 32070317
Site Type	Flint findspot
Period	?Mesolithic
Source	Fieldwalking
Conditions	Good

Small core fragment in brown boulder clay flint, with blade scars, found on the sandy soil of the main Ince Blundell ridge overlooking the Little Crosby peat-filled channel *c*800m to the west.

Me 047	**Ince Blundell**
NGR	SD 32200321
Site Type	Flint findspot
Period	Prehistoric uncertain
Source	Fieldwalking
Conditions	Good

A single small flake from a very small, light blue flint pebble with water-worn cortex from the main Ince Blundell ridge. The peat-filled Little Crosby channel lies *c*950m to the west.

Me 048	**Ince Blundell**
NGR	SD 31600314
Site Type	Flint findspot
Period	Mesolithic
Source	Fieldwalking
Conditions	Good

A single flint blade from the western edge of a slight sandy ridge, which lies *c*20m from an indistinct peaty area with patches of sand and a low sandy ridge visible within it. The peat appears fairly well humified.

Me 049	**Little Crosby**
NGR	SD 3118002
Site Type	Flint Findspot
Period	?Mesolithic
Source	Fieldwalking
Conditions	Good

Two pieces of struck flint were found within *c*25m of each other on a featureless but slightly clayey patch of ground on the edge of an area of peat to the north of Woodham Knoll. The peat nearest the site is quite sandy and a third struck flint (No 3) came from this surface *c*30m to the west. The nearest surviving peat is *c*50m from this flintwork.

1:micro-blade, light brown boulder clay flint, water-worn cortex. 2:small flake, light grey boulder clay flint. 3:bi-polar core, burnt.

Me 050	**Little Crosby**
NGR	SD 31040279
Site Type	Flint concentration
Period	?Mesolithic
Source	Fieldwalking
Conditions	Good

Four pieces of struck flint came from an area of *c*20x15m on a small sandy island *c*40m to the south-east of Me 054. Two of the pieces came from the surrounding organic sand into which the former peat has degenerated in this area. Two further struck piece (Nos 1 and 5) came from this surface *c*20m and 40m respectively to the south. The presence of the cores suggests a flint knapping area on the sand island and possibly on the surface under the former peat.

1-3:very small cores, two single platform, one bi-polar, in brown and grey boulder clay flint, two of which have come from the beach or a river site. 4-5:chips, light grey and whitish grey flint, probably from beach pebbles, one (No 5) is from a blade and has been lightly retouched along one edge. 6:natural beach-pebble, light grey flint, limited retouch on one end.

Me 051	**Little Crosby**
NGR	SD 31010282
Site Type	Flint findspot
Period	Mesolithic
Source	Fieldwalking
Conditions	Good

A single long regular blade, possibly of early Mesolithic date, was found in the humified peat on the south-east of the small sand island adjacent to Me 054, *c*60m from the main flint group. It is made from a light grey opaque boulder clay flint which may have some affinities with that of a large flake which came from *c*50m to the north-east on the sand island. This latter piece is closer to the flint group from Me 054, but the raw material and technology differs from that group of material in addition to it being the only piece in the vicinity to come from the island as opposed to the peat channel or the ridge to the north-east. The cortex on the flake shows that it was not brought from the beach or riverside.

Me 052	**Little Crosby**
NGR	SD 3096083
Site Type	Flint concentration
Period	?Mesolithic
Source	Fieldwalking
Conditions	Good

A thin scatter, consisting of 7 pieces of struck flint, three of which came from an area of *c*50x90m probably represents a focus of Mesolithic activity (Nos 1-3) on the north-eastern edge of a low sandy ridge adjacent to a peat hollow within which low sand islands and ridges are visible. A fourth piece of probably the same date was located *c*110m to the east on a very low sandy ridge in the peat. The peat around this site is fairly well humified producing an organic sand surface which includes a scattering of lumps of dried peat and small pieces of wood. The main ridge was previously covered by peat and the frequency of the humified organic material is greater here than in the hollow to the east. The other three pieces of struck flint (Nos 5-8), which may not be of the same date, come from the same topographic feature but are more widely spread, extending the general area of this scatter to *c*140x75m. It seems very likely that, notwithstanding the large area involved, the slight slope of the ridge has formed a focal point for prehistoric activity.

1:small blade, medium grey boulder clay flint, beach pebble. 2-4:chips, boulder clay flint, one burnt. 5:flake, grey-brown boulder clay flint. 6:edge retouched irregular flake. 7:small bi-polar core, light brown boulder clay flint, beach pebble.

Me 053	**Little Crosby**
NGR	SD 31060290
Site Type	Flint findspot
Period	Prehistoric uncertain
Source	Fieldwalking
Conditions	Good

A single, small bi-polar core on a brown flint beach pebble came from the north-eastern side of the slight sand ridge on which Me 054 was located, *c*50m to the south. The peat to the north-east of the ridge is here the most well-preserved in the field.

Me 054	**Little Crosby**
NGR	SD 31060286
Site Type	Flint concentration
Period	?Mesolithic
Source	Fieldwalking
Conditions	Good

Twelve pieces of struck flint found in an area of *c*20x15m suggests a reasonably well defined site lies around a narrow peat channel, *c*12m wide, between the adjacent low sand ridge to the north-east and a small sand island to the south-west. Eight of the pieces come from the channel where the brown peat is slightly less humified than the peat that surrounds the ridge and the island, where it is almost an organic sand. All the remaining pieces come from the edge of the slight sand ridge, which here is about 35m wide but widens out to about 80m to the north. The peat on the north-eastern side of the ridge is generally fairly well preserved, presumably through being a little deeper than elsewhere in the field. A single flint flake came from *c*15m away on the island but has been placed in a separate site (Me 051). Apart from this piece, the flintwork is generally quite poor and consists mostly of small waste. On other sites such waste is associated with blades, suggesting a Mesolithic date, although there is a little more doubt here in that the one complete flake from this group (No 1) appears to have been struck with a hard hammer technique which rarely occurs in the blade assemblages.

1:small, thick flake, from a light blue flint beach pebble. 2:small thick primary flake, burnt. 3-9:chips, including one white flint, three light blue beach pebble flint and three either medium grey or brown boulder clay flint. 10-12:waste chunks, either light grey or medium grey coloured boulder clay flint two of which are from beach pebbles.

Me 055	**Little Crosby**
NGR	SD 30630315
Site Type	Flint findspot
Period	Mesolithic
Source	Fieldwalking
Conditions	Good

A single corticated blade from the surface of the blown sand on the coastal plain. In this area the sand overlies a marine clay, both of which are likely to post-date the Mesolithic.

Me 056	**Little Crosby**
NGR	SD 30800349
Site Type	Flint concentration
Period	Mesolithic
Source	Fieldwalking
Conditions	Good

Three pieces of struck flint, all cores or related pieces, came from an area *c*65x10m on the north-western edge of a slight ridge of Shirdley Hill Sand which marks the approximate limit of the blown sand covering the coastal plain.

1:small conical single platform core, light grey brown boulder clay flint. 2:small brown flint pebble, prepared platform and three small blade scars. 3:?core trimming flake, medium grey boulder clay flint. 1 and 3 have slightly coarse cortex.

Me 057	**Ince Blundell**
NGR	SD 31300365
Site Type	Findspot
Period	?Neolithic
Source	Fieldwalking
Conditions	Good

A single utilised flake, possibly a hollow scraper, with secondary fine, edge retouch, in a patinated, light grey flint. The implement came from a topographically indistinct area of Shirdley Hill Sand *c*200m to the south-west of the edge of the Orrell Hill peat.

Me 058	**Ince Blundell**
NGR	SD 31430369
Site Type	Flint concentration
Period	Mesolithic
Source	Fieldwalking
Conditions	Good

Five pieces of struck flint were located within an area of *c*30x7m, on the southern edge of a slight sandy ridge. Another single find came from *c*30m to the south of the main group. The finds are located within *c*150m of the present edge of the Orrell hill peat area.

1:blade. 2-3:chips. 4:waste chunk. 5:possible bashed lump. Nos 2-4:yellowy to reddy brown boulder clay flint. No 1:dark grey lustrous flint.

Me 059	**Ince Blundell**
NGR	SD 31400437
Site Type	Flint findspot
Period	Later prehistoric
Source	Fieldwalking
Conditions	Good

A struck flake and a possible flake from the eastern edge of a very low alluvial clay capped ridge, *c*80m to the west of site Me 060.

Me 060	**Ince Blundell**
NGR	SD 31490422
Site Type	Flint findspot
Period	Prehistoric uncertain
Source	Fieldwalking
Conditions	Moderate

Three pieces of struck flint, including a blade fragment (which suggests a possible Mesolithic date), from an area *c*40x12m on a sandy ridge lying between two narrow peat-filled channels. The blade came from the edge of the peat in the channel which may suggest more of the site lies buried by organic material.

The peat is c0.4m deep and overlies at least 0.3m of organic clays on the east of the ridge. To the north-east the sand divides into at least three very slight ridges running parallel to each other, with patches of organic sand adjacent in places.

Me 061	**Ince Blundell**
NGR	SD 31800426
Site Type	Flint Findspots
Period	Mesolithic
Source	Fieldwalking
Conditions	Good

Two pieces of struck flint were found within c50m of each other on the edge of the Shirdley Hill Sand as it starts to fall towards the Alt valley to the north-east. The topography of the natural slope is, however, hidden by a belt of post-medieval alluvial clay reaching to within c12m of the more easterly find. A nearby ditch section shows that the alluvial clay is c80mm thick beneath the topsoil, with a black peaty clay of similar depth beneath that overlying the sand.

1:small point, pale grey/blue flint. 2: Core rejuvenation piece, with blade removals, creamy brown flint.

Me 062	**Ince Blundell**
NGR	SD 31730410
Site Type	Flint Findspot
Period	Mesolithic/?Neolithic
Source	Fieldwalking
Conditions	Good

Two pieces of struck flint found in an area c12x8m on the western edge of a very slight sandy ridge. About 60m to the north-west is a tongue of shallow eroding peat within which two small, low sand islands are just visible. The north-eastern edge of the peat is marked by a series of low, linear sand ridges.

1:blade with partially retouched edge, in a bluey grey flint. 2:flake, broad platform, with light retouch around one edge, heavier inverse retouch along one side, in a light grey, white speckled flint.

Me 063	**Ince Blundell**
NGR	SD 31960405
Site Type	Flint Findspot
Period	?Neolithic
Source	Fieldwalking
Conditions	Moderate

An end/side scraper on a thick blade-like flake in local flint and a burnt fragment of flint were found within c12m of each other. They lie towards the northern edge of the major Ince Blundell sand ridge, c250m from the peat at Orrell Hill.

Me 064	**Ince Blundell**
NGR	SD 31780393
Site Type	Flint concentration
Period	?Neolithic/Mesolithic
Source	Fieldwalking
Conditions	Good

Five pieces of struck flint were found over an area c40x12m, on the north-west edge of a low sandy ridge.

1:broad, truncated blade. 2:broad, broken blade. 3:nodule trimming flake, retouched. 4:small blade. 5:waste chunk.

No 2 is lustrous reddy brown flint, No 1 a patinated light blue/grey flint, the others are dull, medium grey flint. Nos 1, 3 and possibly 4 have waterworn cortex.

Two more pieces were found around this concentration. 6:small, flint single platform blade core, c40m to the south. 7:burnt flint, the same distance to the south east. No 6 is yellowy brown boulder clay flint with waterworn cortex around one side.

Me 065	**Ince Blundell**
NGR	SD 31850382
Site Type	Flint concentration
Period	Neolithic/Mesolithic
Source	Fieldwalking
Conditions	Good

Four pieces of struck flint were found over an area c20x5m with a further piece c45m to the south-west. The site lies on a sandy soil along a very slight ridge.

1:blade. 2:leaf arrowhead. 3:waste chip. 4:thick flake. All yellow-brown boulder clay flint.

Me 066	**Ince Blundell**
NGR	SD 31930387
Site Type	Findspot
Period	Mesolithic
Source	Fieldwalking
Conditions	Good

Eleven pieces of struck flint found spread over an area of c18x20m. The subsoil is Shirdley Hill Sand, but the finds are not associated with any obvious topographic feature.

1:microlith fragment. 2-3:blades. 4:crested blade. 5:blade-like flake, retouched. 6:blade tip. 7-8:flakes. 9:chunk. 10:chip. 11:core rejuvenation piece. 12:burnt split pebble. All are either reddy brown or yellow brown boulder clay flint, except 6:grey, patinated flint, 10:browny grey slightly lustrous chert, and 11:dark bluey grey chert, possibly from a local pebble.

Me 067	**Ince Blundell**
NGR	SD 31980397
Site Type	Findspot
Period	Prehistoric uncertain
Source	Fieldwalking
Conditions	Good

A single flint scraper came from the south-western edge of a low sandy ridge.

Me 068	**Ince Blundell**
NGR	SD 32050374
Site Type	Flint findspot
Period	Bronze Age
Source	Fieldwalking
Conditions	Good

A single flint flake came from the sandy soil c40m to the south of site Me 066 . Reddish brown boulder clay flint.

Me 069	**Ince Blundell**
NGR	SD 32380414
Site Type	Flint Findspots
Period	?Mesolithic
Source	Fieldwalking
Conditions	Good

Three struck flints, including a burnt core fragment, and two chips in an orange and a grey boulder clay flint were found over an area of c24x12m. They lay near the edge of the sandy plain, part of the major Ince Blundell ridge, before it falls gently to the north-east towards the Alt valley. The nature of the original Shirdley Hill Sand relief has been smoothed by a belt of Flandrian deposits, of which the surface deposit, an alluvial clay, lies within *c*120m.

Me 070	**Ince Blundell**
NGR	SD 32540401
Site Type	Flint findspot
Period	Prehistoric uncertain
Source	Fieldwalking
Conditions	Moderate

A small retouched flake in a poor quality boulder clay flint from the sandy soil of the main Ince Blundell ridge.

Me 071	**Ince Blundell**
NGR	SD 32710400
Site Type	Flint findspot
Period	Mesolithic/?Neolithic
Source	Fieldwalking
Conditions	Good

A single honey-coloured flint scraper on a thin flake, found in sandy soil just above the break in slope of the terrace of the river Alt, which is here a little more pronounced than to the south.

Me 072	**Ince Blundell**
NGR	SD 32850405
Site Type	Flint findspot
Period	Prehistoric uncertain
Source	Fieldwalking
Conditions	Moderate

A flint flake and burnt flint were found together beneath the slight break in slope of the terrace of the Alt. About 300m to the north-west in a similar topographic situation was a possible flake of chert. The surface is a mixture of mainly sandy clay, with some patches of sand and peaty sand.

Me 073	**Ince Blundell**
NGR	SD 32800385
Site Type	Flint findspot
Period	?Mesolithic
Source	Fieldwalking
Conditions	Poor

A single blade core, found in sandy soil at the *c*4m OD contour, which here marks a slightly more pronounced terrace to the river valley. The conditions and the nature of the find suggest that it is likely that a more significant site may exist here. Miscellaneous grey flint and rough cortex (used as the platform), suggests local boulder clay flint.

Me 074	**Ince Blundell**
NGR	SD 33150391
Site Type	Flint findspots
Period	?Mesolithic
Source	Fieldwalking
Conditions	Moderate

A core rejuvenation piece in brown, boulder clay flint and a possible flake with two crushed edges from the junction of the floodplain and the slight slope of the River Alt terrace. They come from an area that is a mixture of patches of peat and sandy peat, which must reflect the thin and humified nature of the organic layer.

Me 075	**Ince Blundell**
NGR	SD 33270394
Site Type	Flint findspot
Period	Uncertain
Source	Fieldwalking
Conditions	Moderate

A single flake in brown boulder clay flint, with a possible rudimentary chipped point from the floodplain of the River Alt. The surface here is clayey (alluvial) in nature with a mixture of humified peat and sandy patches.

Me 076	**Ince Blundell**
NGR	SD 33330382
Site Type	Flint concentration
Period	Mesolithic
Source	Fieldwalking
Conditions	Good

Sixteen pieces of struck flint were found on a small sand island, *c*65m at its widest and c80m long, in the floodplain of the River Alt. Ten pieces, including the worked points (1-3), came from an area of *c*15x20m. This group lies on the north-west edge of the island which is surrounded by brown alluvial clay, although the underlying peat is showing through in patches on the northern side. Between *c*40-80m to the south-east, on the southern side of the island three widely separated pieces may indicate a second focus. Two of these pieces, a core and a burnt chip, lie off the sand on the alluvial clay, suggesting part of the site may be more protected from ploughing.

1-3:obliquely blunted points. 4-5:blades. 6-7:flakes. 8-10:crude pebble cores or bashed lumps. 11-13:chips. 14-16:burnt chips. The material is almost all a light grey-brown flint, the cortex of the core suggesting it has come from either the river channel or the beach. The exceptions are two of the points, one of which is a dark grey chert, the other a medium grey banded chert.

Me 077	**Ince Blundell**
NGR	SD 33370379
Site Type	Flint concentration
Period	Mesolithic
Source	Fieldwalking
Conditions	Good

Six pieces of struck flint including 3 small blades, within an area of *c*30x5m, located on the north-west and south-east edges of a small sand island. Brown alluvial clay laps against the sand at a distance of between *c*20m and 10m from the findspots respectively, patches of the underlying peat are showing through the clay on the north-west side. Two further flint pieces came from distances of *c*30m and *c*60m along the sand ridge to the north-east. The latter piece is burnt and lay on the brown alluvial clay, *c*3m from the edge of the sand. This may suggest a largely buried site is starting to be eroded from beneath the clay and peat.

Me 078	**Ince Blundell**
NGR	SD 33290377
Site Type	Flint concentration
Period	Mesolithic
Source	Fieldwalking
Conditions	Good

Six pieces of struck flint and a bashed pebble within an area of *c*12x5m, on the south-west edge of a low sand island, orientated *c*40m east-west. Brown alluvial clay lies within less than *c*10m on three sides, with patches of peat showing through it on the west side. The ridge runs for about 30m northwards and in many places is just emerging from the brown clay. This probably means that there are further elements of the scatter lying under these capping layers.

The microlith (1), which is made from a dark grey chert (similar to material from North Lancashire) may possibly imply an early date within the Mesolithic, although there is rather too little evidence to be sure. The main raw material on the site is a pale grey, probably beach, flint (2, 3, and 5), the final piece (6) is a white patinated flint.

1:obliquely blunted microlith. 2:small blade. 3-4:small flakes (1 burnt). 5-6:chips

Me 079	**Ince Blundell**
NGR	SD 33720374
Site Type	Burnt flint concentration
Period	Prehistoric uncertain
Source	Fieldwalking
Conditions	Good

Three pieces of burnt flint, a small chip, and two possible struck pieces of flint were found across an area of *c*50x60m. They lay on a slight ridge *c*25m wide, one of several which run intermittently across this part of the floodplain of the river Alt. The surface of the ridge consists of patchy sand which shows through the general surface of very humified peat and alluvial clay.

Me 080	**Ince Blundell**
NGR	SD 33160357
Site Type	Flint concentration
Period	Mesolithic
Source	Fieldwalking
Conditions	Good

Thirteen pieces of struck flint and four, possibly associated, pebble fragments. They lay at the *c*5m OD contour, a little above the junction between the Alt floodplain and the very gradual slope to the south-west up to the top of the dryland ridge at *c*12m OD. The scatter covers an area of *c*90x36m, with a slight trend to two *foci* within the material, although this could be a recovery variable. The main part of the scatter lies on the eastern edge of a very slight linear depression in the Shirdley Hill Sand, which may mark a former stream course.

1-3:blade cores, single platform. 4:bi-polar core. 5-7:blades. 8-10:flakes. 11-13:waste chunks. 14:burnt flint chip. 15-18:possible shatter and bashed pebbles. All pieces are of yellowy brown, boulder clay flint, except one blade, two flakes, and a chunk, which are of either dark grey, lustrous flint or a dull, light grey-brown flint.

Me 081	**Ince Blundell**
NGR	SD 33940352
Site Type	Flint findspot
Period	Prehistoric uncertain
Source	Fieldwalking
Conditions	Good

A single struck flint from the sandy clay margins of a low, narrow sand island, *c*90m long. The island is surrounded by alluvial clay and in the vicinity of the island the underlying peat is showing through. A larger site may therefore still await discovery.

Me 082	**Ince Blundell**
NGR	SD 33960350
Site Type	Flint concentration
Period	Mesolithic
Source	Fieldwalking
Conditions	Good

Ten pieces of struck flint from a sandy ridge, measuring *c*85x20m, in the floodplain of the Alt. The flint comes mainly from the south-eastern edge of the sand ridge at the junction with the alluvial clay in a band *c*5m wide over a distance of about 12m. A further piece, a blade (11), lay on the north-western edge of the ridge *c*70m away, on the surface of the alluvial clay.

1:struck nodule. 2-3:micro-blades. 4:blade fragment. 5-6:small chunks. 7-10:burnt chips. 11:blade. The latter is pale brown boulder clay flint the rest consist of a variety of greys and browns. Two pieces, 1 and 2, have a very smooth water-worn cortex.

Me 083	**Ince Blundell**
NGR	SD 33900348
Site Type	Flint findspot
Period	Prehistoric uncertain
Source	Fieldwalking
Conditions	Good

Two struck flints from an area of *c*15x5m on the edge of a low island. In most areas the island consists of a modern alluvial clay surface with the underlying sand only appearing intermittently or as very finely dispersed grains in the clay. This may account for the reason why so few finds have been located and it is to be suspected that a larger site lies on the upstanding part of the island or on that part of it which is still under the surrounding alluvial clay. A number of small patches of the underlying peat are also now becoming visible within this clay around the island.

Me 084	**Ince Blundell**
NGR	SD 33903460
Site Type	Flint findspot
Period	?Mesolithic
Source	Fieldwalking
Conditions	Good

Two pieces of struck flint came from a sand island on the floodplain of the River Alt. This island is slightly higher and measures *c*60x30m. The flint comes from within an area of *c*5x5m on the north-eastern edge of the island, which is mostly marked by the brown alluvial clay, although small patches of black peat occur in its surface on the eastern edge of the island.

1:small flake, light grey flint. 2:chip, boulder clay flint.

Me 085	**Ince Blundell**
NGR	SD 33980345
Site Type	Flint concentration
Period	Mesolithic
Source	Fieldwalking
Conditions	Good

Five pieces of struck flint came from an area of *c*5x8m on the eastern edge of a very slight sandy island, measuring *c*12x10m, lying on the floodplain of the Alt valley. The sand island is mainly surrounded by brown alluvial clay, probably of post-medieval date, although in the vicinity of the flint a narrow, *c*2m wide, band of the underlying peat is visible on the surface. A sixth piece of struck flint came from the alluvial clay on the northern side of the island, *c*8m from the main concentration. All the pieces, except one, are less than 20mm in size and each one is of a different type of flint. Several pieces are either beach or river derived.

1:primary blade. 2:blade chip. 3-6:chips.

Me 086
NGR SD 33960341
Site Type Flint findspot
Period Prehistoric uncertain
Source Fieldwalking
Conditions Good

A single, small, irregular boulder clay nodule with some small flakes taken off was found on a small patch of alluvial clay distinguished by the presence of fine sand, although it was topographically indistinct.

Me 087 **Ince Blundell**
NGR SD 33850334
Site Type Flint findspot
Period Prehistoric uncertain
Source Fieldwalking
Conditions Good

Two possible flakes from within the peat, *c*60m apart, may mark the location of a site starting to erode out.

Me 088 **Ince Blundell**
NGR SD 33600329
Site Type Flint findspot
Period ?Mesolithic
Source Fieldwalking
Conditions Moderate/poor

A small single platform core, with two faces partially shattered through use but with four small blade scars still visible, was found at about the 5m OD contour which approximately marks the edge of the river Alt terrace, although here it consists of an almost continuous shallow gradient slope. The present edge of the peat lies *c*100m to the north-east. About 70m to the south were found a (possibly blade) core fragment, and a rudimentary core on an irregular flint pebble. The less than ideal conditions at the time of walking may have prevented a larger focus of activity from being discovered. May possibly be a similar site to Me 080.

Me 089 **Ince Blundell**
NGR SD 34100330
Site Type Flint findspot
Period Prehistoric uncertain
Source Fieldwalking
Conditions Good

A single flake in an opaque creamy-grey flint found in the floodplain of the River Alt, close to the junction of the surface upper peat (post *c*5000 Cal BC) and a clayey ridge which has resulted from cleaning out the adjacent ditch. This suggests there is a possibility that it has come from the ditch and therefore may have lain under the peat originally.

Me 090 **Ince Blundell**
NGR SD 34320335
Site Type Flint findspot
Period Prehistoric uncertain
Source Fieldwalking
Conditions Good

A single flake in a poor quality, medium grey boulder clay flint came from the alluvial clay surface of the floodplain close to the line of the River Alt. The upper peat (post *c*5000 Cal BC) lies at the surface within *c*20m of the findspot.

Me 91 **Thornton**
NGR SD 34420273
Period Mesolithic
Site Type Flint concentration
Source Fieldwalking
Conditions Moderate

Seven pieces of struck flint, including a blade core and a blade were found at approximately the 4m OD contour on the east facing terrace slope of the Alt valley. They came from an area of *c*40mx50m *c*70m from the present edge of the peat in the valley.

Me 092 **Thornton**
NGR SD 34090250
Period Prehistoric uncertain
Site Type Flint findspot
Source Fieldwalking
Conditions Moderate

A single flake of light grey flint was found on the peat surface close to a ditch on the flood plain of the Alt. This may mark the existence of a larger site in the vicinity. A narrow ridge running along the northern part of the field in the vicinity of the findspot suggests the Shirdley Hill Sand may lie close to the present ground surface and could mark the potential location for such a site.

Me 093 **Lunt**
NGR SD 34880230
Site Type Flint concentration
Period Mesolithic
Source Fieldwalking
Conditions Moderate

Sixty-nine pieces of struck flint were found mainly in two groups separated by a space of *c*12m (Fig 20 A *this vol*), a distance which was walked at 2m intervals. There is, however, little difference between the flintwork of the two groups and a continuous scatter of material, partially obscured by post-depositional factors, could be represented at this location. To the north-west a small group of struck flint were located over a distance of *c*12m along a narrow (c10m wide) Shirdley Hill Sand ridge marking the north-facing terrace of the Alt flood plain. The flint was mainly concentrated on the ridge, but four pieces were more widely spread over a distance of *c*25m to the foot of the gentle terrace slope. Another piece was located *c*30m from the foot of the slope on a second (very slight) sand ridge. This is partially composed of an organic sand and partly of a peaty ploughsoil, with small fragments of natural wood quite common.

The second group of material is found over an area of *c*18x10m to the south-east where the ridge merges into a scarp and the slope becomes more pronounced and extensive, although its foot is marked by an almost imperceptible ridge which is a feature of the entire terrace in the field. The flint (*above*) from the main ridge is found on Shirdley Hill Sand, while the scarp slope is mainly a dark (presumably organic) sand. The incidence of humic sand and small pieces of driftwood south-west of the scarp edge shows that peat had formerly lain in the narrow hollow behind the ridge. It would seem unlikely that much if any organic material will survive beneath the ploughsoil in these areas. The artefacts from the slight ridge at the foot of the scarp may have better potential as some parts of the ridge are covered by a woody, peaty soil and a number of artefacts are found at the boundary with the peat, suggesting other elements of the site(s) are relatively protected from erosion in a possible organic context.

1:scraper. 2:retouched flake. 3-9:blades. 10-19:flakes. 20-26:primary flakes. 27-54:chips (2 burnt). 55-66:waste chunks (3 burnt) 67:core. 68:core fragment. 69:bashed pebble.

Me 094	**Lunt**
NGR	SD 35390240
Site Type	Flint concentration
Period	Mesolithic
Source	Fieldwalking
Conditions	Moderate

Ninety struck pieces of flint and chert were found over an area of *c*50x45m (Fig 20.B, *this vol*). The distribution of the material is slightly greater than the area of a low sand island across which most of the material was found. The island lies in the floodplain of the river Alt. To the north the surface is formed by a post-medieval alluvial clay and to the south the ploughsoil surface is mainly a humified peat. A second very small island (*c*12x8m) is located *c*12m to the north of the main island, within the clay, although the underlying peat is also visible in patches around the edge of the island. Two pieces of struck flint were found on this island, suggesting another, better protected, focus of prehistoric activity. The flintwork also occurs patchily in the clay surface to the north-east of the main island, where there is an almost imperceptible rise in ground level. This must signify that part of the occupation on the island lies in a less eroded area and may have limited survival of organic material in the sand beneath the clay and peat. To the south-west of the main island a few small, apparently plank-shaped pieces of wood, one with a pointed tip lay in and on the peat, although their humified and unstratified nature makes it difficult to be sure about their age or function.

The flint material has three main characteristics: a group of 7 dark grey chert pieces, found towards the north-west of the island, may point to a possible early Mesolithic focus, by analogy with sites on the Wirral. It is this material, including cores, trimming flakes, and blades, that shows the most controlled knapping techniques and curation of raw material. This group could be augmented on grounds of raw material by four pieces of possible dark grey/blue chert. In the east of the island a small group of flints (12) includes most of the other controlled knapped blades and the microlith, which suggests a late Mesolithic date. The final group, largely found in the centre of the sand island contrasts with the former two, showing less controlled knapping techniques, probably as a reflection of the poorer quality material involved. Most of the primary flakes in this group are from very small (<20mm) flint pebbles. Nearly all of the burnt pieces occur in this area. It is difficult to know if this represents a separate chronological phase of activity or represents technological or functional differences within the Mesolithic occupation of the island.

Apart from the chert, which may have been brought in from outside the area, most of the other material has been collected from the beach or river, most probably the former. If the material is being brought to the site to be knapped there as is suggested by the debris, then it would seem likely that flint knapping would form only one aspect of the activity on the site, which must have been positioned to exploit the floodplain of the river. There is little of the brown or honey-coloured boulder clay-derived flint (although two of the best blades are in this material), the main emphasis being on medium grey or bluey/grey flint, which is of quite poor quality.

1:microlith. 2-4:retouched flakes. 5-20:blades. 21-30:flakes (2 burnt). 31-35:primary flakes. 36-65:chips (5 burnt). 66:blade fragment. 67-80:chunks (1 burnt). 81-84:core trimming flakes. 85-89:small blade cores. 90:core fragment.

Central mosslands

Me 095	**Knowsley**
NGR	SJ 44259664
Site Type	Flint findspot
Period	Prehistoric uncertain
Source	Fieldwalking
Conditions	Moderate

A single flake was found on the Shirdley Hill Sand plain between Knowsley Park Moss and Kirkby Moss. This field has been largely quarried of its sand, but the find came from a narrow elevated ridge along its western edge which may not have been destroyed.

Me 096	**Knowsley**
NGR	SJ 44889650
Site Type	Flint concentration
Period	Mesolithic
Source	Fieldwalking
Conditions	Good

A small concentration of four struck flints came from an area of *c*10x10m during fieldwalking in 1982. It lay on the south-west facing upper slopes of a large sandy ridge which falls gently down to the edge of Knowsley Park Moss *c*400m away.

1-3:small blades, boulder clay flint patinated white. 4:flake, with central blade scar on back, reddy brown boulder clay flint.

Me 097	**Kirkby**
NGR	SJ 43949742
Site Type	Flint findspot
Period	Prehistoric uncertain
Source	Fieldwalking
Conditions	Moderate

A single burnt pebble fragment with *c*40% cortex came from the Shirdley Hill Sand plain to the west of Kirkby Moss. A flake (lost) was also recorded *c*100m to the east of this findspot, from the foot of the slope of a low, localised sandy ridge.

Me 098	**Kirkby**
NGR	SJ 44909757
Site Type	Flint findspot
Period	Mesolithic/Neolithic
Source	Fieldwalking
Conditions	Good

A single, large broken blade with oblique truncation and inverse backing on the opposite edge was found on a sandy ridge *c*400m to the south of Kirkby Moss. It lay adjacent to a slight hollow containing humified peat. It is made from a dark grey flint with light grey patination.

Me 099	**Kirkby**
NGR	SJ 45169770
Site Type	Flint findspot
Period	?Prehistoric uncertain
Source	Fieldwalking
Conditions	Good

A single piece of broken and burnt flint came from a thin sandy soil overlying boulder clay, within *c*70m to the south of the present edge of the peat of Kirkby Moss. The soil contains extremely humified peat in the vicinity of the findspot.

Me 100	**Knowsley**
NGR	SJ 46109596
Site Type	Flint concentration
Period	Mesolithic
Source	Fieldwalking
Conditions	Good

A small concentration of struck flint, over an area of *c*12x15m, was located on the lower slopes of the ridge which marks the eastern edge of Knowsley Park Moss. The subsoil is Shirdley Hill Sand in this part of the ridge which changes to boulder clay *c*250m to the south. The present wetland edge lies *c*110m to the west.

1: micro-blade in dark grey flint. 2: small burnt natural flake with scraper retouch on one edge. 3-4:waste chips in light grey/ blue and light brown flint. A boulder clay pebble, with water-worn cortex, and possible signs of working, also came from the same area.

Me 101	**Knowsley**
NGR	SJ 46049582
Site Type	Flint findspot
Period	?Neolithic
Source	Fieldwalking
Conditions	Good

A single flint implement (a knife?) found *c*15m into the humified peat on the eastern edge of Knowsley Park Moss. A low sand ridge is just visible in the organic soil at the foot of the main ridge bounding the moss.

Me 102	**Knowsley**
NGR	SJ 46319566
Site Type	Flint findspot
Period	Prehistoric uncertain
Source	Fieldwalking
Conditions	Moderate

A single bi-polar flake, in browny-grey boulder clay flint, was found on the upper part of the west-facing boulder clay ridge skirting Knowsley Park Moss, the present edge of which lies *c*330m away.

Me 103	**Windle**
NGR	SJ 48199729
Site Type	Flint findspot
Period	?Mesolithic
Source	Fieldwalking
Conditions	Good

A small burnt flint chip was found on the Shirdley Hill Sand plain lying between Windle Moss to the north and Catchdale Moss to the south.

Me 104	**Rainford**
NGR	SJ 46909959
Site Type	Flint findspot
Period	?Mesolithic
Source	Fieldwalking
Conditions	Moderate

A single struck piece came from the brow of the boulder clay ridge overlooking Reeds Moss to the south-east. It is a core-trimming flake with micro blade scars, on a small yellowy-brown boulder clay flint pebble.

Me 105	**Bickerstaffe**
NGR	SD 45540317
Site Type	Flint findspot
Period	Mesolithic
Source	Fieldwalking
Conditions	Poor

A single core, with three blade scars from one platform and a flake from the opposing platform, was found on the lower slope of the north-east-facing ridge of the Rainford Brook valley which here contains a belt of surface peat. It is made from a grey flint pebble of boulder clay origin.

Me 106	**Rainford**
NGR	SD 48570237
Site Type	Flint Findspot
Period	Mesolithic
Source	Fieldwalking
Conditions	Good

Two finds were located within *c*10m of each other on the southern brow of the Shirdley Hill Sand ridge lying between Holiday Moss and Holland Moss, at a height of *c*62m OD. The main piece is a long bladed microlithic point in an opaque light grey flint, which is probably early Mesolithic in date. The second piece, which came from a re-walk of the area at a 2m interval, is a small chip of micro-blade proportions, which could be of similar date, in a whiter opaque flint.

Me 107	**Billinge**
NGR	SJ 50140230
Site Type	Flint findspot
Period	?Neolithic
Source	Fieldwalking
Conditions	Good

A single small scraper in pale grey flint came from the lower slopes of the high, sandstone ridge overlooking Holiday Moss, *c*500m to the south. The lower slopes of the ridge here have a Shirdley Hill Sand capping.

Me 108	**Billinge**
NGR	SJ 50770234
Site Type	Flint findspot
Period	?Neolithic
Source	Fieldwalking
Conditions	Good

A single, large flake in deep brown boulder clay flint was found on the middle slopes of the high, sandstone ridge overlooking Holiday Moss, *c*1100m to the south. The soil in the vicinity was clayey.

Me 109	**Billinge**
NGR	SD 51080170
Site Type	Flint findspot
Period	Mesolithic
Source	Fieldwalking
Conditions	Moderate

A struck lithic piece was found on the lower slopes of the high, sandstone ridge overlooking Kings Moss, whose present edge lies *c*700m to the south. It is a broken flake with dorsal blade scars, made from medium grey chert. It came from a small ridge alongside a dried-up stream channel.

Me 110
NGR SD 50800132
Site Type Flint findspot
Period Prehistoric uncertain
Source Fieldwalking
Conditions Moderate

A small burnt chip was found near the foot of the high Billinge sandstone ridge, *c*500m away from Me 109. The findspot is located on a belt of Shirdley Hill Sand, which stretches from the foot of the ridge to the mossland *c*500m to the south. Nearly all the sand has been commercially extracted to the north of the road for a distance of *c*1km parallel to the roadway, from where the geology becomes clayier. The find came from a thin strip which may have escaped extraction, as the sand is probably thinning out here since it dies out completely a little higher up the ridge, where it is replaced by a clayey soil.

Me 111 **Bickerstaffe**
NGR SD 448028
Site Type Cropmarks
Period Uncertain
Source Aerial Photography

A double ditched lane with attached ditched features, which may be enclosures were recognised from the air. They may predate the post medieval pattern of fields seen on the 1st edition OS map. Field conditions did not allow the site to be checked on the ground.

Me 112 **Billinge**
NGR SD 503017
Site Type Cropmarks
Period Uncertain
Source Aerial Photography

A possible banked and multi-ditched rectilinear enclosure with linear ditches and field boundaries was recognised from the air. The field lies on the boulder clay slopes above Holiday Moss and was fieldwalked but provided no artefactual evidence.

Palaeoecological sampling points

P1 **Great Meols, Wirral**
NGR SJ 227882
Site name Newton Carr 11 (NC11)

P2 **Great Meols, Wirral**
NGR SJ 221883
Site name Newton Carr 7 (NC7)

P3 **Great Meols, Wirral**
NGR SJ 228888
Site name Newton Carr 8 (NC8)

P4 **Little Meols, Wirral**
NGR SJ 239901
Site name Park Road, Meols (PRM)

P5 **Bidston, Wirral**
NGR SJ 285909
Site name Bidston Moss 2 (BM2)

P6 **Little Crosby, Sefton**
NGR SJ 307015
Site name Sniggery Wood (SW)

P7 **Little Crosby, Sefton**
NGR SJ 311027
Site name Flea Moss Wood (FMW)

P8 **Ince Blundell, Sefton**
NGR SJ 342031
Site name Alt Valley, Ince Blundell (IB)

P9 **Simonswood, Knowsley**
NGR SJ 445996
Site name Simonswood Moss B (SMB)

P10 **Simonswood, Knowsley**
NGR SD 448010
Site name Simonswood Moss C (SMC)

P11 **Knowsley, Knowsley**
NGR SJ 455961
Site name Knowsley Park (KP)

P12 **Parr, St Helens**
NGR SD 545940
Site name Parr Moss (PMB)

P13 **Prescot, Knowsley**
NGR SJ 468929
Site name Prescot Moss (PM)

Appendix 2: Radiocarbon dates

Site Name and Context	Lab Code	Uncalibrated Date (BP)	Calibrated* Date
Wirral mosslands			
BIDSTON MOSS 2			
196-200cm	SRR-2926	7360±60	6400-6080 Cal BC
98-105cm	SRR-2925	5840±70	4900-4530 Cal BC
32-36cm	SRR-2924	4740±70	3690-3360 Cal BC
PARK ROAD, MEOLS			
121-125cm	SRR-2694	5250±50	4234-3980 Cal BC
81-84cm	SRR-2929	5120±50	4034-3790 Cal BC
76-79cm	SRR-2693	4620±50	3560-3140 Cal BC
60-64cm	GU-1312	4315±70	3100-2704 Cal BC
Sefton mosslands			
FLEA MOSS WOOD			
290-298cm	SRR-2697	7230±70	6215-5960 Cal BC
116-124cm	SRR-2696	5920±50	4939-4720 Cal BC
28-36cm	SRR-2695	4670±50	3622-3340 Cal BC
SNIGGERY WOOD			
170-174cm	SRR-2699	5770±50	4780-4510 Cal BC
10-14cm	SRR-2698	4510±50	3370-3080 Cal BC
ALT VALLEY, INCE BLUNDELL			
80-82cm	GU-5210	7130±110	4186-5740 Cal BC
30-32cm	GU-5211	6040±60	5204-4791 Cal BC
18-20cm	GU-5212	6080±90	5230-4790 Cal BC
Central mosslands			
KNOWSLEY PARK			
42-48cm	Birm-1177	1680±50	Cal AD 240-440
67-73cm	Birm-1176	2670±60	930-790 Cal BC
93-97cm	Birm-1175	2860±60	1258-900 Cal BC
120-125cm	Birm-1174	3490±70	2028-1670 Cal BC
180-185cm	Birm-1173	4370±70	3325-2788 Cal BC
232-238cm	Birm-1172	4600±60	3510-3100 Cal BC
320-327cm	Birm-1191	5290±80	4340-3970 Cal BC
336-344cm	GU-5239	5440±80	4480-4044 Cal BC
407-415cm	Birm-1223	5750±160	4998-4330 Cal BC
441-449cm	GU-5237	6190±80	5322-4906 Cal BC
516-524cm	GU-5240	7700±80	6690-6410 Cal BC
551-559cm	GU-5241	8060±80	
581-589cm	GU-5242	8650±80	
616-624cm	GU-5243	8880±90	
636-644cm	GU-5238	9030±100	Beyond Calibration range
651-659cm	GU-5245	9160±80	
676-684cm	GU-5246	9280±80	
706-712cm	SRR-4515	9305±65	

Site Name and Context	Lab Code	Uncalibrated Date (BP)	Calibrated* Date
SIMONSWOOD MOSS B			
343-351cm	Birm-1217	4090±140	3023-2280 Cal BC
293-301cm	Birm-1218	3790±140	2600-1829 Cal BC
268-276cm	Birm-1219	3390±100	1950-1450 Cal BC
163-172cm	Birm-1220	2730±100	1154-780 Cal BC
114-122cm	Birm-1221	2380±80	790-257 Cal BC
SIMONSWOOD MOSS C			
388-386cm	Birm-1222	5440±160	4669-3970 Cal BC
Southern mosslands			
PRESCOT MOSS			
50-52cm	GU-2723	6850±80	5957-5560 Cal BC
20-22cm	GU-2725	4650±80	3630-3106 Cal BC
8-10cm	GU-2726	4520±140	3630-2890 Cal BC
38-42cm	GU-2724	7980±60	7051-6709(1σ) Cal BC
(on the wood stump and seems to be intrusive)			
PARR MOSS			
327-335cm	Birm-1224	4740±140	3898-3100 Cal BC
230-240cm	Birm-1225	3910±140	2878-1986 Cal BC
195-205cm	Birm-1226	3620±90	2279-1750 Cal BC
115-120cm	Birm-1227	2590±100	930-410 Cal BC

* These dates have been calibrated using the CALIB computer programme (Stuiver and Reimer 1986) using a 20 year (bi-decadal) radiocarbon age dataset which has been derived from the calibration curve published in:

Stuiver and Pearson (1986)	AD 1950-500 BC
Pearson and Stuiver (1986)	500 BC-2500 BC
Pearson *et al* (1986)	2500 BC-5210 BC
Bi-decadal weighted average of data from:	
Linick *et al* (1985)	5210 BC-7210 BC
Stuiver *et al* (1986)	
Kromer *et al* (1986)	

The calibrated dates are shown to 1 (upper) and 2 (lower) standard deviations (σ). There is a 66% chance the date will lie between the 1σ limits and a 99% chance between the 2σ limits.

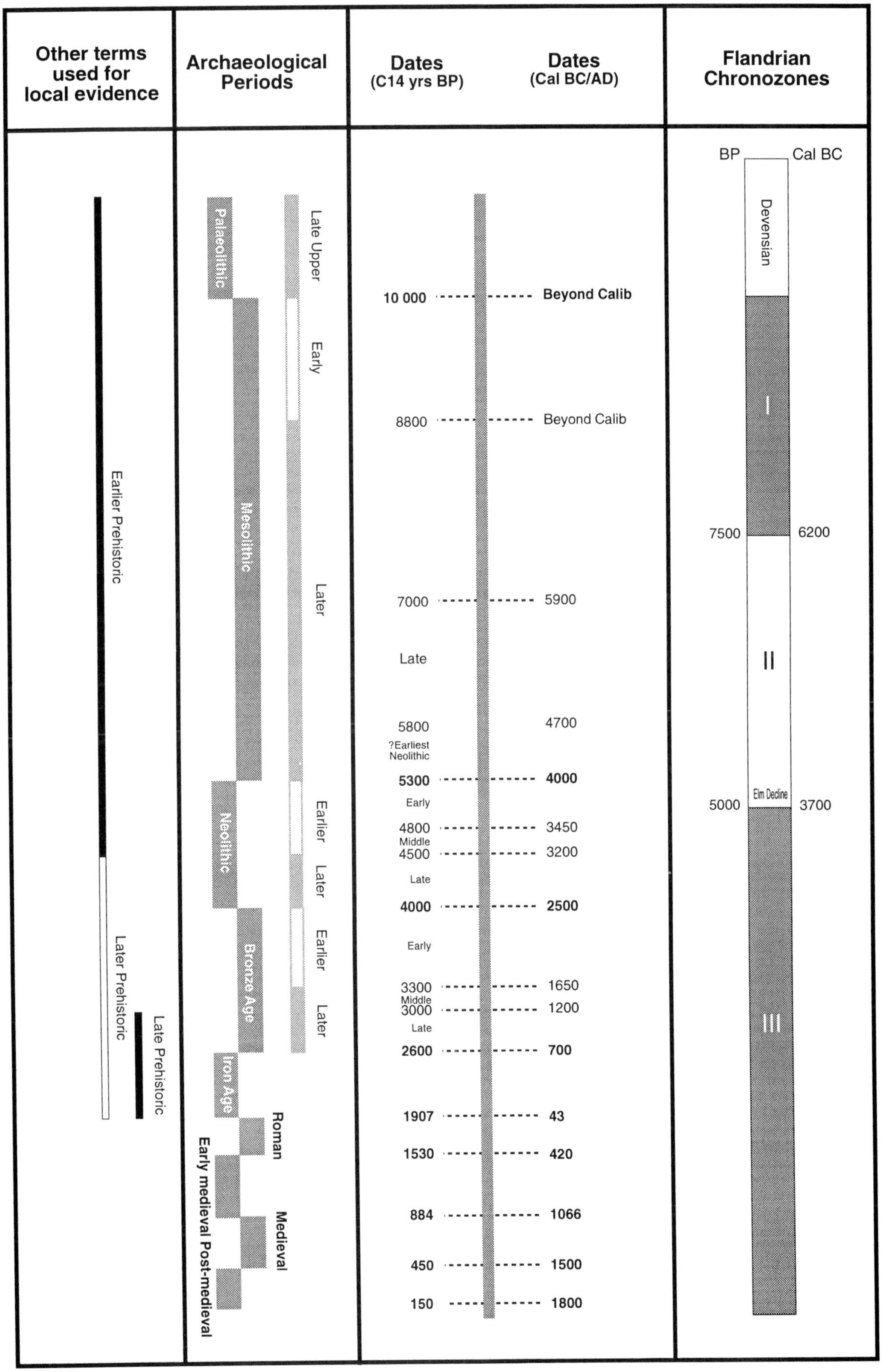
Other terms used for local evidence
Archaeological Periods
Dates (C14 yrs BP)
Dates (Cal BC/AD)
Flandrian Chronozones
Earlier Prehistoric
Later Prehistoric
Late Prehistoric
Palaeolithic
Mesolithic
Neolithic
Bronze Age
Iron Age
Roman
Early medieval
Medieval
Post-medieval
Late Upper
Early
Later
Earlier
Later
Earlier
Later
10 000
Beyond Calib
8800
Beyond Calib
7000
5900
Late
5800
4700
?Earliest Neolithic
5300
4000
Early
4800
3450
Middle
4500
3200
Late
4000
2500
Early
3300
1650
Middle
3000
1200
Late
2600
700
1907
43
1530
420
884
1066
450
1500
150
1800
BP
Cal BC
Devensian
I
7500
6200
II
Elm Decline
5000
3700
III

Appendix 3: Botanical names used in the text and pollen diagrams

Acer	Sycamore or Common maple
Agropyron	Couch-grass
Alnus	Alder
Anemone	Anemone
Armeria	Thrift
Artemisia	Mugwort
Aster	includes Aster
Althaea	Marsh mallow
Betula	Birch
Botrychium	Moonwort
Calluna vulgaris	Heather or ling
Calystegia	Bindweed
Campanula	Bellflower
Caltha	Marsh marigold
Cannabis	Hemp
Carpinus	Hornbeam
Caryophyllaceae	Pink family
Castanea	Chestnut
Centaurea cyanus	Cornflower
Centaurea nigra	Lesser knapweed
Chenopodiaceae	Goosefoot family
Circaea	Enchanter's nightshade
Cirsium	Thistle
Coryloid	Hazel or Bog myrtle
Crataegus	Hawthorn
Cruciferae	Brassica family
Cynoglossum	Houndstongue
Cyperaceae	Sedge family
Drosera	Sundew
Elymus	Lyme-grass
Empetrum	Crowberry
Epilobium	Willowherbs
Equisetum	Horsetails
Endymion-type	Bluebell
Erica-type	Heaths or Heathers
Ericaceae	Heather family
Eriophorum	Cotton-grasses
Eriophorum angustifolium	Common cotton-grass
Euonymus	Spindle tree
Fagus	Beech
Filicales	Ferns
Filipendula	Meadowsweet
Frangula	Alder buckthorn
Fraxinus	Ash
Galium-type	Bedstraws
Genista-type	Greenweed or Needle furze
Gentiana	Gentians
Geranium	Cranesbill
Geum	Avens
Glaux maritima	Sea milkwort
Glyceria	Flote-grass or Reed-grass
Gramineae	Grasses
Hedera	Ivy
Helianthemum	Rockrose
Hippophaë	Sea buckthorn

Hordeum	Barley
Hydrocotyle	Pennyworts
Ilex	Holly
Iris	Yellow flag
Isoetes	Quill-wort
Juniperus	Juniper
Leguminosae or Papilionaceae	Pea family
Lemna	Duckweed
Linum catharticum	Purging flax
Lobelia	Lobelia
Lonicera	Honeysuckle
Lotus	Birdsfoot trefoil
Lycopodium selago	Fir clubmoss
Lysimachia	Loosestrifes
Lythrum	Purple loosestrife
Malva	Mallow
Matricaria	Mayweeds
Melampyrum	Cow-wheats
Mentha	Mints
Menyanthes trifoliata	Bogbean
Mercurialis-type	includes Dog's mercury
Molinia caerulea	Purple moor-grass
Myosotis-type	Forget-me-nots
Myriophyllum alterniflorum	Alternate-flowered Water-milfoil
Myriophyllum spicatum	Spiked water-milfoil
Nuphar	Yellow water-lily
Nymphaea	White water-lily
Oenanthe-type	Dropwort
Onobrychis-type	Sainfoin
Ononis-type	Restharrow
Osmunda	Royal fern
Oxyria-type	Mountain sorrel
Peplis	Water purslane
Phragmites	Common reed
Picea	Spruce
Pinus	Pine
Plantago coronopus	Buck's-horn plantain
Plantago lanceolata	Ribwort plantain
Plantago major	Great plantain
Plantago media	Hoary plantain
Plantago maritima	Sea plantain
Polygonum	Knotweed and Bistorts
Polygonum aviculare	Knotgrass
Polygonum convolvulus-type	Black bindweed
Polypodium	Polypody
Potamogeton	Pondweed
Potentilla	Cinquefoils
Poterium	Salad burnet
Prunus	Cherry and Blackthorn
Pteridium	Bracken
Quercus	Oak
Ranunculaceae	Buttercup family
Ranunculus	Buttercups
Rhamnus	Buckthorn
Rhinanthus-type	Yellow rattle

Rhododendron	Rhododendron
Rosaceae	Rose family
Rubus	includes Brambles and Raspberry
Rubus chamaemorus-type	Cloudberry
Rumex	Sorrels and Docks
Rumex acetosa-type	Sorrel
Sagittaria	Arrowhead
Salix	Willow
Sanguisorba	Burnets
Saxifraga oppositifolia	Purple saxifrage
Saxifraga stellaris	Starry saxifrage
Scleranthus-type	includes Knawels
Scrophulariaceae	Figwort family
Senecio	Ragworts
Silene-type	Campions
Solanum	Nightshade
Sorbus	Mountain ash (Rowan) and Whitebeam
Spergula arvensis	Corn spurrey
Sphagnum	Spaghnum moss
Stellaria	Stitchworts
Stellaria-type	Includes Chickweeds and Stitchworts
Succisa	Devil's bit scabious
Taxus	Yew
Taraxacum	Dandelions
Teucrium	Germanders
Thalictrum	Meadow rue
Thelypteris	Marsh fern
Tilia	Lime
Triticum sp	Wheat
Typha angustifolia	Lesser bulrush
Typha latifolia	Bulrush
Ulex-type	Furze or Gorse
Ulmus	Elm
Umbelliferae	Carrot family
Urtica	Nettle
Vaccinium-type	Bilberries
Valeriana	Valerian
Viburnum	Viburnums
Vicia	Vetches (Tares)
Vicia sativa-type	Common vetch
Vicia sylvatica	Wood vetch

Bibliography

Andersen, S T, 1979 Identification of wild grass and cereal pollen, *Danmarks Geologiske Undersøgelse, Årbog*, **1978**, 69-92

Anon, 1796 Remains of a large forest not far from Liverpool, *Gentleman's Magazine*, **66 (2)**, 549-51

Aston, M, 1985 *Interpreting the landscape. Landscape archaeology in local studies*, London

Atkinson, D, and Houston, J (eds), 1993 *The sand dunes of the Sefton coast*, Liverpool

Bailey, F A (ed), 1937 Prescot Court Leet and other records 1447-1600, *Rec Soc Lancashire Cheshire*, **89**

Barker, T C, and Harris, J R, 1959 *A Merseyside town in the Industrial Revolution, St Helens 1750-1900*, Liverpool

Barnes, B, 1982 *Man and the changing landscape*, Liverpool

Baxter, J, 1983 *Vegetation history of the Shirdley Hill Sand in South West Lancashire*, unpubl PhD Thesis, UCW Aberystwyth

Baxter, J, and Taylor, J A, 1981 *Some preliminary notes on the vegetation history of the South West Lancashire plain*, unpubl report, UCW Aberystwyth

Bedlington, D J, Sea-level changes and coastal movements in North Wales and the Wirral, unpubl PhD thesis, Univ Durham

Behre, K E, 1981 The interpretation of anthropogenic indicators in pollen diagrams, *Pollen et Spores*, **23**, 225-45

Bewley, R H, 1994 *Prehistoric and Romano-British settlement in the Solway Plain, Cumbria*, Oxford

Binford, L R, 1983 *In pursuit of the past*, London

Binney, E N, and Talbot, J H, 1843 On the petroleum found in Downholland Moss, near Ormskirk, *Trans Manchester Geol Soc*, **8 (3)**, 41-8

Birks, H J B, 1964 Chat Moss, Lancashire, *Mem Proc Manchester Lit Phil Soc*, **106**, 22-45

Birks, H J B, 1965 Late glacial deposits at Bagmere, Cheshire, and Chat Moss, Lancashire, *New Phytol*, **64**, 270-85

Bonsall, C, 1981 The coastal factor in the Mesolithic settlement of North West England, in *The Mesolithic in Europe* (ed B Gramsch), Potsdam, 451-72

Bonsall, C, Sunderland, D, Tipping, R, and Cherry, J, 1986 The Eskmeals Project 1981-5: an interim report, *Northern Archaeol*, **7 (1)**, 1-30

Bradley, R, 1984a *The social foundations of prehistoric Britain*, London

Bradley, R, 1984b Regional studies in Neolithic Britain, in R Bradley and J Gardiner (eds), *Neolithic Studies: A review of some current research*, Brit Archaeol Rep, **133**, Oxford, 5-14

Bradley, R, and Edmonds, M, 1993 *Interpreting the axe trade. Production and exchange in Neolithic Britain*, Cambridge

Brownbill, J, 1936 A history of the old parish of Bidston, Cheshire, *Trans Hist Soc Lancashire Cheshire*, **87**, 133-99

Bu'Lock, J D, 1959 The Pikestones: A chambered long cairn of Neolithic type on Anglezarke Moor, Lancashire, *Trans Lancashire Cheshire Antiq Soc*, **68**, 143-5

Bu'Lock, J D, 1960 The Celtic, Saxon, and Scandinavian settlement at Meols in Wirral, *Trans Hist Soc Lancashire Cheshire*, **112**, 1-28

Burgess, C, 1974 The Bronze Age, in C Renfrew (ed), *British Prehistory*, London, 165-233

Burgess, C, 1980 *The Age of Stonehenge*, London

Burton, R G O, and Hodgeson, J M, 1987 *Lowland peat in England and Wales*, Soil Survey Special Publication, **15**, Harpenden

Camden, W, 1610 *Britain vol 1: England*, London

Cantor, L M (ed), 1982 *The English medieval landscape*, Oxford

Chitty, G S, 1978 Wirral rural fringes survey report, *J Merseyside Archaeol Soc*, **2**, 1-25

Chitty, G S, 1981 *St Helens rural fringe survey report*, unpubl report, Liverpool Museum

Chitty, G S, and Warhurst, M, 1977 Ancient Meols: finds from the Cheshire shore, *J Merseyside Archaeol Soc*, **1**, 19-42

Clapham, A R, Tutin, A G, and Warburg, E F, 1962 *Flora of the British Isles*, London

Clare, T, 1979 Rayset Pike long cairn in the Machel Mss, *Trans Cumberland Westmorland Antiq Archaeol Soc*, **75**, 1-16

Clark, R H, and Schofield, A J, 1991 By experiment and calibration: an integrated approach to the archaeology of the ploughsoil, in A J Schofield (ed), *Interpreting artefact scatters*, Oxford, 93-106

Cleary, R, and Philpott, R A, 1993 *An archaeological evaluation at Church Farm, Bidston*, unpubl report, Liverpool Museum

Clymo, R S, 1984 The limits of bog growth, *Phil Trans R Soc Lond B*, **303**, 605-654

Coles, B, and Coles, J, 1986 *Sweet Track to Glastonbury*, London

Coles, J, 1978 The Somerset Levels: a concave landscape, in H C Bowen and P J Fowler (eds), *Early land allotment*, British Archaeol Rep, **48**, Oxford, 147-8

Coles, J, 1991 *From the waters of oblivion*, Stichting voor de Nederlandse Archeologie, Assen

Corser, T (ed), 1845 James' Iter Lancastrense 1636, *Chet Soc*, 1 ser, **7**

Cowell, R W, 1981 *Knowsley rural fringes survey*, unpubl report, Liverpool Museum

Cowell, R W, 1982 *Liverpool urban fringe survey*, unpubl report, Liverpool Museum

Cowell, R W, 1991a The prehistory of Merseyside, *J Merseyside Archaeol Soc*, **7**, 21-61

Cowell, R W, 1991b Interim report on the survey of the wetlands in Merseyside, winter 1990-91, *NWWS Ann Rep 1991*, Lancaster, 13-20

Cowell, R W, 1992a Wetland survey in Merseyside, winter 1991-2: aspects of Mesolithic and Neolithic settlement, *NWWS Ann Rep 1992*, Lancaster, 29-36

Cowell, R W, 1992b Prehistoric survey in North Cheshire, *Cheshire Past*, **1**, 6-7

Cowell, R W, 1992c Greasby, North Wirral, Merseyside: interim report on the excavation of an early Mesolithic site, *Archaeology North West*, **4**, 7-15

Cowell, R W, in press Neolithic and Bronze Age finds from Merseyside and North West England, *J Merseyside Archaeol Soc*, **9**

Cowell, R W, in prep a *The nature of prehistoric settlement in the lowlands of North West England: field survey in Merseyside and Cheshire*

Cowell, R W, in prep b *The lowland Mesolithic in North West England: excavations in Merseyside*

Cowell, R W, in prep c *The Liverpool Museum local prehistory collections*

Cowell, R W, and Chitty, G S, 1989 A timber framed building at 21-23 Eccleston St, Prescot (Site 30), *J Merseyside Archaeol Soc*, **5**, 23-33

Cowell, R W, Milles, A, and Roberts, G, 1993 Prehistoric footprints on Formby Point beach, Merseyside, *NWWS Ann Rep 1993*, Lancaster, 43-48

Cowell, R W, and Nicholson, S M, 1990 *An archaeological implications study of the Rimrose Valley, Sefton*, unpubl report, Liverpool Museum

Cowell, R W, and Philpott, R A, 1994 *Excavations along the route of the M57-A562 link road, Merseyside, in 1993*, unpubl report, Liverpool Museum

Cowell, R W, and Warhurst, M H, 1984 *The Calderstones: a prehistoric tomb on Merseyside*, Liverpool

Cox, E W, 1895 Recent antiquarian 'finds' in Cheshire, *Trans Hist Soc Lancashire Cheshire*, **46** (3 ser **10**), 175-179

Cox, C, 1992 *Aerial archaeology in the North West Wetlands: Merseyside 1992*, unpubl report, Liverpool Museum

Cundill, P R, 1981 The history of vegetation and land use of two peat mosses in South West Lancashire, *Manchester Geog*, n ser **2 (2)**, 35-44

Cundill, P R, 1984 Palaeobotany and archaeology on Merseyside: additional evidence, *Circaea*, **2**, 129-131

Cunliffe Shaw, R, 1956 *The Royal Forest of Lancaster*, Preston

Davey, P J, 1976 The distribution of Bronze Age metalwork from Lancashire and Cheshire, *J Chester Archaeol Soc*, **59**, 1-13

Davey, P J, 1978 *Prescot Action Area: an archaeological view*, Liverpool

Davey, P J, and McNeil, R, 1985 Excavations at South Castle Street, Liverpool, 1976 and 1977, *J Merseyside Archaeol Soc*, **4**, 1-158

Davies, J, and McMillan, N F, 1956 Notes on post-glacial deposits in the Wirral, Cheshire, *Liverpool Manchester Geol J*, **1.6**, 507-511

de Rance, C E, 1870 On the post-glacial deposits of Western Lancashire and Cheshire, *Quarterly J Geol Soc (Lond)*, **26**, 655-668

de Rance, C E, 1872 *The geology of the country around Southport, Lytham, and South Shore*, Mem Geol Surv UK, London

de Rance, C E, 1877 *The superficial geology of the county adjoining the coasts of South West Lancashire*, Mem Geol Surv UK, London

Dodgson, J McN, 1972 *The place names of Cheshire IV*, Eng Place Name Soc, **47**

Edwards, B J N, 1991 Bonds Farm, Pilling, *NWWS Ann Rep 1991*, Lancaster, 42-43

Edwards, K J, and Hirons, K R, 1984 Cereal pollen grains, in pre-elm decline deposits: implications for the earliest agriculture in Britain and Ireland, *J Archaeol Science*, **11**, 71-80

Edwards, K J, and Ralston I B M, 1984 Post-glacial hunters, gatherers and vegetational history in Scotland, *Proc Soc Antiq Scotland*, **114**, 15-34

Ekwall, E, 1922 The place names of Lancashire, *Chet Soc*, **81**

Erdtman, G, 1928 Studies in the post-arctic history of the forests of north-western Europe. I Investigations in the British Isles, *Geologiska Föreningens i Stockholm Förhandlingar*, **50**, 123-192

Faegri, K, and Iversen, J, 1989 *Textbook of pollen analysis*, Chichester, 4 edn

Farrer, W, 1899 Notes on the Domesday Survey of the land between Ribble and Mersey, *Trans Lancashire Cheshire Antiq Soc*, **16**, 1-38

Farrer, W, 1900 The chartulary of Cockersand Abbey, *Chet Soc*, **43 (2.2)**

Farrer, W, 1906 Domesday Survey, in W Farrer and J Brownbill (eds), *A History of the County of Lancaster*, **1**, The Victoria History of the Counties of England, London, 260-290

Farrer, W, 1910 Final Concords of the County of Lancaster, IV, 1510-1558, *Rec Soc Lancashire Cheshire*, **60**

Farrer, W, and Brownbill, J (eds), 1907 *A History of the County of Lancaster*, **3**, The Victoria History of the Counties of England, London

Fleming, A, 1971 Territorial patterns in Bronze Age Wessex, *Proc Prehist Soc*, **37 (1)**, 138-66

Foley, R, 1981 A model of regional archaeological structure, *Proc Prehist Soc*, **47**, 1-17

Ford, S, Bradley, R, Hawkes, J, and Fisher, P, 1984 Flint working in the Metal Age, *Oxford J Archaeol*, **3**, 157-73

Freke, D J, 1981 Bromborough Court House moated site excavations 1979, *J Merseyside Archaeol Soc*, **2**, 47-52

Freke, D J, and Holgate, R, 1990 Excavations at Winwick, Cheshire in 1980: 1 excavation of two second millennium BC mounds, *J Chester Archaeol Soc*, **70**, 9-30

Garstang, J, 1906 Early Man, in W Farrer and J Brownbill (eds), *A History of the County of Lancaster*, **1**, The Victoria History of the Counties of England, London, 211-56

Girling, M A, and Greig, J, 1985 A first fossil record for *Scolytus scolytus* (F) (elm bark beetle); its occurrence in elm decline deposits from London and its implications for Neolithic elm disease, *J Archaeol Science*, **12**, 347-51

Glenn, T A, 1914 Flint flakes found at Hilbre Point, *Trans Lancashire Cheshire Antiq Soc*, **32**, 245-46

Godwin, H, 1959 Studies in the post-glacial history of British vegetation. XIV, Late-glacial deposits at Moss Lake, Liverpool, *Phil Trans R Soc B*, **242**, 127-49

Godwin, H, 1975 *History of the British Flora*, Cambridge, 2 edn

Godwin, H, and Willis, E H, 1962 Cambridge University radiocarbon measurements, V, *Radiocarbon*, **4**, 57

Green, H S, 1974 Early Bronze Age burial, territory, and population in Milton Keynes, Buckinghamshire and the Great Ouse Valley, *Arch J*, **131**, 75-139

Green, H S, 1980 *The flint arrowheads of the British Isles*, British Archaeol Rep Brit Ser, **75**, Oxford

Green, H S, 1984 Flint arrowheads: typology and interpretation, *Lithics*, **5**, 19-39

Greenwood, H W, 1910 Exposures of the old beach surface at Leasowe, *Proc Liverpool Geol Ass*, n ser **4**, 49-50

Gresswell, R K, 1953 *Sandy shores in South Lancashire*, Liverpool

Gresswell, R K, 1958 Hillhouse coastal deposits in South Lancashire, *Liverpool Manchester Geol J*, **2**, 60-78

Grimm, E, 1991 *TILIA and TILIA-GRAPH*, Illinois

Hageman, B P, 1969 Development of the western parts of the Netherlands during the Holocene, *Geologie Mijnb*, **48 (4)**, 373-88

Hall, B R, 1955 *Borehole records from the mosses of South West Lancashire*, Soil Survey of England and Wales, MS **65**

Hall, D N, 1987 *The Fenland Project, 2: fenland landscapes and settlement between Peterborough and March*, E Anglian Archaeol, **35**, Norwich

Hall, D N, and Coles, J M, 1994 *Fenland Survey: an essay in landscape and persistence*, London

Hall, B R, and Folland, C J, 1970 *Soils of Lancashire*, Soil Survey of England and Wales, Bulletin **5**, Harpenden

Hawkes, A J, 1936 Some thirteenth and fourteenth century deeds relating to Parbold, Ince and Sutton, *Trans Lancashire Cheshire Antiq Soc*, **51**, 60-74

Healey, E, 1987 Lithic technology, in C A Smith and F M Lynch, *Trefignath and Din Dryfol, the excavation of two megalithic tombs in Anglesey*, Cambrian Archaeol Monogr, **3**, 50-59

Healy, F, 1984 Lithic assemblage variation in the late third and early second millennia in eastern England, *Lithics*, **5**, 10-18

Healy, F, 1987 Prediction or prejudice? The relationship between field survey and excavation, in A G Brown and M R Edmonds (eds), *Lithic analysis in later British prehistory: some problems and approaches*, British Archaeol Rep Brit Ser, **162**, Oxford, 9-17

Hibbert, F A, 1977 *Risley Moss, a palaeoecological study*, unpubl report, Liverpool Polytechnic

Hibbert, F A, Switsur, V R, and West, R G, 1971 Radiocarbon dating of Flandrian pollen zones at Red Moss, Lancashire, *Proc Royal Soc London B*, **177**, 161-76

Hibbert, F A, and Switsur, V R, 1976 Radiocarbon dating of Flandrian pollen zones in Wales and northern England, *New Phytol*, **77**, 793-807

Holgate, R, 1989 Excavations at Prescot, December 1980-January 1981 (Sites A-E), *J Merseyside Archaeol Soc*, **5**, 11-22

Holt, J, 1795 *A General view of the agriculture of the county of Lancaster*, London

Hoskins, W G, 1955 *The making of the English landscape*, London

Howard-Davis, C, Stocks, C, and Innes, J B, 1988 *Peat and the past*, Lancaster

Huddart, D, 1992 Coastal environmental changes and morphostratigraphy in South West Lancashire, England, *Proc Geol Ass*, **103** (3), 217-36

Huddart, D, Tooley, M J, and Carter, P A, 1977 The coast of North West England, in C Kidson and M J Tooley (eds), *The Quaternary history of the Irish Sea*, Liverpool, 119-54

Hume, A, 1863 *Ancient Meols*, London

Innes, J B, 1982 Notes on the vegetation history of Sefton District, Merseyside, in J Lewis, *Sefton rural fringes survey report*, Appendix 1, unpubl report, Liverpool Museum

Innes, J B, 1986 The history of the Shirdley Hill Sand revealed by examination of associated organic deposits, *Proc North of England Soils Discussion Group*, **21**, 31-43

Innes, J B, and Tomlinson, P R, 1983a An approach to palaeobotany and survey archaeology in Merseyside, *Circaea*, **1**, 83-93

Innes, J B, and Tomlinson, P R, 1983b Cultural implications of Holocene landscape evolution in the Merseyside region, *Amat Geol*, **10**, 2-17

Innes, J B, and Tomlinson, P R, 1991 Environmental archaeology in Merseyside, *J Merseyside Archaeol Soc*, **7**, 1-20

Innes, J B, and Tooley M, 1993 The age and vegetational history of the Sefton coast dunes, in D Atkinson and J Houston (eds), *The sand dunes of the Sefton coast*, Liverpool, 35-40

Innes, J B, Tooley, M J, and Tomlinson, P R, 1989 A comparison of the age and palaeoecology of some sub-Shirdley Hill Sand peat deposits from Merseyside and South West Lancashire, *Naturalist*, **114**, 65-69

Innes, J B, Bedlington, D J, Kenna, R J B, and Cowell, R W, 1990 A preliminary investigation of coastal deposits at Newton Carr, Wirral, Merseyside, *Quaternary Newsletter*, **62**, 5-12

Iversen, J, 1941 Land occupation in Denmark's Stone Age, *Damn Geol Unders Serial ii*, **66**, 1-68

Jackson, J W, 1919 Shell-pockets on sand-dunes on the Wirral coast, Cheshire, and notes on ancient land surfaces, *Lancashire Cheshire Naturalist*, **12**, 9-14, 39-44

Jackson, J W, 1936 The prehistoric archaeology of Lancashire and Cheshire, *Trans Lancashire Cheshire Antiq Soc*, **50**, 65-106

Jacobi, R M, 1978 Northern England in the eighth millennium BC: an essay, in P Mellars (ed), *The early post-glacial settlement of Northern Europe*, London, 243-94

Jacobi, R M, 1980 The early Holocene settlement of Wales, in A J Taylor (ed), *Culture and environment in prehistoric Wales*, British Archaeol Rep Brit Ser, **76**, Oxford, 131-206

Jacobi, R M, 1987 Misanthropic miscellany: musings on British early Flandrian archaeology and other flights of fancy, in P Rowley-Conwy, M Zvelebil, and H P Blankholm (eds), *Mesolithic North West Europe: recent trends*, Sheffield, 163-68

Jelgersma, S, de Jong, J, Zagwijn, W H, and van Regteren Altena, J F, 1970 The coastal dunes of western Netherlands; geology, vegetation history, and archaeology, *Meded Rijks geol Dienst*, n ser **21**, 93-167

Johnson, R H, 1985 The geomorphology of the regions around Manchester: an introductory review, in R H Johnson (ed), *The geomorphology of North West England*, Manchester, 1-23

Jones, T A, 1919 Notes on some borings recently made in Wirral and in Liverpool, *Proc Liverpool Geol Soc*, **12 (4)**, 339-45

Jones, G D B, 1980 Archaeology and coastal change in North-West England, in F H Thompson (ed), *Archaeology and coastal change*, London, 87-102

Jowsey, P C, 1966 An improved peat sampler, *New Phytol*, **65**, 245-48

Keen, L, and Hough, P, 1993 *Beeston Castle, Cheshire, a report on the excavations 1968-85*, HBMC Archaeol Rep, **23**, London

Kenna, R J B, 1978 Early settlement on the North Wirral coastal area, *J Merseyside Archaeol Soc*, **2**, 27-34

Kenna, R J B, 1979 Coastal change and early settlement on the North Wirral, *J Merseyside Archaeol Soc*, **3**, 41-53

Kenna, R J B, 1986 The Flandrian sequence of North Wirral (North West England), *Geol J*, **21**, 1-27

Kidson, C, and Heyworth, A, 1973 The Flandrian sea-level rise in the Bristol Channel, *Proc Ussher Soc*, **2**, 565-584

Kinnes, I A, 1988 The Cattleship Potemkin: reflections on the early Neolithic in Britain, in J C Barret and I A Kinnes (eds), *The archaeology of context in the Neolithic and Bronze Age: recent trends*, Sheffield, 2-8

Kromer, B, Rhein, M, Bruns, Z M, Schoch-Fischer, H, Munnich, K O, Stuiver, M, and Becker, B, 1986 Radiocarbon calibration data for the sixth and eighth millennia BC, *Radiocarbon*, **28**, 954-60

Lamb, H H, 1981 Climate from 1000 BC to 1000 AD, in M Jones and G Dimbleby (eds), *The environment of Man: the Iron Age to the Anglo-Saxon period*, Brit Archaeol Rep Brit Ser, **87**, 53-65

Lees A M, 1979 *Merseyside Structure Plan: draft written statement*, Liverpool

Lewis, J M, 1982 *Sefton rural fringes survey report*, unpubl report, Liverpool Museum

Lewis J M, 1991 Medieval Merseyside, *J Merseyside Archaeol Soc*, **7**, 21-61

Lewis, J M, in prep *Excavations at Micklehead Green, Barrow Old Hall, and Castle Hill, St Helens*

Lindsay, R A, 1989 Bogs, in *Guidelines for the selection of biological SSSIs*, Peterborough, 159-73

Linick, T W, Suess, H E, and Becker, B, 1985 La jolla measurements of radiocarbon in south German oak tree-ring chronologies, *Radiocarbon*, **27**, 20-32

Longley, D M T, 1987 Prehistory, in B E Harris and A T Thacker (eds), *A History of the County of Chester*, **I**, The Victoria History of the Counties of England, London, 36-114

Maidwell, F T, 1920 Some borings through the marshes bordering the southern shore of the Mersey estuary, *Proc Liverpool Geol Soc*, **12 (4)**, 417-28

Mayer, J, 1849 On the British urns found at West Kirby, *Trans Hist Soc Lancashire Cheshire*, **1**, 153-55

McNeil, R, 1989 Excavation of an eighteenth century pottery in Eccleston St, Prescot (Site F), *J Merseyside Archaeol Soc*, **5**, 49-94

Mellars, P, 1974 The Palaeolithic and Mesolithic, in C Renfrew (ed), *British Prehistory*, London, 41-92

Mellars, P, 1976 Fire ecology, animal populations and man: a study of some ecological relationships in prehistory, *Proc Prehist Soc*, **42**, 15-45

Middleton, R, 1990 Archaeological fieldwork in Cumbria and Lancashire 1989-90, *NWWS Ann Rep 1990*, Lancaster, 35-40

Middleton, R, 1993 Landscape archaeology in the North West and the definition of surface lithic scatter sites, *NWWS Ann Rep 1993*, Lancaster, 1-8

Middleton, R, and Wells, C, 1990 A research design for an archaeological survey of the wetlands of North West England, *NWWS Ann Rep 1990*, 1-6

Middleton, R, Wells, C, and Huckerby, E, forthcoming *The Wetlands of North Lancashire*

Moffatt, M, 1977 *The prehistory of the Wirral peninsula*, unpubl thesis, Univ Liverpool

Moore, P D, 1975 The origin of blanket mires, *Nature London*, **256**, 267-69

Moore, P D, 1986 Hydrological changes in mires, in B E Berglund (ed), *Handbook of Holocene palaeoecology and palaeohydrology*, Chichester

Moore, P D, and Bellamy, D J, 1974 *Peatlands*, London

Moore, P D, Evans, A T, and Chater, M, 1986 Palynological and stratigraphic evidence for hydrological change in mires associated with human activity, in K E Behre (ed), *Anthropogenic indicators in pollen diagrams*, Rotterdam

Moore, P D, and Webb, J A, 1978 *An illustrated guide to pollen analysis*, London

Moorhouse, S A, 1981 Field systems in West Yorkshire, in M L Faull and S A Moorhouse (eds), *An archaeological survey to AD 1500*, **3**, Wakefield, 656-681

Morton, G H, 1897 *The geology of the country around Liverpool, including the north of Flintshire*, London

Muirburn Working Party, 1977 *A guide to good muirburn practice*, Dept of Agriculture and Fisheries for Scotland / Nature Conservancy Council, Edinburgh

Myers, A, 1987 All shot to pieces? Inter-assemblage variability, lithic analysis, and Mesolithic assemblage types:

some preliminary observations, in A G Brown and M Edmonds (eds), *Lithic analysis and later British prehistory*, Brit Archaeol Rep, **162**, Oxford, 137-54

Nevell, M D, 1989 Great Woolden Hall Farm: excavations on a late prehistoric/Romano-British native site, *Greater Manchester Archaeol J*, **3**, 35-44

Nevell, M D, 1991 A field survey of High Legh Parish: Part one, prehistoric and Roman evidence, *Archaeology North West*, **2**, 16-19

Newstead, R, 1927 The excavations at Hilbre, 1926, *Trans Hist Soc Lancashire Cheshire*, **78**, 136-143

O'Hanlon, D M, 1991 Dark Age Merseyside, *J Merseyside Archaeol Soc*, **7**, 75-86

Pearson, G W, 1987 How to cope with calibration, *Antiquity*, **61**, 98-103

Pearson, G W, Pilcher, J R, Baillie, M G, Corbett, D M, and Qua, F, 1986 High-precision C14 measurement of the Irish oaks to show the natural C14 variations from AD 1840-5210 BC, *Radiocarbon*, **28**, 911-934

Pearson, G W, and Stuiver, M, 1986 High-precision calibration of the radiocarbon time-scale, 500-2500 BC, *Radiocarbon*, **28**, 839-62

Perry, I and Moore, P D, 1987 Dutch elm disease as an analogue of Neolithic elm decline, *Nature London*, **326**, 72-73

Petch, D F, 1976 Stray finds: Little Hilbre Island, *Cheshire Archaeol Bull*, **4**, 37

Philpott, R A, 1988 *Historic towns of the Merseyside area: a survey of urban settlement to c1800*, Liverpool Museum Occas Papers, **3**, Liverpool

Philpott, R A, 1989 Prescot and its archaeology, *J Merseyside Archaeol Soc*, **5**, 107-9

Philpott, R A, 1991 Roman Merseyside, *J Merseyside Archaeol Soc*, **7**, 61-81

Philpott, R A, 1993 A Romano-British farmstead at Irby, Wirral, and its place in the landscape: an interim statement, *Archaeol North West*, **5**, 19-25

Philpott, R A, 1994 New light on Roman settlement; recent aerial photography in Cheshire, *Cheshire Past*, **3**, 6-7

Philpott, R A, in prep, *Medieval excavations in Merseyside*, Liverpool

Philpott, R A, and Cowell, R W, 1992 *An archaeological assessment of land at Irby for a golf course development*, unpubl report, Liverpool Museum

Philpott, R A, and Davey, P J, 1984 *Prescot sampling project 1983-4*, Liverpool

Philpott, R A, Simmons, P, and Cowell, R W, 1993 *Excavations on a Romano-British farmstead and prehistoric sites in Southworth-with-Croft, Cheshire*, unpubl report, Liverpool Museum

Pilcher, J R, and Smith, A G, 1979 Palaeoecological investigations at Ballynagilly, a Neolithic and Bronze Age settlement in County Tyrone, Northern Ireland, *Phil Trans Royal Soc B*, **268**, 345-69

Pitts, M W, 1978 Towards an understanding of flint industries in post-glacial England, *Bull Institute Archaeol Univ Lond*, **15**, 179-97

Pitts, M, and Jacobi, R M, 1979 Some aspects of change in flaked stone industries of the Mesolithic and Neolithic in southern Britain, *J Archaeol Science*, **6**, 163-77

Plater, A J, Huddard, D, Innes, J B, Pye, K, Smith, A J, and Tooley, M J, 1993 Coastal and sea-level changes, in D Atkinson and J A Houston (eds), *The sand dunes of the Sefton coast*, Liverpool, 23-34

Potter, C, 1876 Observations on the geology and archaeology of the Cheshire shore, *Trans Hist Soc Lancashire Cheshire*, **4**, 121-42

Potter, C, 1894 On the so-called forest beds of Leasowe, *J Liverpool Geol Ass*, **14**, 5-7

Pryor, F, 1976 Fen-edge land management in the Bronze Age: an interim report on excavations at Fengate, Peterborough, 1971-5, in C Burgess and R Miket (eds), *Settlement and economy in the third and second millennium BC*, British Archaeol Rep, **33**, Oxford, 29-49

Pryor, F, 1978 *Excavation at Fengate, Peterborough, England: the second report*, Arch Monogr ROM, **5**, Toronto

Pryor, F, 1980 *Excavation at Fengate, Peterborough, England: the third report*, Northants Arch Soc Monograph, **1**/ Arch Monogr ROM, **6**, Toronto and Northampton

Pryor, F, 1991 *Flag Fen: prehistoric Fenland centre*, London

Pryor, F M M, and French, C A I, 1985 *The Fenland Project, 1: archaeology and environment in the lower Welland valley*, E Anglian Archaeol, **27**, Cambridge

Pye, K, 1990 Physical and human influences on coastal dune development between the Ribble and the Mersey estuaries, North West England, in K F Nordstrom, N Psuty, and R W G Carter (eds), *Coastal dunes: form and process*, Chichester, 339-359

Pye, K, and Neal, A, 1993a Stratigraphy and age structure of the Sefton dune complex: preliminary results of field drilling investigations, in D Atkinson and J Houston (eds), *The sand dunes of the Sefton coast*, Liverpool, 41-44

Pye, K, and Neal A, 1993b Late Holocene dune formation on the Sefton coast, North West England, in K Pye (ed), *The dynamics and environmental context of aeolian sedimentary systems*, Geol Soc Special Publication, **72**, London, 201-217

Rackham, O, 1980 *Ancient woodland; its history, vegetation and uses in England*, London

Reade, T M, 1872a The geology and physics of the post-glacial period, as shown in deposits and organic remains in Lancashire and Cheshire, *Proc Liverpool Geol Soc*, **2**, 36-88

Reade, T M, 1872b The post-glacial geology and physiography of West Lancashire and the Mersey estuary, *Geol Mag*, **9**, 111-119

Reade, T M, 1878a The trees of the post-glacial forest beds in the neighbourhood of Liverpool, *Trans Hist Soc Lancashire Cheshire*, **6**, 27-28

Reade, T M, 1878b Some further notes on the submarine forest of the Alt mouth, *Proc Liverpool Geol Soc*, **3 (4)**, 362-369

Reade, T M, 1908 Post-glacial beds at Great Crosby as disclosed by the new outfall sewer, *Proc Liverpool Geol Soc*, **10 (4)**, 249-61

Richards, J, 1990 *The Stonehenge environs project*, Engl Heritage Archaeol Rep, **16**, London

Ricketts, C, 1866 A wooden implement found in Bidston Moss, *Proc Birkenhead Literary and Scientific Soc*, Session **9**, 138-45

Roe, F, 1979 Typology of stone implements with shaft-holes, in T H McK Clough and W A Cummins (eds), *Stone axe studies*, CBA Res Rep, **23**, London, 23-48

Roeder, C, 1900 On a newly discovered Neolithic settlement at the Red Noses, New Brighton, near Liverpool, *Trans Hist Soc Lancashire Cheshire*, **50**, 123-130

Rylands, J P, 1887 Lancashire Inquisitions: Stuart Period Part II, *Rec Soc Lancashire Cheshire*, **31**, 1-102

Schofield, A J, 1991a Interpreting artefact scatters; an introduction, in A J Schofield (ed), *Interpreting artefact scatters; contributions to ploughsoil archaeology*, Oxbow Monogr, **4**, Oxford, 3-8

Schofield, A J, 1991b Artefact distributions as activity areas; examples from South East Hampshire, in A J Schofield (ed), *Interpreting artefact scatters; contributions to ploughsoil archaeology*, (Oxbow Monogr, **4**, Oxford, 117-128

Sheppard, B, Tomlinson, P, and Innes, J, 1978 Towards an understanding of the environmental history of Merseyside, *J Merseyside Archaeol Soc*, **2**, 91-92

Shimwell, D W, 1985 The distribution and origins of the lowland mosses, in J H Johnson (ed), *The geomorphology of North West England*, Manchester, 299-312

Shone, W, 1911 *Prehistoric man in Cheshire*, London and Chester

Silvester, R J, 1991 *The Fenland Project*, **4**: *Norfolk Survey, The Wissey Embayment and Fen Causeway*, E Anglian Archaeol, **52**, Norwich

Simmons, I G, 1975 The ecological setting of Mesolithic man in the Highland Zone, in J G Evans, S Limbrey, and H Cleere (eds), *The effect of man on the landscape; the Highland Zone*, CBA Res Rep, **11**, London 57-63

Simmons, I G, Dimbleby, G W, and Grigson, C, 1981 The Mesolithic, in I G Simmons and M J Tooley (eds), *The environment in British prehistory*, London, 82-124

Simmons, I G, and Innes, J B, 1987 Mid-Holocene adaptations and later Mesolithic forest disturbance in northern England, *J Archaeol Science*, **14**, 385-403

Simmons, P, and Cowell, R, 1993 *A watching brief at Mason Sports Ground, Bidston Moss, Wirral*, unpubl report, Liverpool Museum

Smith, A G, Grigson, C, Hillman, G, and Tooley, M J, 1981 The Neolithic, in I G Simmons and M J Tooley (eds), *The environment in British prehistory*, London, 125-209

Smith, C, 1992 *Late Stone Age hunters of the British Isles*, London

Smith, H E, 1866a Notabilia of the archaeology and natural history of the Mersey district during the years 1863-4-5, *Trans Hist Soc Lancashire Cheshire*, **18** (n ser **6**), 195-266

Smith, H E, 1866b Notice of a recent disruption of soil by Rimrose Brook, Bootle, *Trans Hist Soc Lancashire Cheshire*, **18** (n ser **6**), 267-70

Smith, H E, 1867 Archaeology of the Mersey district, 1866, *Trans Hist Soc Lancashire Cheshire*, **19 (**n ser **7)**, 169-188

Smith, H E, 1868 An ancient British cemetery at Wavertree, *Trans Hist Soc Lancashire Cheshire*, **20** (n ser **8**), 131-146

Smith, H E, 1869 Archaeology of the Mersey district, 1868, *Trans Hist Soc Lancashire Cheshire*, **21** (n ser **9**), 199-218

Smith, H E, 1871 Archaeology of the Mersey district, 1870, *Trans Hist Soc Lancashire Cheshire*, **23** (n ser **11**), 119-152

Smith, H E, 1872 Archaeology of the Mersey district, 1871, *Trans Hist Soc Lancashire Cheshire*, **24** (n ser **12**), 131-150

Smith, H E, 1873 Archaeology of the Mersey district, 1872, and Liverpool notabilia, *Trans Hist Soc Lancashire Cheshire*, **25** (n ser **13**, 3 ser **1**), 113-152

Smith, H E, 1874 Archaeology of the Mersey district, 1873, and Liverpool notabilia, *Trans Hist Soc Lancashire Cheshire*, **26** (3 ser **2**), 83-112

Smith, H E, 1875 Archaeology of the Mersey district, 1874, and Liverpool notabilia, *Trans Hist Soc Lancashire Cheshire*, **27** (3 ser **3**), 85-108

Smith, I F, 1979 The chronology of British stone implements, in T H McK Clough and W A Cummins (eds), *Stone axe studies*, CBA Res Rep, **11**, London, 13-22

Smith, J P G, 1849 On the Bidston Marsh, *Proc Lit Phil Soc Liverpool*, **5**, 169-172

Smith, R W, 1984 The ecology of Neolithic farming systems as exemplified by the Avebury region of Wiltshire, *Proc Prehist Soc*, **50**, 99-120

Stanning, J H, 1892 The Royalist Composition Papers 1643-1660, 2, *Rec Soc Lancashire Cheshire*, **26**

Stewart-Brown, R, 1925 *Birkenhead Priory and the Mersey Ferry*, Liverpool

Stuiver, M, and Pearson, G W, 1986 High precision of the radiocarbon time-scale, AD 1950-500 BC, *Radiocarbon*, **28**, 805-838

Stuiver, M, Pearson, G W, and Braziunas, T, 1986 Radiocarbon age calibration of marine samples back to 9000 Cal 1/r BP, *Radiocarbon*, **28**, 980-1021

Switsur, V R, 1986 884 BP and all that!, *Antiquity*, **61**, 214-216

Taylor, J A, 1952 The relation of crop distribution to the drift pattern in South West Lancashire, *Trans Inst Brit Geog*, **18**, 77-91

Terret, I B, 1962 Lancashire, in H C Darby and I S Maxwell (eds), *The Domesday geography of northern England*, 392-414

Thomas, J, 1991 *Rethinking the Neolithic*, Cambridge

Tomlinson, P R, and Innes, J B, 1989 Pollen analysis of the Warrington Road, Prescot, peat bed, *J Merseyside Archaeol Soc*, **5**, 7-10

Tooley, M J, 1974 Sea-level changes during the last 9000 years in North West England *Geog J*, **140**, 18-42

Tooley, M J, 1976 Flandrian sea-level changes in West Lancashire and their implications for the Hillhouse coastline, *Geol J*, **11**, 137-152

Tooley, M J, 1977 The Altmouth, in M J Tooley (ed), *The Isle of Man, Lancashire Coast, and Lake District, 10th INQUA Excursion Guide*, Geoabstracts, **8**, Norwich, 32

Tooley, M J, 1978a *Sea-level changes in North West England during the Flandrian stage*, Oxford

Tooley, M J, 1978b The interpretation of Holocene sea-level changes, *Geologiske Foreningens i Stockholm Forhandlingar*, **100**, 203-212

Tooley, M J, 1980 Theories of coastal change in North West England, in F H Thompson (ed), *Archaeology and coastal change*, London, 74-86

Tooley, M J, 1982 Sea-level changes in northern England, *Proc Geol Assoc*, **93 (1)**, 43-51

Tooley, M J, 1985 Sea-level changes and coastal morphology in North West England, in J H Johnson (ed), *The geomorphology of North West England*, Manchester, 94-121

Tooley, M J, 1990 The chronology of coastal dune development in the United Kingdom, *Catena Supplement*, **18**, 81-88

Tooley, M J, and Kear, B S, 1977 Shirdley Hill Sand formation, in M J Tooley (ed), *The Isle of Man, Lancashire Coast, and Lake District, 10th INQUA Excursion Guide*, Geoabstracts **8**, Norwich, 9-12

Travis, C B, 1908 On some borings near Hightown, *Proc Liverpool Geol Ass*, **3**, 11-13

Travis, C B, 1926 The peat and forest beds of the South West Lancashire coast, *Proc Liverpool Geol Ass*, **14 (3)**, 263-77

Travis, C B, 1929 The peat and forest beds at Leasowe, Cheshire, *Proc Liverpool Geol Ass*, **15 (2)**, 157-78

Travis, W G, 1922 On peat bands in the Wallasey sandhills, *Proc Liverpool Geol Ass*, **13 (3)**, 207-14

Troels-Smith, J, 1954 Ertebollekultur-Bodekultur, *Aarb Nord Oldkynd Hist 1953*, 5-62

Troels-Smith, J, 1955 Characterisation of unconsolidated sediments, *Danmarks Geologiske Undersøgelse*, 4 ser, **3 (10)**

Tyrer, F, 1968 *The Great Diurnall of Nicholas Blundell of Little Crosby*, Rec Soc of Lancashire and Cheshire, **110**

Varley, W J, 1932 Early man in the Cheshire Plain, *J Chester North Wales Archit Archaeol Hist Soc*, **29**, 50-65

Varley, W J, 1964 *Cheshire before the Romans*, Chester

Varley, W J, and Jackson, J W, 1940 *Prehistoric Cheshire*, Chester

Walker, F, 1939 Historical geography of South West Lancashire before the Industrial Revolution, *Chet Soc*, **51**, 86-110

Warhurst, M, 1977 Stray finds; Prehistoric, *Cheshire Archaeol Bull*, **5**, 49-50

Warhurst, M, 1982 *Merseyside sylloge of coins of the British Isles*, **29**, New York

Wells, C, 1992 Stratigraphic survey in Merseyside and West Lancashire borderlands, *NWWS Ann Rep 1992*, 43-47

Williams, C T, 1985 *Mesolithic exploitation patterns in the Central Pennines. A palynological study of Soyland Moor*, Brit Archaeol Rep Brit Ser, **139**, Oxford

Williams, M (ed), 1990 *Wetlands: a threatened landscape*, Oxford

Wilson, P, 1985 The mere sands of Lancashire – a forgotten Flandrian deposit, *Quaternary Newsletter*, **45**, 23-6

Wilson, P, Bateman, R M, and Catt, J A, 1981 Petrography, origin, and environment of deposition of the Shirdley Hill Sand of South West Lancashire, England, *Proc Geol Assoc*, **92**, 211-229

Wiltshire, P E J, and Edwards, K J, 1993 Mesolithic, early Neolithic, and later prehistoric impacts on vegetation at a riverine site in Derbyshire, England, in F M Chambers (ed), *Climate change and human impact on the landscape*, London, 157-168

Wood, K W, 1993 Animal and plant groups; birds, in D Atkinson and J Houston (eds), *The sand dunes of the Sefton coast*, Liverpool, 94-96

Wray, D A, and Cope, F W, 1948 *Geology of Southport and Formby, One inch Geological Sheets 74 and 83*, Mem Geol Surv UK, London

Zvelebil, M, 1986 Mesolithic prelude and Neolithic revolution, in M Zvelebil (ed), *Hunters in transition*, Cambridge, 5-16

Zvelebil, M, and Rowley-Conwy, P, 1986 Foragers and farmers in Atlantic Europe, in M Zvelebil (ed), *Hunters in Transition*, Cambridge, 67-94

Index